FOR REFERENCE ONLY

DO NOT REMOVE FROM THE LIBRARY.

THE ENCYCLOPEDIA OF THE ENVIRONMENT

THE ENCYCLOPEDIA OF THE ENVIRONMENT

Edited by Colin Tudge

CHRISTOPHER HELM
London

Editor Lionel Bender
Designers Ayala Kingsley, Niki Overy
Picture Researcher Alison Renney

Design Consultant John Ridgeway
Project Editor Lawrence Clarke

Advisor
Professor Richard Lewontin, Harvard University

Contributing Editor Colin Tudge

Contributors
Michael Allaby (16, 17, 19)
Robin Dunbar (4, 5, 6, 7)
Professor N.J. Mackintosh (1, 2, 3)
Professor D.M. Moore (15)
Dr Peter D. Moore (8, 9, 10)
Colin Tudge (11, 12, 13, 14, 18, 20, 21)

AN EQUINOX BOOK

Planned and produced by:
Equinox (Oxford) Ltd,
Littlegate House,
St Ebbe's Street,
Oxford OX1 1SQ

This edition published 1988 by:
Christopher Helm (Publishers) Ltd,
Imperial House, 21-25 North Street,
Bromley, Kent BR1 1SD

ISBN 0-7470-3204-1

British Library Cataloguing Data applied for

Introductory pictures (pages 1-8)
1 Lumber mill (➧ page 178)
2-3 Nomad Lapps (➧ page 80)
4-5 Flowers bloom after rain (➧ page 108)
7 Oil burn-off (➧ page 223)
8 Monarch butterflies (➧ page 74)

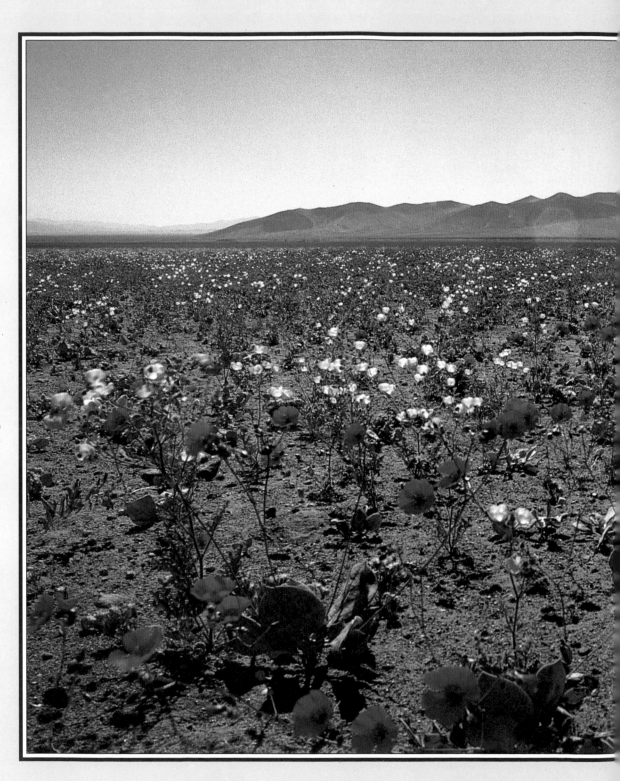

Contents

Introduction

When we first evolved as hunter gatherers on the plains of Africa, we were entirely reliant upon nature, and therefore we had to respect it. But we were also in competition with our fellow creatures – the antelope we contrived to hunt, and leopards that tried to beat us to the prey. When we became farmers we sought to tame nature. We looked after the plants and animals that were useful to us, but tended to regard all other living things, from wolves to greenfly, as pests or arch enemies. Finally, on becoming town-dwellers we largely cut ourselves off from nature – we became indifferent to it.

At different stages of our development, then, we have had a different relationship with nature. But human evolution has now reached a point when we can and should frame a totally new relationship with the planet's flora and fauna, one that is more sophisticated and more subtle than anything before.

Our power over nature

It is vital that we form such a new relationship, both for our own welfare and for the survival of all living things. Perhaps the most important single difference between ourselves and our ancestors is that we now have stupendous power over nature. Indeed, we now have the power to destroy it totally. The first reason for this, and the most significant, is that our population is now vastly greater than ever could have been conceived – hundreds of times greater than at the time of Christ. What makes this worse is that human beings are big animals – we consume a great deal. We are also, to a far greater extent than any other animal, able to convert the environment to suit ourselves. For example, we not only take over land to grow food on, we also convert river mouths into marinas, just for our own pleasure. We build vast roads between cities simply in order to go a little faster, and so divide woods and hills into small patches of land that may individually be too small to support more than a few species of plant or animal. So even if we do not destroy our planet with nuclear war, which is possible, we could destroy it simply by being too successful.

On the other hand, our power over nature is such that we no longer have any good reason to fear it. We do not need to kill wolves in order to protect our sheep; we can arrange for them to live apart from the sheep. We do not need to try to destroy all insects in order to grow crops; we can find out which insects actually cause damage, and arrange to make life difficult just for them. In short, although we are able to destroy life on Earth, we have no excuse to do so. We should now start using our great powers of control not for the elimination of our fellow species, but for their salvation.

The need for planet management

The recognition that we need to conserve is not entirely for the benefit of other animals or plants. Modern ecologists have shown that although we feel we are "above" nature, we are at least as dependent upon a healthy environment, well stocked with wildlife, as were our most primitive ancestors. And by the "environment", ecologists mean the assemblages of all the external factors or conditions that influence each and every living thing in any way. This includes all living creatures and nonliving things such as the air, land, and water. If we pollute the seas we may destroy the microscopic plants that supply most of the oxygen in the atmosphere. If we destroy the forests, we may forever eliminate plants and animals that could, in the future, have served us as new sources of food or drugs.

Lastly, but very importantly, biologists have realized in recent years that other animals are far more intelligent, sensitive, and varied in their behavior than had ever been realized in the past. Once, people felt it was perfectly reasonable to treat animals simply as chattels, to be cooped up or killed as was convenient. Indeed, such an attitude unfortunately is still widespread. But more and more people are coming to appreciate that we should treat animals humanely and with respect, because although they may be less intelligent and versatile than us, they cannot be regarded simply as inanimate machines.

Put all these things together – our power, the control that should remove fear, the realization that we need nature and that we should treat our fellow creatures with respect – and you have the basis of this book. In it, we describe what we are doing to nature, what we should be doing in the name of conservation, and why our attitudes must change.

We begin with an examination of our changing view of animals, looking at the modern, difficult research that has shown that their lives and mental processes are so much richer than was appreciated in the past. In practice the research can be considered in two sections. First, the work on individual animals, which is beginning to show how they "think" – this can be classed animal psychology. Second, the research on animals operating in groups – social behavior. Underlying both subjects, in modern biology, is the theme of evolution. Biologists no longer ask simply what animals do. They ask why animals have evolved each particular kind of behavior. What advantage does each form of behavior confer?

Next, we look at the strangest and, for the time being, the most successful large animal of all: our own species. Clearly, we are different from our fellow creatures. We walk upright with our feet flat. We are relatively hairless. But which of the discernable differences account for our astonishing success? Modern research is showing that the significant differences between ourselves and other animals may be far more subtle than was previously appreciated.

The fact that human beings are successful, however, is undeniable. We commence the second half of this book with a look at the ways in which humans use their success. We explore the ways in which they exploit nature – through agriculture, forestry and fishing. We show that each of these forms of exploitation is in practice highly destructive, but we also show how they might be carried out in ways that would reduce their impact on the environment, and increase the security of this planet. In particular, we describe ways of controlling pests without destroying the environment, highlight specific ways of combining food production with conservation, and of raising livestock without cruelty.

The way forward

Finally, we look specifically at conservation, showing how we might put our new attitude towards nature into practice. If matters go on as they are, then by the middle of the next century we seem bound to have destroyed 90 percent of all of the 1.6 million or so named species of animal and plant that are now on Earth. Yet this is not quite inevitable, and if we work at the problem, we should at least save those species not already endangered, which could include many of the large vertebrates. Furthermore, we could as individuals gain much from nature, by encouraging wildlife back into our cities. Urban foxes and kestrels are already common in many north European cities. In New England, in the United States, the osprey, symbol of wildness, is beginning to look at home among the holiday houses and pleasure yachts. But the task of conservation is urgent and difficult. As we stress, animals need to be kept in breeding colonies of several hundreds if they are to be considered safe from extinction. And scores of species, including many of the largest mammals and birds, are now below the minimum "safe" number.

Though we have not underestimated the enormity of the problems facing this planet, the message of this book is hopeful. Human beings may seem to be a long way down the road to their own destruction, and so to the destruction of all life on Earth. But there is still time to change direction. What is required is that all human beings, and particularly governments, must take the issues seriously. This means a change in moral attitude. It also requires that we should try to find out a great deal more about our fellow species, and about ourselves, than we generally do. The purpose of this book is to provide the essential insights.

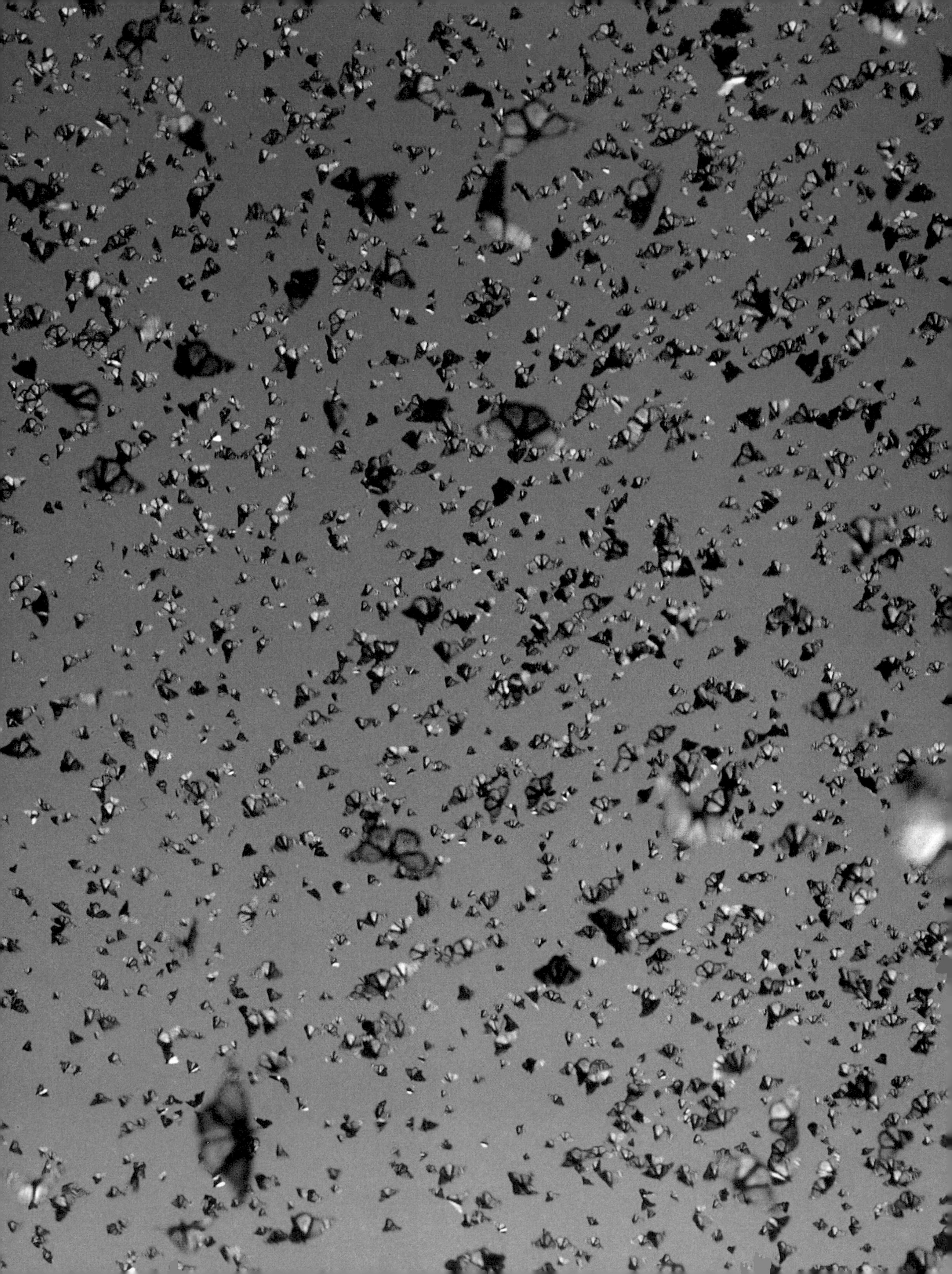

Looking at Animals

1

Why study animal behavior?...Methods of observation and analysis...Neurosurgery as a way of examining behavioral mechanisms...Field studies and laboratory experiments...The role of learning in young animals... PERSPECTIVE...Sound spectrographs and bird ringing ...How animals find their way in the dark...Imprinting ...Young rats learn from their mother how to avoid poisons

We are interested in animals because we hope that understanding them may help us to understand ourselves and for the practical contribution their study may make to human welfare. It was common enough once to use animals to point a human moral; Aesop's fables and medieval bestiaries are the best-known examples. Many animals are of considerable economic importance to us, some because they can be domesticated and made to work (➧ page 163), or provide food or clothing, others because they destroy crops or carry disease. Successful domestication requires some knowledge of the animal's habits and way of life, and since it is often impossible or self-defeating to control pests by eradicating them, successful control may require more subtle techniques which will depend on knowing more about them. Medical research has benefited from the use of animal models of the human condition. Although some knowledge of basic anatomy can be acquired by dissecting human cadavers, the anatomy of structures such as the human brain may be so complex that it makes more sense to start with simpler animal models (➧ page 23), while understanding the function or physiology of a given structure depends on observing it in action in a living animal. Both points must apply even more strongly to the study of behavior.

The modern student of animal behavior wants the answers to several different questions: What do animals do? How and why do they do it? And like other scientists, he will devise hypotheses to explain his data and then seek to test and refine his hypotheses in the light of further data. He starts with observations, then seeks to explain those observations, and finally tests his explanations by further observation or by devising experiments. What is that animal doing? What are the mechanisms that allow it to do this? And why is it doing this rather than something else?

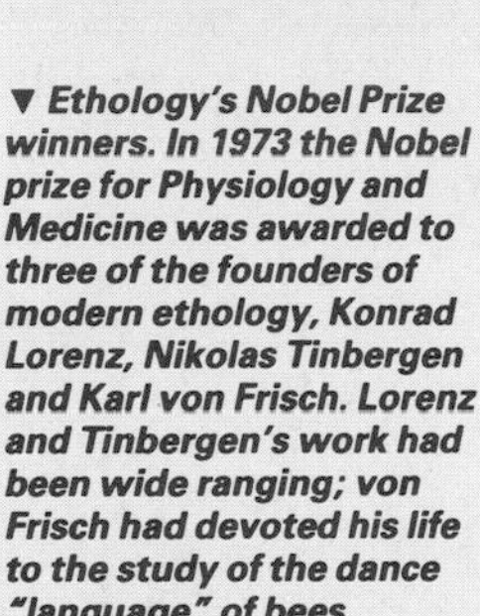

▼ Ethology's Nobel Prize winners. In 1973 the Nobel prize for Physiology and Medicine was awarded to three of the founders of modern ethology, Konrad Lorenz, Nikolas Tinbergen and Karl von Frisch. Lorenz and Tinbergen's work had been wide ranging; von Frisch had devoted his life to the study of the dance "language" of bees.

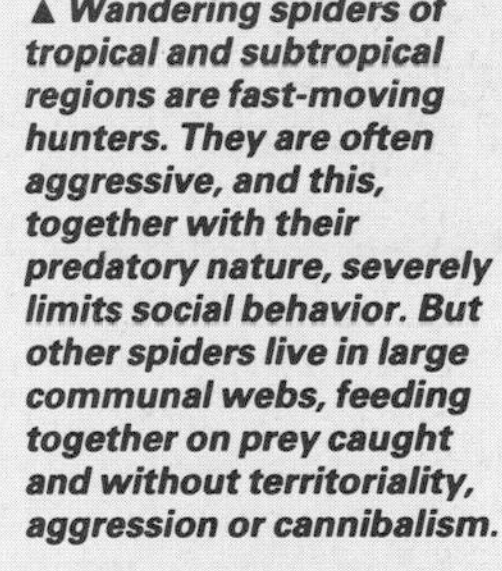

▲ Wandering spiders of tropical and subtropical regions are fast-moving hunters. They are often aggressive, and this, together with their predatory nature, severely limits social behavior. But other spiders live in large communal webs, feeding together on prey caught and without territoriality, aggression or cannibalism.

The science of ethology

"Why" questions in biology are often the trickiest because they can be answered in so many different ways. One answer is in terms of the animal's immediate goal. Why is the lion lying in the grass intently watching a herd of zebra? Because he is hunting. Another answer is in terms of the function or adoptive value of the activity. Because hunting provides food and animals (and their young) need food to survive.

One of the leading animal behaviorists of this century, Dutchman Nikolas Tinbergen (b. 1907), summarized his subject of ethology by saying that there are four questions about the behavior of animals. These are the development of behavior, its immediate causes, how it evolved and the function it now serves. Ethologists seek answers to these questions by studying animals in the wild.

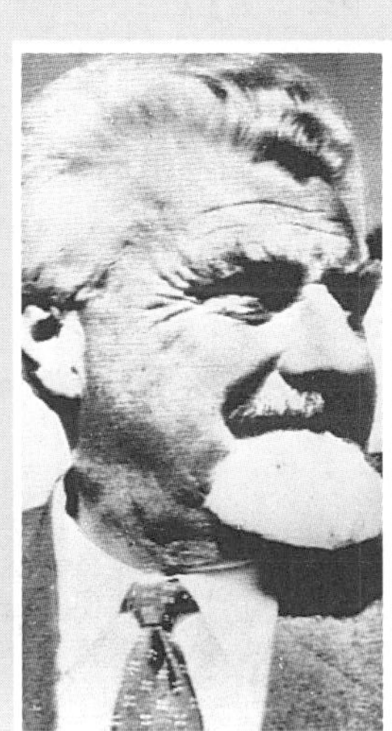

Radar sets are sensitive enough to pick up a single bird 100km away, flying at 8,000m

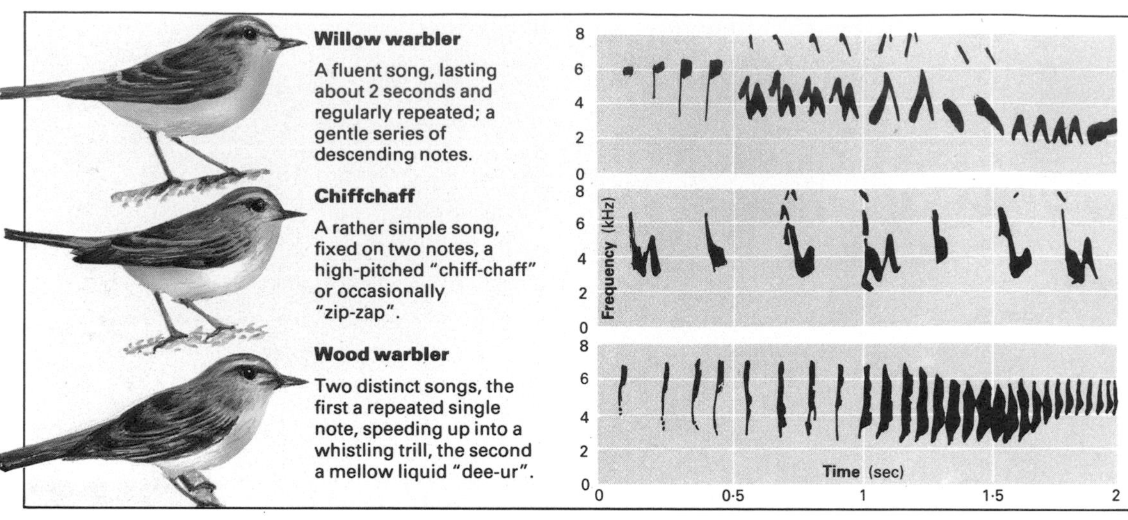

Willow warbler
A fluent song, lasting about 2 seconds and regularly repeated; a gentle series of descending notes.
Chiffchaff
A rather simple song, fixed on two notes, a high-pitched "chiff-chaff" or occasionally "zip-zap".
Wood warbler
Two distinct songs, the first a repeated single note, speeding up into a whistling trill, the second a mellow liquid "dee-ur".
Frequency (kHz)
Time (sec)
0
0·5
1
1·5
2

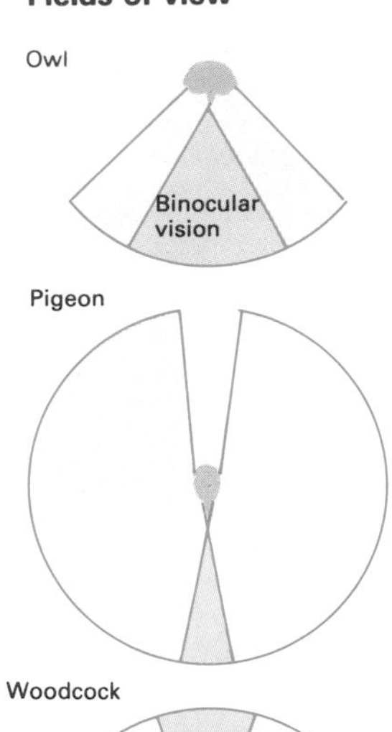

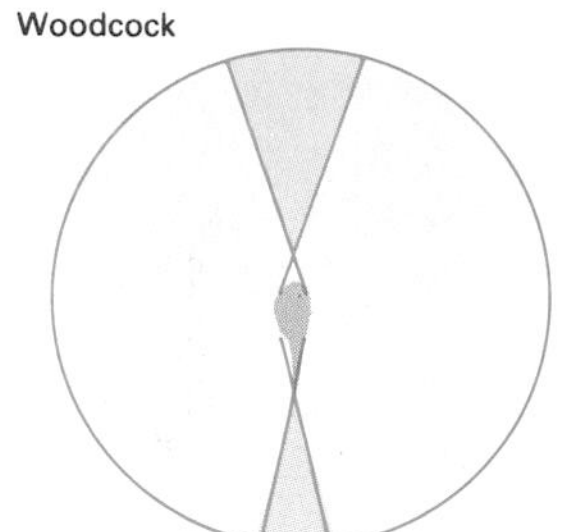

◀ How does one keep track of an animal's movements underwater? If all we want to know is how deep an animal dives, it is a simple matter to fit it with a depth recorder, seen here strapped onto a fur seal. Seals have been recorded diving to depths of 600m, and whales to 1200m.

▲ Birdsong often provides the best way of telling closely related species apart. Many species of warblers or of North American sparrows look alike, but have distinctly different calls and songs. Presumably it is not only humans that identify species in this way.

Analyzing bird songs
Using sound spectrographs scientists have discovered that different groups of white crowned sparrows have different "dialects" of the same basic species song, that young chaffinches denied experience of their parents' song will still produce a song that, although grossly impoverished, contains structural elements of normal chaffinch song, and that marsh warblers develop an idiosyncratic individual song that contains elements of up to 100 other species' songs that they happen to have heard, all depended on the use of sound spectrograms.

Monitoring animals' movements
*A much-used method is to mark individual animals so that they can be identified again wherever they appear. Bird ringing, or banding, is one of the best known examples and has provided a vast amount of information about patterns of migration. But it is inevitably a hit or miss affair, for it depends on the ringed bird being found by someone willing to report the fact. And even then, all one knows is the bird's final destination and not necessarily much about the route it took to get there (**➧ page 78**).*

Observation and investigation

Detailed and accurate observation of animals often requires special techniques for specifying and recording what has been seen or heard. Filming a sequence of behavior not only provides a permanent record, it also allows the ethologist to see details that would otherwise escape the eye. By slowing down or speeding up the film one can study more closely the pattern of the sequence. An experienced and gifted ornithologist might be able to describe and reproduce the songs of several dozen birds, but only a tape recorder can provide a permanent and accurate record, while a sound spectrograph will reveal details of the basic pattern and repetitive nature of the song simply not detectable to the normal human ear.

In other instances, the unaided observer simply cannot record what an animal is doing. Provided that the subject remains in a relatively small home range, it can in principle be watched continuously. But the length of time required to make systematic observations of free-living groups of animals demands heroic patience. Similarly, many mammals, such as cats and foxes, are nocturnal and tracking their movements at night may require special aids. Radio telemetry provides one, albeit expensive, solution. Migrating fish, birds and mammals can cover thousands of kilometers every year on seasonal migrations, and it would be a daunting task to follow them in person (➧ page 72).

Observation can only tell what an animal does, however. If we want to know how it does it, we must go further. A parent bird flies off from the nest to obtain food and returns to feed the young. How does it recognize food and find its way back to the nest? In colonial breeders such as many gulls, several thousands of nests will be perched along a series of cliff ledges, and the colony itself can hardly be missed. But how does the parent recognize its own nest and own young from the many others around it? How do predators detect and capture their prey, and how do the prey detect and avoid the predator?

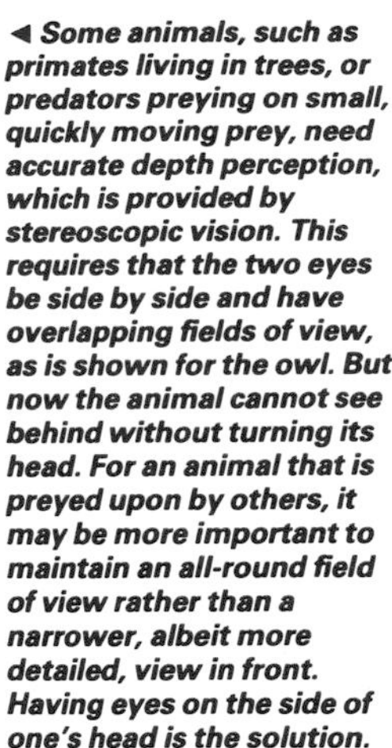

◄ *Some animals, such as primates living in trees, or predators preying on small, quickly moving prey, need accurate depth perception, which is provided by stereoscopic vision. This requires that the two eyes be side by side and have overlapping fields of view, as is shown for the owl. But now the animal cannot see behind without turning its head. For an animal that is preyed upon by others, it may be more important to maintain an all-round field of view rather than a narrower, albeit more detailed, view in front. Having eyes on the side of one's head is the solution.*

► *Ears are often more useful than eyes, both for communicating with one's conspecifics and for finding one's way about. Animals living in dense jungle, or, like these humpback whales, deep in the ocean, where vision is obscured, rely extensively on vocal communication and sometimes on echolocation to detect prey.*

Sensory structures

The mechanism of an animal's behavior raises questions about its sensory capacities and a first answer is to study its anatomy. The basic visual system of an eye with lens and retina connected to the brain by an optic nerve, is common to virtually all vertebrates. But the variations on this theme are legion (◆ page 21). Nocturnal animals have eyes designed to maximize sensitivity to low levels of light, and most diurnal animals have instead detailed color vision.

For some nocturnal animals, and others such as burrowing moles, cave-living salamanders, and very deep-sea fish that live in permanent darkness (◆ page 122), the visual system has degenerated, sometimes, as in the salamander, being present at the larval stage but disappearing by adulthood. All these animals must rely on other senses. Hearing can be astonishingly effective. Barn owls have good night vision but catch mice in total darkness by hearing them move or breathe.

Perhaps the most interesting use of hearing is for echolocation, for example by bats whose large forward-facing ears are designed to pick up the echoes of the very high-frequency sounds they emit.

Lesions

Anatomy alone, as several of these examples show, is not always enough to uncover how animals find their way about. We may be confident that a mole whose eyes are covered with skin or the Mexican cave fish whose eyes completely degenerate by adulthood are not relying on vision. But barn owls have large eyes and excellent night vision and it needs experimental intervention to establish that they also rely on hearing. One approach is to mechanically obstruct or surgically remove the particular sense organ. Since blinded bats have no difficulty flying around, bats do not need vision to navigate. Similarly, plugging up their ears does destroy their ability to fly, so they must, somehow, rely on hearing.

The mystery of bat navigation

The Italian scientist Lazzaro Spallanzani (1729-1799) had become interested in how various nocturnal animals made their way about. The owls he tested refused to fly in total darkness, but bats had no difficulty. In order to prove that the bats were not relying on vision, he resorted to blinding them. He found they could still fly and avoid obstacles. When he released some blinded bats outdoors and several days later went to look for them where they roosted, the blinded bats were there as fit and healthy as the others. And when he cut open the stomach of one of them, it was as crammed with recently eaten insects as any other bat's. Attempts to interfere with the sense of touch or taste seemed similarly without effect, but when Spallanzani plugged their ears, the bats started colliding with obstacles as they flew. To make sure that this was not caused by mechanical injury to the ear canal, he built tiny hollow brass tubes that he could insert into the bats' ears but which had no effect until he filled them with wax.

Since the sounds that bats generate to produce echoes are mostly well outside the range of human hearing, Spallanzani had no notion how the animals relied on sound to navigate. It was not until the 1930s that the critical discovery was made. A Dutch zoologist Sven Dijkgraaf relied on bats' faintly audible clicks and noted that their frequency increased as the animal approached an obstacle. Indepently, the American psychologist Donald Griffin used electronic equipment capable of detecting sounds outside the range of human hearing. Subsequent experiments established that blocking the bat's mouth was as effective a way of disrupting navigation as blocking its ears and the mystery was solved.

If a mother goat is denied contact with her newly born kid in the first hour following birth, she will not be imprinted with its smell and will reject it

The sensitive period

It is so unsurprising to observe a flotilla of baby ducklings swimming after the mother duck or a young lamb running bleating to the mother ewe that we might suppose such behavior depends only on instinct. It took the Englishman Douglas Spalding and the Austrian ethologist Konrad Lorenz to establish the role of learning in young animals' identification of their mother. Chickens, ducklings and goslings will follow the first conspicuous moving object they see during the critical two or three days after they hatch, whether this be their mother, or a bright red ball. It is partly this "imprinting" that makes it possible to domesticate some animals: by coming into contact with humans during this critical or sensitive period, the young of many domestic species become partially imprinted on people and tolerate their presence much better than they otherwise would.

Making the right conclusions

*How does one decide that a particular part of the brain is involved in the storage of memories? It sounds simple. First, train an animal to perform a particular task, for example, a laboratory rat to press a lever whenever a buzzer is turned on in order to obtain a piece of food (**▸ page 27**). Second, interfere with some part of the brain. Finally, test for retention of this behavior some weeks later. If the lesioned rat performs less well than an unoperated animal, the scientist may think that he has discovered the locus of memory in the rat's brain. But he must be sure the operation is not interfering with the rat's ability to hear the buzzer or to execute the response; that it has not reduced its hunger or increased some other motivational state that interferes with feeding; or that it has not made the rat so hyperactive or so comatose that it is unable to press levers. It is thus astonishingly difficult to pinpoint the precise process or processes impaired by a particular injury. Lesions to the hippocampus, a brain structure lying deep inside the cerebral hemispheres, have been thought, at one time or another, to impair animals' ability to inhibit responses, initiate voluntary actions, distinguish between novel and familiar stimuli, learn conditional discriminations, hold information in short-term memory, transfer information to long-term memory, span temporal intervals, or construct a spatial map of their environment. A wondrous structure indeed.*

Maturation or learning?

The water beetle Dytiscus *has good vision but does not use vision to capture its prey. Tadpoles placed in a glass test tube in the beetle's tank were ignored, whereas tadpoles in an opaque muslin bag were immediately attacked. The beetle seems to rely on chemical stimuli to detect tadpoles. By contrast, among cuttlefish, which feed on shrimp, young inexperienced individuals that appear to need to learn to attack shrimp under natural conditions, attack a shrimp in a glass test tube as readily as one in the water. Simple behavioral maturation, rather than learning dependent on successful practice, is involved.*

◂ The Austrian ethologist Konrad Lorenz (b. 1903) may not have been the first person to discover imprinting, but pictures such as this, of young greylag geese imprinted on Lorenz and faithfully following him, did much to bring it to popular attention. In species such as ducks, geese, sheep and cattle, the newborn infant can run around within a few minutes of birth. Imprinting ensures that they form a strong attachment to the first conspicuous moving object they see, normally the mother, and this reduces the chances of their getting lost.

▾ Field studies often involve capturing animals to measure and weigh them to determine their age, monitor their growth, and check their state of health. Here a lion, sedated using a dart gun, is being weighed.

Using visual landmarks

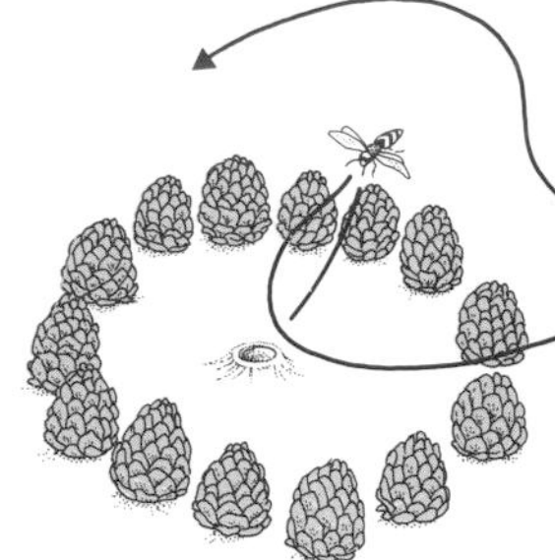

Digger wasp

Nest

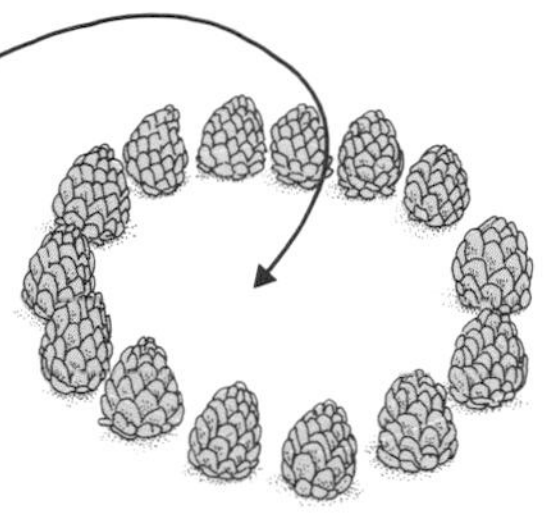

◀ *The female digger wasp provisions her nest with food for which she must go and search. How does she find her way back to the barely visible nest entrance? The Dutch ethologist Niko Tinbergen showed that she used neighboring visual landmarks. He placed a ring of pine cones round the nest while the wasp was laying her eggs, but then moved the entire ring a little way after she left the nest in search of food. The wasp returned from this foraging trip to the center of the pine-cone circle rather than to the actual nest entrance.*

▶ *Field studies of mute swans have revealed that a territory-owning male will furiously attack another, intruding, mute swan but will tolerate other swans. The intruder's orange beak elicits the attack.*

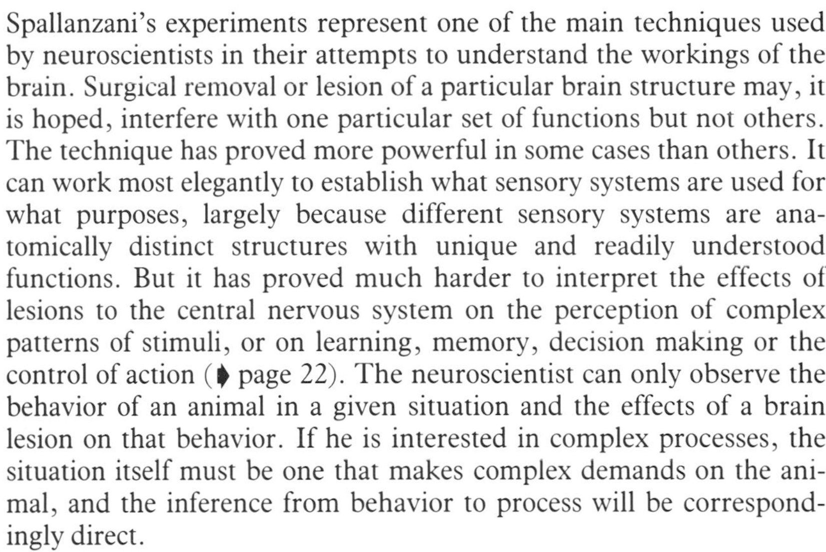

Spallanzani's experiments represent one of the main techniques used by neuroscientists in their attempts to understand the workings of the brain. Surgical removal or lesion of a particular brain structure may, it is hoped, interfere with one particular set of functions but not others. The technique has proved more powerful in some cases than others. It can work most elegantly to establish what sensory systems are used for what purposes, largely because different sensory systems are anatomically distinct structures with unique and readily understood functions. But it has proved much harder to interpret the effects of lesions to the central nervous system on the perception of complex patterns of stimuli, or on learning, memory, decision making or the control of action (➧ page 22). The neuroscientist can only observe the behavior of an animal in a given situation and the effects of a brain lesion on that behavior. If he is interested in complex processes, the situation itself must be one that makes complex demands on the animal, and the inference from behavior to process will be correspondingly direct.

Simple behavioral experiments can often tell one as much as this more drastic form of intervention and are a necessary prerequisite to understanding the functions of the brain. They can often be conducted in the field, as has been done in classic experiments by Tinbergen with digger wasps and by British ornithologist David Lack with European robins. However, they are more commonly done in the laboratory, where it is easier to establish controls (➧ page 21).

Role of experience

Often, as in the case of many young birds that appear to learn to fly by assiduous practice, we believe that a particular skill depends on learning when in fact it seems to develop with normal maturation (➧ page 79). In other instances, we miss the crucial role of early experience. It took the experiments of the Englishman Douglas Spalding and the Austrian Konrad Lorenz to establish the role of learning in young animals' identification of their mother. This is the phenomenon of imprinting.

Games Theory 1

Games animals play

Naturalists seek to describe what animals do and modern biologists to explain why they do it. But the question "why" has a very particular meaning in biology. It means: What evolutionary advantage is there in a particular form of behavior?

Recent decades have seen two major changes in the way in which biologists approach this issue. The first has to do with a significant shift in the level at which they perceive evolution to be operating. The second involves the increasing application, throughout biology, of mathematics. Putting the two together, we find biologists such as Professor John Maynard Smith, formerly of Sussex University, England, applying a branch of mathematics – games theory – to animal behavior, and coming up with novel explanations.

Evolution: a matter of level

Charles Darwin proposed that evolution proceeded by means of natural selection. That is, more organisms were born in each generation than could possibly survive. Some of these would inevitably be better adapted to the environment than others. Those that were better adapted, or fitter, would be the ones that survived and reproduced others of their own kind. Darwin pointed out that natural selection would help to shape behavior, just as it would lead to adaptive changes in body size, color, or any other character of an organism.

Most modern biologists still accept that natural selection as defined by Darwin is an important guiding force in evolution, which does indeed tend to produce organisms that are progressively fitter as the generations pass.

However, there have been some important modifications to the basic idea. One is that Darwin supposed that natural selection acted upon individuals. Thus, he suggested that in a litter of kittens, say, the biggest and the cleverest might survive, while the small less alert ones died.

*Nowadays, biologists know that the characters of organisms are each determined by genes (**▸ page 202**). Instead of regarding natural selection as a force that tends to weed out the less fit individuals, they urge us to look upon it as a force that tends to eliminate genes that produce less desirable characteristics. Of course the slow-footed or slow-witted kittens are more likely to die than the swift and bright. However, the important point is not that some individual kittens survive and some die, but that the genes that produce slow-footedness or slow-wittedness tend to be eliminated, while genes that produce swift and bright kittens become more common because the animals that possess them live and reproduce.*

*At first sight, it may seem that the question of whether natural selection really operates upon individuals or upon genes is academic. After all, in practice some animals live and some die, even if biologists choose to think that the genes they contain are more important than the animals themselves. However, this new way of thinking about evolution does affect the way biologists interpret animal behavior (**▸ page 16**).*

*First, however, we should look at the other revolutionary changes in biology (**▸ page 15**).*

◄ *Among Northern elephant seals, during the breeding season the huge bulls haul themselves out onto the beaches, where they remain without food for 4 months. There they battle with one another, at great energetic cost, and with occasional severe wounding, until the winners have established dominion over certain stretches of beach. The bulls mate only with females landing on their stretch.*

► *An adult lion plays with his offspring. When a group of males take over a pride they are liable to kill cubs sired by ousted males because the young bear the competitors' genes. The new pride males soon produce their own offspring, to which they are extremely tolerant.*

The rise of mathematics

One important way in which modern biology differs from traditional natural history is that biology is quantitative: naturalists observe, but biologists observe and measure. Over the past 100 years, however, biologists have begun to do a great deal more than simply measure what they see. They have also begun to apply many different kinds of mathematical theories to their observations, and thus have gained quite new insights into a whole range of biological problems. In fact, biology has begun to be underpinned by mathematical theory in precisely the same way as has always been the case in physics.

We can now start to put these two ideas together. In the early 1970s Professor Maynard Smith and his colleagues began to apply the particular branch of mathematics known as games theory to the behavior of animals. Games theory, as the term implies, enables us to predict what lines of action are liable to produce victory, and which are liable to end in defeat.

They began with the observation that animals adopt various behavioral strategies; they show innate patterns of behavior that are programmed by particular genes. The first question to ask, then, was how can we decide which strategy is better than another, and therefore liable to be favored by natural selection?

◄ Gannets breed in vast colonies on islands off the coast of South Africa. Here food supplies are generally abundant, reliable and easily exploited so that adults are able to feed broods of up to four chicks. However, often only two eggs are laid. Within a few days of hatching one chick always kills the other, regardless of the availability of food. This behavioral strategy has evolved to ensure that one stronger chick, rather than two weaker ones, has the opportunity to become independent. The laying of the second egg also ensures that a chick will always survive should the first offspring to hatch die.

▲ In social animals such as olive baboons, each individual must discover its place in the interlocking web of relationships. This is achieved progressively during a variety of interactions involving threats, displays, contests and grooming. For example, young baboons play rough-and-tumble games during which they learn not only combat skills useful for adult life, but also about the existence of a dominance hierarchy within the colony, with some playmates behaving as bullies others as cowards. Males often participate in aggressive encounters, fighting over females or establishing their position.

Games Theory 2

Fighting strategies: doves and hawks

Games theory provides a means of measuring the likely success of a particular strategy. Suppose, for example, that an animal always fought others of its own kind, or at least of its own sex, whenever it met them. That would be a behavioral strategy (♦ page 44). Would it be a good strategy? If the animal wins all its fights then it would be good: the winner would gain more food and more mates. But not every animal can win all its fights; in each contest there is a winner and a loser. Even if it wins a fight, an animal loses time in fighting, and while it is doing battle it cannot look for food. Fighting is dangerous, too, because the rival fights as well. So each combatant risks serious injury. In addition, whenever a fighting animal defeats one rival, it also benefits another rival by eliminating one of the animals that the other rival might otherwise have had to fight.

Games theory makes sense of all this musing by giving each possible outcome a points value: winning a fight might gain 10 points, serious injury might lose 20, and wasting time, winning or losing, might lose 3. In general, such a points system would confirm what common sense suggests; that animals that simply fought all they met would be most unlikely to succeed. In genetic terms, the gene that promotes indiscriminate aggression would tend to be eliminated.

Suppose now that we had a population in which no animal fought any other: that each adopted what we might call a "dove" strategy. Such animals might begin to suffer from overcrowding, and so might start going hungry, but at least they would have nothing to fear from each other. Games theory would suggest that "doves" are better off than compulsive fighters, the "hawks". On this basis, then, we might suppose that animals would evolve peaceful behavior; the genes that produce peacefulness would prevail over the genes that promote fighting.

However, this is clearly not what happens in nature. In most animal societies, animals do not fight with each other all the time, but they certainly quarrel over territory, food, and mates (♦ page 68). But games theory, combined with the notion that natural selection operates in a way that tends to eliminate some genes and promote others, provides the explanation. For suppose that in a population of doves an animal was born with a mutant hawk gene. That animal could gain all the advantages of fighting, but suffer none of the drawbacks – because the other animals in the population would not strike back. The hawk would grow up to have many offspring, who would inherit the fighting gene and therefore in turn do well.

Such success contains the seeds of its own downfall. After a short time, the animals with the fighting gene would become more common than the peaceful animals. So when a hawk picked a fight it would be more likely to find itself up against another natural fighter than against a dove. So the hawks would start to suffer the disadvantages of fighting – time wasted, possible injury, and perhaps untimely death. Then the fighting animals would start to become less common and the peaceful ones would again begin to prevail.

The Hawk-Dove game

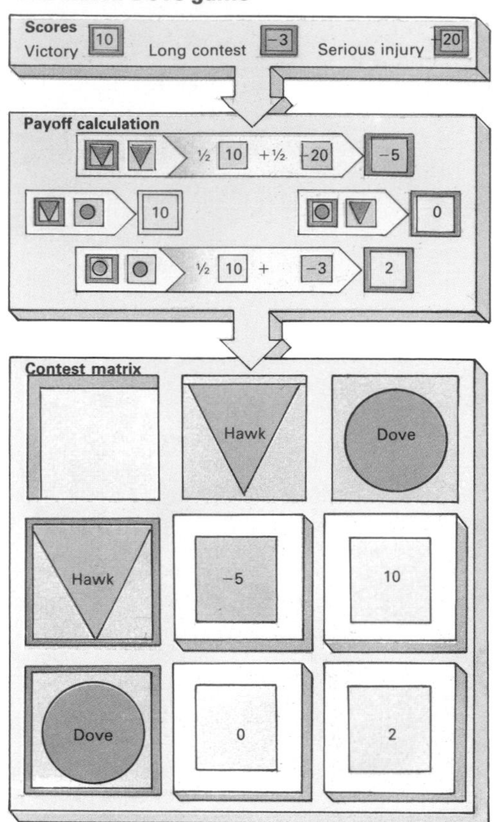

◄ ***Following the game theory, it is possible to compare contests and fighting strategy scores in order to determine the evolutionary stable behavior for animals in conflicts with their own kind. In a simple "hawk" and "dove" situation, an individual who has a high average pay-off will leave many genes behind in the gene pool (♦ page 202). By drawing out a contest matrix, left, it is possible to see that with a population of all doves, the average pay-off for a mutant hawk (+10) is greater than for a dove (+2). So hawk genes will rapidly spread through the population. Now each hawk may encounter in each contest either a dove or a hawk. If the whole population comprised hawks, a hawk's average payoff will now be considerably less (−5). Here, the ESS is to play hawk in 8 of every 13 contests and dove in the other 5. This model predicts that in real animal contests either different animals adopt hawk or dove tactics or individuals within a population vary their fighting strategy.***

The Hawk-Dove-Bourgeois game

► ***Contestants differ not only in fighting strategy, such as hawk or dove, but also in size, strength and need for the resource they are fighting for. There are also differences between individuals that do not always affect the outcome or pay-offs of contests. For example, in a contest an individual may adopt a hawk tactic if it is the owner of the resource in question and a dove tactic if it is not. Such a strategy is called a bourgeois tactic. The pay-offs for contests involving hawks and doves are unchanged but new pay-offs arise for contests involving bourgeois individuals (see table and matrix right). Each contest is between an owner and an interloper; each individual can be in either role; and each individual knows which role it is playing. The pure bourgeois strategy turns out to be the evolutionary stable behavior. Examples of bourgeois strategy have been found in hamadryas baboon, fiddler crab and speckled wood butterfly populations (♦ page 17).***

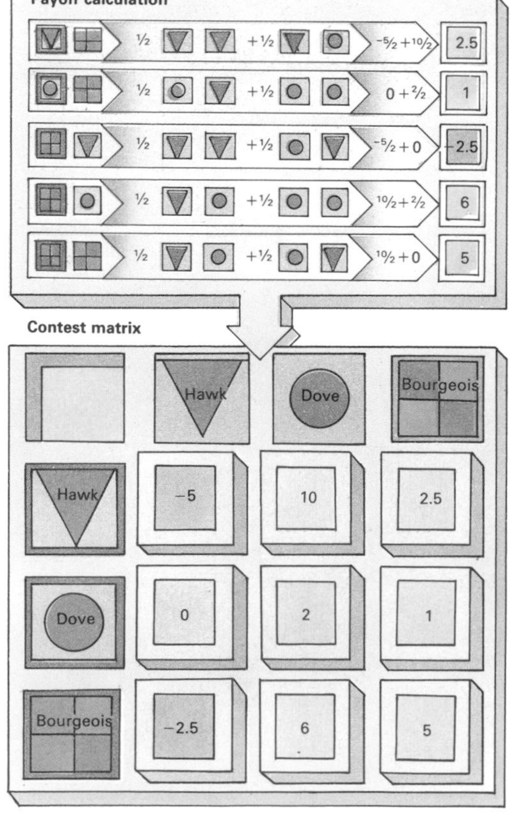

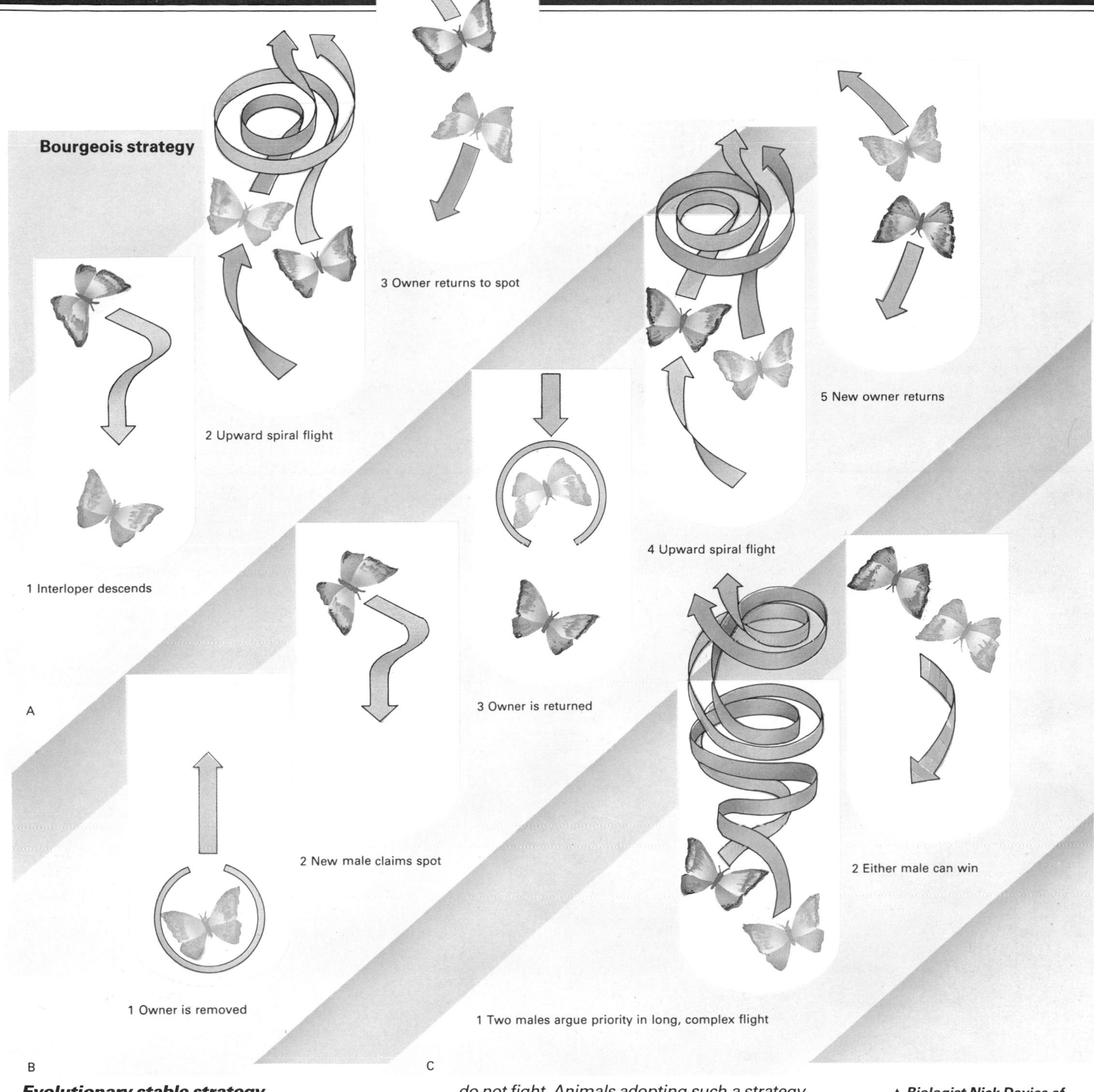
Bourgeois strategy
1 Interloper descends
2 Upward spiral flight
3 Owner returns to spot
4 Upward spiral flight
5 New owner returns
A
1 Owner is removed
2 New male claims spot
3 Owner is returned
B
1 Two males argue priority in long, complex flight
2 Either male can win
C

Evolutionary stable strategy

Thoughts on hawk and dove strategies led Professor Smith to introduce the notion of evolutionarily stable strategy, or ESS. An ESS is a mode of behavior that is the best possible such that animals that practiced that behavior could not be routed by some mutant practicing some other kind of behavior. Genes that produce ESS's are the ones that in the end must prevail. Simply to fight at each encounter is not an ESS because in the end hawks would lose out to peaceful animals that did not waste time or risk injury in fighting. But pure dove behavior is not an ESS either because in the short term the peaceful lose out to mutant fighters.

What does emerge as an ESS is a compromise form of behavior – "retaliator". Retaliators begin an encounter with another animal by threatening to fight. However, only if the rival attacks them do they fight back. If it does not attack them, then they do not fight. Animals adopting such a strategy would always threaten, but would generally in the end refrain from fighting. They would not lose out to hawks because they would fight back. But they would not be ousted by doves either, because if another animal treated them dovishly they would respond dovishly.

This idea of how animals ought to behave is only a mathematical model. But in practice it seems to apply to most animal societies. Most social animals do have codes of threatening behavior, but few fight with their own kind unless they really have to. The great Austrian ethologist Konrad Lorenz observed this fact, and ascribed it to a kind of chivalry among animals; he thought that individuals do not hurt each other more than they really have to. Games theory suggests, rather, that if animals refrain from fighting it is primarily in order to avoid being hurt themselves.

▲ Biologist Nick Davies of Oxford University, England, has shown a bourgeois behavior among speckled wood butterflies. Males follow this strategy in territorial disputes over sunlit spots on the forest floor. Contests involve an upward spiral flight, after which the original owner returns (A, 1-3). The flight seems to inform the interloper that the spot is occupied. Davies removed the owner and introduced another male (B, 1-2), who then acted as the new owner (3-5). If two males both consider themselves the owner, a long spiral flight follows (C, 1-2).

See also
Animals as Machines 19-24
The Limits of Animals 25-34
Sex and Species 35-46
Groups and Troops 47-62
Territory 63-70
Animal Migration 71-80

Imprinting is not entirely one way. Not only does the infant animal need to recognize its parents but the adults, if they are to pass on their genes to future generations, must both breed successfully and rear their young. In colonial nesting birds and other group-living animals, there is a premium on recognizing and caring for one's own offspring rather than one's genetically unrelated neighbor's. In many species, experimental intervention suggests that such recognition depends on contact during a very brief critical period after hatching or birth.

In order to understand whether a particular pattern of behavior depends on learning, and if so what the animal has learned and how, there is no alternative to careful experimental analysis. In some cases, an apparent improvement with practice is not due to learning at all but to maturation of the nervous system and musculature. In others, such as a young rat's developing preference for some kinds of food over others, parental influence is paramount, but this does not necessarily imply that the young rat is learning anything. Even where we are confident that behavior is modified by learning, precisely what is learned can be established only by varying the experience given to different animals (⧫ page 29). Experimental psychologists have been the major investigators of the ways in which animals learn, and their research forms the basis of the science of ethology.

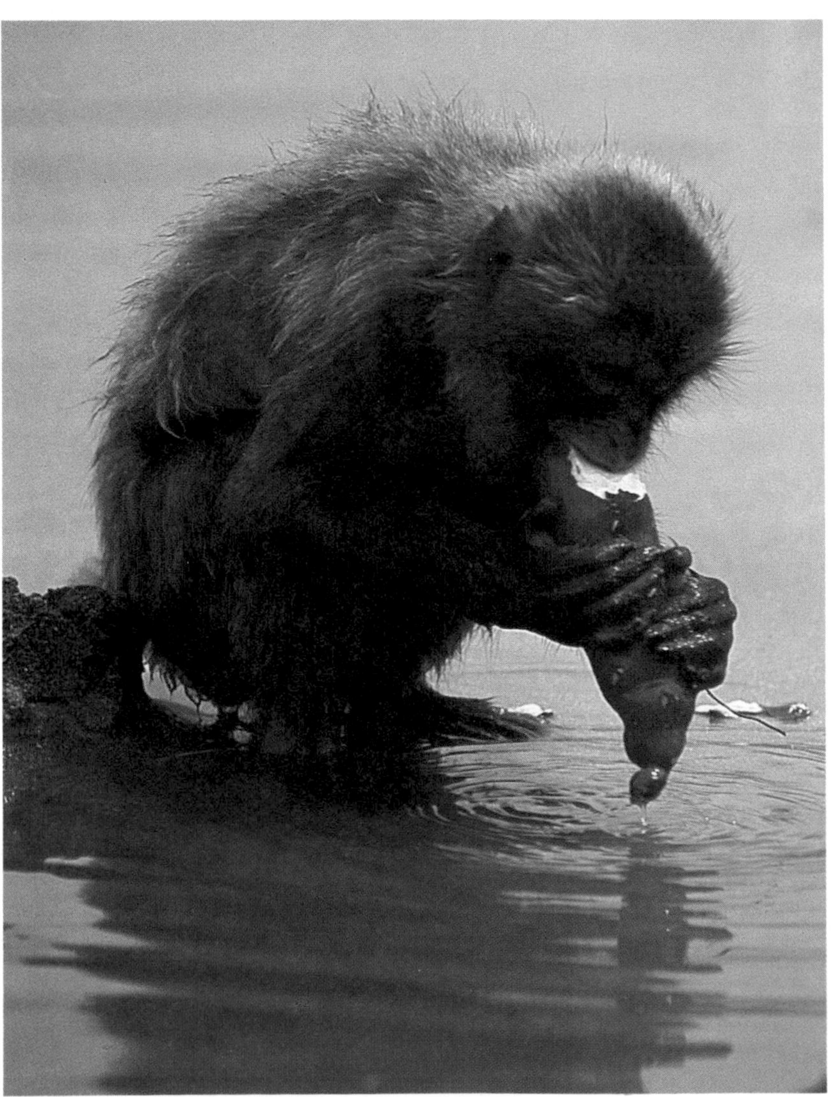

Social transmission of food preferences in rats

One reason why it is so difficult to control wild rats by poison is that however effective the initial onslaught, a few rats will almost certainly survive by refusing to eat the poisoned bait, and in subsequent generations most of the rats will never even touch the bait.

What has happened here that allows some rats to transmit their apparent immunity to their offspring? One possibility is that rats differ in how attractive they find the bait, and if these differences are genetically determined, the survivors' descendants will also be protected by an innate dislike of the bait. In fact, experimental analysis suggests that the effect is socially rather than genetically transmitted. But not all social transmission depends on individual learning.

Young rats, when they first start feeding for themselves, will invariably feed in the company of their mother or another adult. If the mother never approaches the bait, then neither will her offspring. When later they start feeding alone without an adult, a second mechanism will keep them from the bait: young rats show a marked preference for eating food in surroundings that smell of other rats. If no adult has approached the bait, the young will not either. Finally, now that they have started eating other food, the young will continue eating it since like most animals (and people) rats prefer familiar to novel foods. Thus social transmission operates by taking advantage of the young rat's preference for feeding in the company of adults or, failing that, where other rats have been in the past, and the preference is maintained by a preference for familiar over novel food.

So far there is no need to suppose that young rats learn anything directly from their mother. But there is one final factor. If given a choice between two equally palatable foods, each equally associated with the presence of an adult, young rats will choose the one which their mother was eating before they were weaned, even though, in the artificial conditions of the laboratory experiment, they have never seen their mother eating or directly seen, smelled or tasted the food before. How is this done? It turns out that the food that she eats affects the taste of the mother's milk. The infant rat detects this taste and thus regards this food as familiar even though it has never eaten it itself. Because it prefers familiar to novel food, it will prefer its mother's diet to any alternative.

In the real world, therefore, rats are nicely protected from our ceaseless attempts to exterminate them. However, these findings have made possible a more efficient method of poison baiting. First, small amounts of plain food are put in an area infested with rats. These are eventually eaten. Poison is then added to the now familiar food and is taken by the rats in lethal quantities.

◀ A young Japanese macaque monkey carefully washing a sweet potato in the sea to remove the sand (and perhaps add salt) before eating it. The technique was "invented" by one juvenile female, was soon copied, and gradually spread through the entire troop – a nice example of social learning. The monkeys were given food on the beach by scientists investigating their behavior away from the forest.

Animals as Machines

2

Challenging Descartes' theories...The work of Pavlov...The stimulus-response theory...Rats in mazes...Behaviorism and Cognitive Revolution... PERSPECTIVE...The mechanistic view of Man...On six legs...Responses to light...Conditioning experiments... Reinforcement shapes behavior

▲ The French philosopher and scientist René Descartes (1596-1650), who applied the mechanistic philosophy of the new science of physics to the problems of biology. In his Discourse on Method, published in 1637, Descartes argued that all animal action could be explained in physical terms, and that there were no grounds for ascribing an immaterial mind or immortal soul to any animal other than man.

For medieval scholastic philosophers, the critical difference between us and other animals was that we have an immortal soul and they, probably, do not. But even if they have no soul, animals could still have purposes, intentions and goals (➧ page 31). The central characteristic of the 17th-century scientific revolution, however, was the steady attempt to provide mechanical explanations for natural phenomena. If the inanimate world was a great machine, set into motion by God the Creator but thereafter running like clockwork, why should not the animate world be similarly described? Our possession of a reasoning soul might exclude us from this process, but if animals have no soul and cannot reason, then they too are mere machines.

The most explicit statement of the view that animals are automata was by the French philosopher Descartes. Two of his arguments were by analogy. If we can manufacture clockwork machines that can do amazing, lifelike things, how much more impressive must be the machinery created by God. Then, Descartes noted, many of our own actions are performed without conscious thought or control: we blink when a friend aims a blow at us even though we know he will not hit us, and we reach out for a support to save ourselves falling without stopping to think.

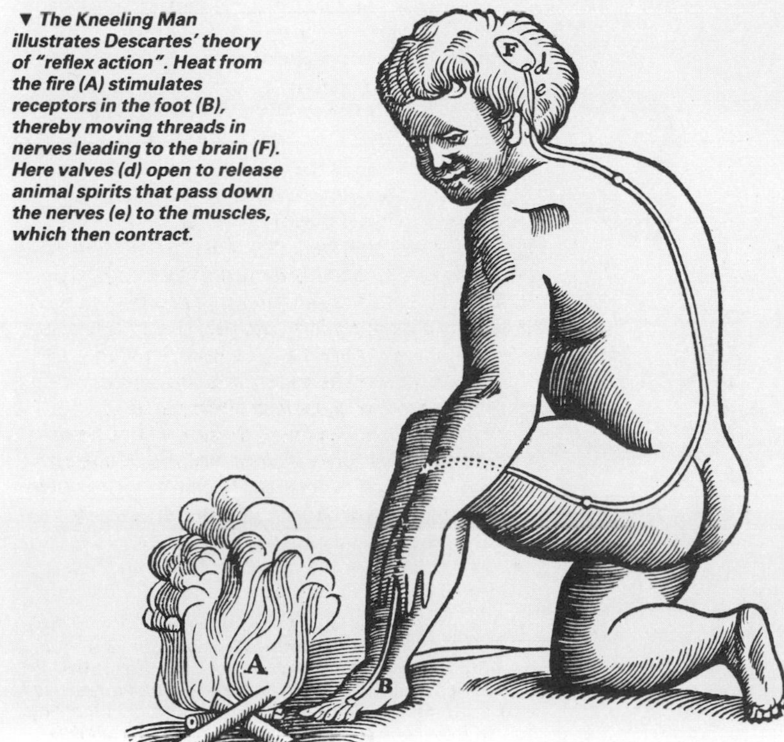

▼ The Kneeling Man illustrates Descartes' theory of "reflex action". Heat from the fire (A) stimulates receptors in the foot (B), thereby moving threads in nerves leading to the brain (F). Here valves (d) open to release animal spirits that pass down the nerves (e) to the muscles, which then contract.

Are we different from other animals?

René Descartes tried to show that the body was simply a complicated piece of machinery: the heart was a pump, the veins and arteries hollow tubes, the nerves also tubes, connecting brain to limbs.

Descartes' arguments did not persuade everyone but for some they were even more persuasive than he would have wished. If purely mechanical principles explain not only such bodily functions as circulation of the blood, respiration or digestion, but also an animal's behavior, might they not also explain human behavior? And if not, where are we to draw the line? Between humans and all other animals? But is there really a greater divide between us and apes than there is between apes and worms or snails? It did not need Darwin's theory of evolution to suggest to many writers that some animals are more like humans than others, and therefore that there may be no hard and fast distinction between us and the rest of the animal kingdom. Either we must allow that some animals can think and reason in a way that no piece of machinery can, or we must extend the process of mechanization to Man, as implied by the title, L'Homme Machine, of the book by the 19th-century French physiologist Julien de la Mettrie. The conflict between these two views is still alive today.

A dog's reflex machinery is sufficiently sophisticated that if the animal is prevented from scratching an itch with one leg, it will endeavor to do so with another

How a cockroach walks

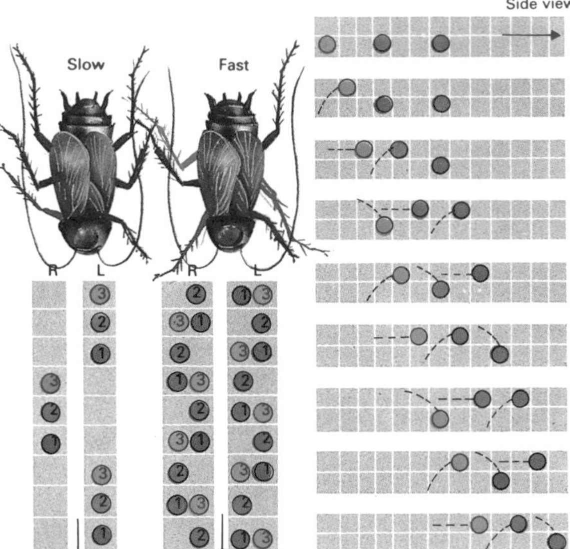

▲ ***Insects have six legs and most, including cockroaches, use all six to walk. The basic pattern is the legs on each side are moved in sequence from back to front, and the right and left legs alternate in phase. In slow locomotion, all three legs on one side are moved and then all three on the other. As the cockroach speeds up, it shortens the interval between finishing a cycle on one side and starting it again on that side. Ultimately, the cycle starts again before the previous cycle has finished, so that R_3 is moving at the same time as R_1 (and L_2).***

How animals walk

The basic principle in walking is not to fall over. Four-footed animals ensure this by following a simple rule: move one leg when your center of gravity lies over the triangle formed by the other three. The regular progression is: move right forefoot, left hindfoot, left forefoot, right hindfoot and so on. This is slow and, in practice, the animals start moving one foot before the preceding one has reached the ground. The order in which they lift their feet remains the same. They simply shorten the interval between successive movements. With a further increase in speed, the two diagonal feet may be moved simultaneously.

Clearly, some relatively simple and fixed actions, and therefore a basic neural machinery, can encompass what seems to be a bewildering variety of gaits. Animals can change speed of movement by varying only one appropriate parameter.

The spinal cord acting alone

The British physiologist Charles Sherrington (1857-1952) showed that a "spinal" dog (one whose spinal cord had been surgically severed from its brain) would show exactly the same reflex behavior as a normal dog. The spinal cord alone, then, contains machinery capable of coordinating an extraordinarily complex sequence of muscular contractions and extensions, and of adapting the animal's behavior to changing circumstances. Although Sherrington himself was a Cartesian dualist, believing that not all mental life can be explained in mechanical terms, his account of the integrative action of the spinal cord provided critical support for the mechanical philosophy.

The discovery of reflexes

The success of the mechanistic view depended on substantiating Descartes' abstract idea that behavior is a product of the machinery of the body. It took nearly 200 years before physiologists made much significant progress (➧ page 23).

Descartes had argued that the effects of a sensory stimulus were transmitted to the brain where they were "reflected" to a motor nerve, whose activity caused a muscle to expand. Here is the familiar notion of the reflex. However, it was not until the early 19th century that the electrical basis of nervous action gradually became clear and that two surgeons, Charles Bell in London and Francois Magendie in Paris, established that there were distinct sensory and motor nerves in the spinal cord. By 1850, the German physiologist Hermann von Helmholtz demonstrated that nerve impulses travel from sense organ to central nervous system to muscle at the speed of 25-45 m/sec, and by the late 19th century the concept of the reflex arc was so widely accepted that it was hailed by some, and denounced by others, as a model of all behavior.

Yet scientists did not understand the full complexity of the machinery actually underlying even rather simple reflexes. When a dog scratches an itch with its hind paw, rhythmic contraction of certain sets of muscles in that limb must be combined with extension of others, while lifting the limb off the ground requires a whole set of postural adjustments in the remaining limbs. Unless we believe that the spinal cord has a soul, we must accept that neural machinery is capable of generating integrated and apparently purposeful behavior.

▲ ***A golden-mantled squirrel scratches its neck in a reflex action.***

Directed movements

However elaborate the model of the reflex, it hardly provides an adequate account of all behavior. Animals respond to a specific stimulus in quite different ways depending upon their current mood or past experience.

Several developments were necessary to extend the mechanical analysis. One was the work of the German physiologist Jacques Loeb. Even the simplest animals do not just move at random, they move in certain directions rather than others. For instance, some (moths) move towards the light while others (cockroaches, woodlice) move away from it. Further analysis suggested a distinction between responses that are triggered or elicited by a releasing stimulus as in the classical model of the reflex, and those which are continuously guided by a stimulus. The former are ballistic responses which once set off by the eliciting stimulus runs their course regardless. In the latter, the response serves to modify the stimulus that guides it; there is a servo-mechanism incorporating feedback.

Intelligence and learning

The most serious challenge to reflex theory, however, came from the growing acceptance that animals will profit from experience and adapt their behavior to changing circumstances (➧ page 26). Descartes himself had argued that one of the infallible proofs that we are not mere machines is that we can act appropriately and rationally when confronted by a novel situation. But is it impossible to provide a mechanistic explanation of any form of learning?

◀ Once a chameleon has initiated the response of shooting out its tongue at a prey dead ahead, it will complete the action even if it is suddenly prevented from seeing the target.

▼ An animal's orientation to a source of light may be based on a set of sensory receptors (as in a fly), simultaneous comparison between two receptors (flatworm), or by making successive comparisons by moving its body from side to side (maggot). Similar receptors enable animals to orientate to gradients of stimulation.

Goal-directed actions

Many long behavioral sequences consist of a mixture of the ballistic and feedback responses. Commonly, early guided movements are under feedback control and bring the animal into the right position where the appropriate stimulus then elicits a ballistic response. A chameleon that detects out of the corner of its eye a potential prey will turn eyes and head towards it, track it until it is centrally fixated, then shoot out its tongue at the beetle or fly. The earlier sequence is under constant feedback control aimed at reducing the discrepancy between the current position of the prey and the desired position (dead ahead). The final movement is ballistic: if the beetle moves after the chameleon has started to shoot out its tongue it will escape.

Tropisms and taxes

Moths and flies which burn themselves on the flame of a candle or, more commonly these days, get trapped inside an illuminated glass lampshade provide a familiar example of directed movement. They are flying towards the light. But how are such movements directed?

There are several different ways an animal can come to approach (or move directly away from) a source of light. The simplest is illustrated by the behavior of various species of maggot, which crawl away from the light. The maggot waves its head from side to side as it crawls along and turns away from the direction in which the light was brighter. All it is doing is comparing the level of illumination received when its head is pointing in various directions and turning appropriately. This is demonstrated by placing the maggot on a dimly illuminated platform and briefly turning on an overhead light whenever the maggot's head points to the left. The maggot will crawl round in a circle to the right indefinitely.

Most animals are "bilaterally symmetrical": they have one eye on either side of their head, two arms, legs or whatever, and so on. It was this fact that inspired Loeb's theory of tropisms. Jacques Loeb (1859-1924), an American physiologist, supposed that an animal such as a moth compared the amount of light falling onto the two eyes. The more strongly illuminated eye would produce more neural activity and this would be translated into more motor activity on the opposite side of the body. The animal would turn until both eyes were equally stimulated. Although the neural machinery is certainly less simple than Loeb supposed, some such process of comparison between the two eyes certainly occurs in many animals. If a pillbug is placed some distance from two sources of light, it will initially crawl in a straight line between them rather than directly towards one or the other. Only as the bug approaches a point directly between the two lights will it suddenly change direction and approach one of them. At first, the path that maintains equal illumination on the two eyes is the bisector of the two lights. Only when the bug is directly between the lights will a chance swing towards one of them eliminate the influence of the other and ensure the animal approaches the former.

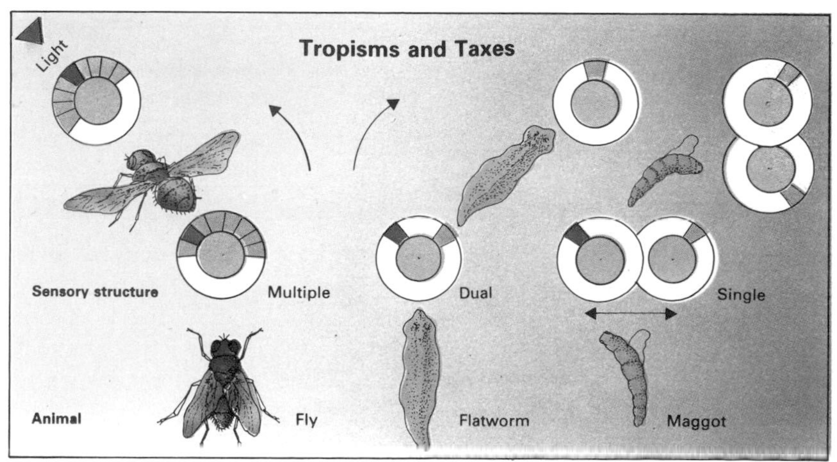

Using stimuli of a tone of 0.7kHz sounded for 8 secs and a squirt in the mouth of lemon juice, humans have shown the same conditional responses to food as Pavlov's dogs

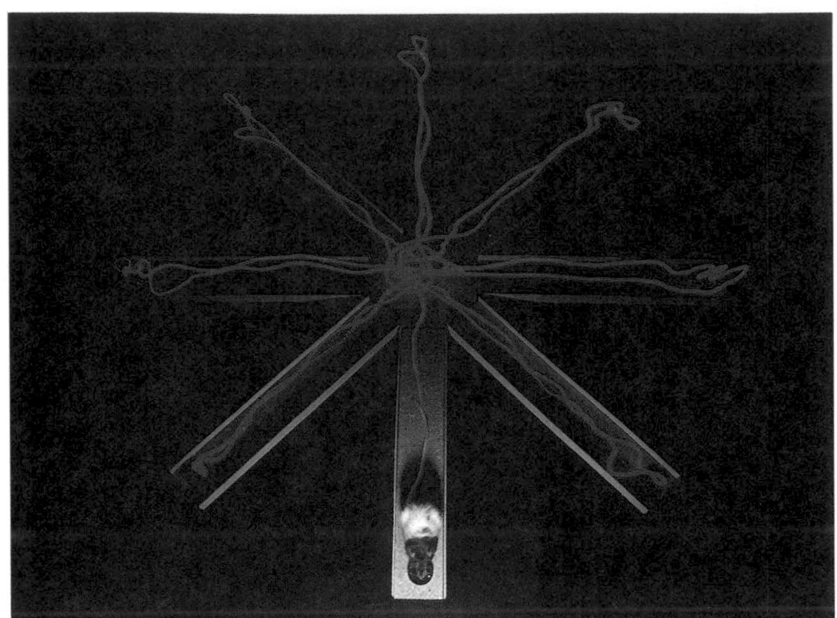

The Law of Effect
Behaviorist Clark Hull set out to show Pavlov's S-R theory was too simplistic. A rat would learn to turn right, rather than left, at the choice-point of a maze because a right turn was followed by food in the goal box and this strengthened the connection between the stimuli at the choice-point and the response of turning right. If both paths led to food but the right-hand path was shorter than the left, the rat would still learn to turn right rather than left because the shorter delay before the receipt of food would strengthen this connection more effectively.

◄ A rat placed in a maze with food at the end of each arm has to learn to collect all the food in the most efficient way. The best strategy for this is to go just once into each of the eight arms, and rats prove very adept at doing this. They must therefore remember which of the arms they have already visited.

▼ How do animals avoid predators? One strategy is to be inconspicuous, relying on camouflage to escape detection. A quite different strategy is to be extremely conspicuous, but also repellent. The striking colors of the fire salamander advertises the fact that it is poisonous. It secretes a venom from pores just behind the eyes.

Pavlov and conditioned reflexes

In his early animal behavior studies, Pavlov had noticed that gastric secretion might be triggered not just by the presence of food, but by the sight of the food bowl or even of the attendant who normally provided the food. His experiments (➧ page 23) consisted in providing some specified, arbitrary stimulus in place of the attendant; when this initially neutral stimulus regularly and immediately preceded the delivery of food, the dog would start salivating as soon as it was turned on. A new, conditioned reflex had been established.

Pavlov's own physiological theory of conditioning, which talked of waves of excitation and inhibition spreading across the cerebral cortex, never found much favor outside Russia. Contrary to his beliefs, indeed, conditioning proceeds perfectly normally in animals with no cortex at all, and the cellular changes underlying conditioning have been most successfully studied in invertebrates. However, Pavlov's greatest influence was on psychologists, who gratefully accepted both his demonstration that it was possible to study the mental life of animals objectively and his assumption that the conditioned reflex was the basic building block from which this mental life was constructed. The conditioned reflex was a new connection between stimulus and response established as a consequence of their association, and stimulus-response, or S-R, theory asserted that all learning was a matter of establishing new S-R connections. Reflex theory could not make good its claim to explain all behavior, however complex.

American behaviorism

Although American S-R theorists such as J. B. Watson and Clark Hull adopted Pavlov's terminology of conditioning, extinction and reinforcement, they did not use his experimental paradigm. They preferred laboratory rats in mazes or Skinner boxes (➧ page 27), where the animal had to perform some designated response to obtain food. The analysis of the experiments suggested that new S-R connections were strengthened not merely, as in Pavlov's experiments, because the stimulus and response repeatedly occurred together, but because the response was followed by a reinforcing consequence.

◄ *A predator attacking this death's head hawkmoth larva will reject it because of its foul taste. It will then associate the prey's markings with repellent consequences and will avoid it in future. This is aposematism. Other, non-repulsive, prey may escape predation because they resemble the larva.*

▲ *The Russian physiologist, Ivan Pavlov, was awarded the Nobel Prize for his work on digestion at the age of 55. But it is for his work on conditioned reflexes in dogs, started only a few years earlier and continuing until he died at the age of 80, that he is best known. The medal depicts the apparatus used in his experiments.*

Pavlovian conditioning

Pavlov's experiments involved first training dogs to stand still on a platform where they were loosely restrained. (Although the restraints were sufficient to prevent the dog escaping, a dog that spent the entire experiment struggling would show little sign of conditioning). A minor operation diverted the salivary duct through a small hole in the dog's cheek so that drops of saliva could now be counted. The hungry dog's salivation to dry meat powder squirted into its mouth was recorded. Because food inevitably elicited a strong response, it was termed an unconditional stimulus (US) which elicited an unconditional response (UR) of salivation. A neutral stimulus (flashing light or ticking metronome) was now presented and elicited no salivary response, but rather an investigatory or orienting reflex. But if the flashing light was repeatedly followed, a few seconds later, by the delivery of food, the dog would soon start salivating to the light before the arrival of food. The light was termed a conditional stimulus (CS) because its ability to elicit the conditional response (CR) of salivation was dependent (conditional) on this specific temporal pairing of light and food.

Further evidence of this was provided by the phenomenon of extinction. If the light was presented on its own for several trials without food, the salivary CR to the light would gradually extinguish, to be restored only by re-pairing the light with food. Other stimuli would elicit a salivary CR to the extent that they were physically similar to the original CS. A regular gradient of generalization could be recorded by conditioning to a metronome ticking 100 times per minute and testing the animal to metronomes set at 80, 60 or 40 times per minute.

Watching a dog in a conditioning experiment, one could assume the animal was anticipating the imminent delivery of food or failing to see the difference between the CS paired with food and another, similar, stimulus. Pavlov forbade such speculation in his laboratory.

See also
Looking at Animals 9-18
The Limits of Animals 25-34
Sex and Species 35-46
An Ape Called Man 129-140

Hull's influence on American psychology was at its peak in the early 1950s, but by 1960 it was declining rapidly. His mantle passed to B. F. Skinner, whose radical behaviorism insisted that the behavior of any animal, including humans, was to be understood by looking at the contingencies of reinforcement that had shaped it.

Skinner showed the extremes to which rigorous antimentalism could be taken, but the extravagance of his claims was his undoing. Clearly there are fundamental differences between a student learning how to solve differential equations and a rat in a Skinner box learning to press a lever for occasional pellets of food. Starting by rejecting S-R theory's pretensions to account for anything interesting about human intelligence, thinking or reasoning, this cognitive revolution soon argued that the theory was equally inadequate to account for the behavior of other animals. Simple conditioning might not be the only way by which animals could adapt their behavior to changing circumstances and conditioning itself might be a more complicated process than S-R theory had imagined.

Conditioning reinforcement in humans
Psychologist B. F. Skinner (b. 1904) used teaching machines that took students through a series of carefully graded steps to produce the desired behavior – for example, giving the correct answers to differential equations – and, in a theoretical analysis of language appropriately entitled Verbal Behavior, *argued that both the form and content of a speaker's utterance had been shaped by past contingencies of reinforcement provided by the verbal community.*

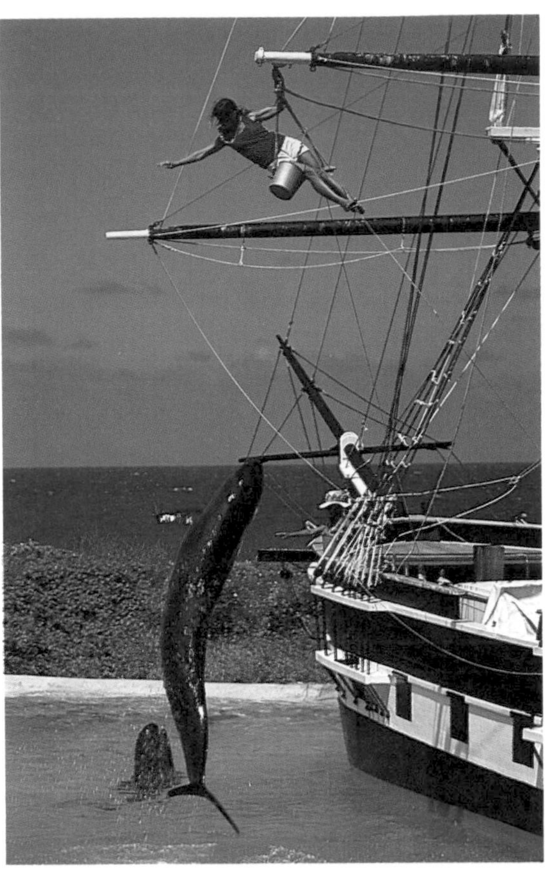

▶ Animals can be taught elaborate circus tricks by standard conditioning – carefully reinforcing successive approximations to the desired end product.

▼ Associative learning, or conditioning, is not confined to vertebrates. Some invertebrates, such as slugs, show evidence of this behavior.

The Limits of Animals

3

How animals solve problems...Using landmarks to locate food resources and shelter...Discrimination tests...Can animals perceive abstract relationships?... Are animals conscious?...Animals and language... PERSPECTIVE...Pigeons and rats in Skinner boxes... Finding solutions to novel problems...Indications of our superior intelligence...Chimps using sign language

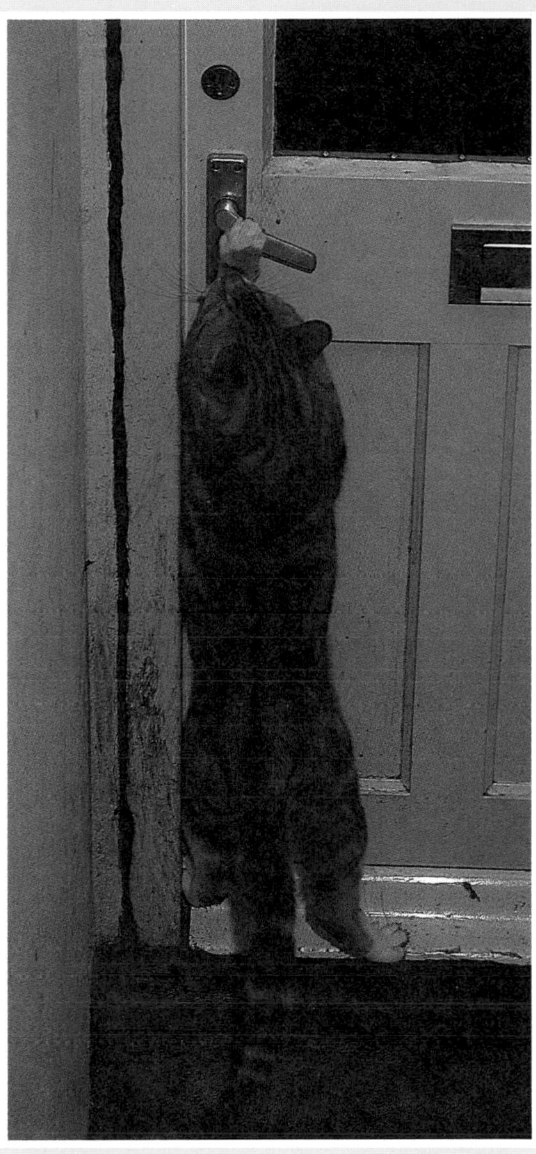

Descartes' view that animals are automata (◊ page 19) did not pass unchallenged. It was in the 18th century, following observation of the behavior of monkeys and apes, and with anatomical dissection of occasional specimens, that many scientists first believed that, even if they were not our close relatives, these animals certainly shared some human characteristics. Almost a century later, in 1863, T. H. Huxley's *Evidence as to Man's Place in Nature* argued on the basis of anatomy that Man and other primates must be descended from common ancestors. And in 1871, Darwin's *Descent of Man* laid out the argument for mental continuity between Man and other animals. Unlike Huxley, Darwin had virtually no serious scientific evidence to support his position. He relied on casual observation and anecdotes of the supposed amazing feats of animals. And although his young disciple, George Romanes, conscientiously sifted through the stories sent to him, accepting only those vouched for by reputable observers, or by several different observers, Romanes' book *Animal Intelligence* remained little more than a string of anecdotes.

It was against this background that psychologist Edward Thorndike (1874-1949), in 1898, set out to establish by careful observation and experiment just how animals did solve problems and whether there was any truth to the extravagant claims made on their behalf. He concluded that there was none. According to Thorndike, animals learned only by trial and error; more formally by the gradual strengthening of appropriate, and weakening of inappropriate, stimulus-response, or S-R, connections (◊ page 22).

▲ If they have to, most pet cats and dogs will learn to open doors. The proud owner is only too eager to be impressed by this evidence of his pet's ingenuity, but the learning involved may not be very elaborate. Thorndike's cats learned to operate catches and levers to escape from his puzzle boxes by little more than random trial and error (or simple conditioning) – errors were followed by failure.

Is a chimpanzee cleverer than a rat?

Thorndike argued that all the animals he tested, cats, dogs, monkeys or chickens, solved the problems he gave them by a process of blind trial and error; they could neither reason nor solve problems by observation alone. Of course, we believe apes and monkeys are more intelligent than, say, rats, chickens or goldfish. If Thorndike could not see this, it was because he set his animals ridiculously simple tasks that made no demands on their intelligence.

However plausible these beliefs may be, it is worth insisting that the processes of simple conditioning (which are a great deal less simple than Thorndike supposed) are probably much the same in animals ranging from goldfish or chickens to apes. All these animals, for example, will learn to approach and contact a stimulus that signals the delivery of food or withdraw from one signaling an adverse consequence. This conditioned approach or withdrawal will generalize to similar stimuli, and extinguish if the original stimulus is presented with no further consequence (◊ page 23).

But what of more complicated problems and the more complex processes their solution requires? Provided that the experimenter takes the trouble to devise appropriate tests, even quite simple animals can solve what appear to be difficult problems. The lowly rat can be placed on a central platform from which radiate a dozen or more arms like the spokes of a wheel. At the end of each arm is a small piece of food and the rat's task is to obtain each piece of food as quickly and economically as possible. To do this, it must enter each arm in turn without reentering one already visited on that trial. To solve the problem, rats must be able to remember which arms they have already entered on a given trial in the absence of any external stimulus distinguishing entered from unentered arms. They can succeed when there are as many as 17 arms.

Such performance is not confined to mammals. Pigeons seem to be just about as proficient as rats. No doubt there are tasks that apes and monkeys can solve and other animals cannot, but they are hard to find. Only better experimental procedures will reveal the existence of unsuspected abilities.

There is little, if any, evidence to show that monkeys and apes can solve simple discrimination problems any faster than rats, chicken or fish

Thorndike's pessimistic views of the limits to animal intelligence form part of the mechanistic tradition. But other psychologists mistrusted both his experimental evidence and his interpretation. The German psychologist Wolfgang Kohler (1887-1967), studying a group of captive chimpanzees on the island of Tenerife during World War I, concluded that if they were set a suitable problem they would solve it intelligently by a process he termed insight. The objection to Thorndike's puzzle boxes was that there was no way that the animal could escape for the first time except by blind trial and error. Evidence of intelligence depended on the elements of the problem being apparent to the animal from the outset. The American psychologist, Edward Tolman, on the other hand, believed that rats were perfectly capable of displaying intelligence even in such unpromising surroundings as the mazes in which they had been studied so long. Both Kohler and Tolman were respected figures, but their influence remained small. It is only within the past 20 years that the long-dominant mechanistic tradition has apparently been superseded.

A new look at conditioning

Stimulus-response theory regarded conditioning not as the acquisition of new knowledge about the world, but as the acquisition of new behavior. Pavlov's dog salivated whenever the metronome started ticking because it had been conditioned to do so (◀ page 23). The rat threading its way through a maze to the goal box had simply learned a new sequence of stimulus-response connections. Neither dog nor rat was credited with any knowledge of what was going to happen next, that the ticking of the metronome signified imminent delivery of food, or that turning one way rather than another would lead to a goalbox with food in it rather than to one that was empty. But there is excellent evidence that animals in conditioning experiments do acquire precisely this sort of knowledge.

The modern view of conditioning, then, is that animals in conditioning experiments are discovering what leads to what, what predicts the occurrence of food (or illness), very much as Tolman suggested. But a second source of the recent reappraisal of conditioning has been the discovery that conditioning does not simply depend on repeatedly presenting two events in close temporal contiguity. A conditional stimulus (CS) will not be associated with an unconditional stimulus (US) unless it is the best available predictor of that US. For example, rats can associate the flavor of a novel food with illness, so as to condition an aversion to that food, after a single trial when there is an interval of an hour or more between their eating the food and actually getting ill. What is important is that there should be no other plausible cause of their illness. If the rats are given a second novel food to eat in the interval between eating the first and becoming ill, they will attribute their illness to this second food rather than the first. The rule is: what you ate most recently caused your illness. But this can be overridden: if you've eaten a particular food before without falling ill, then it probably will not make you ill next time. So if the second food the rats eat is a familiar one, which they have eaten before with no adverse consequence, they will attribute their illness to the first food again.This suggests that the mechanism underlying the process of conditioning is a sophisticated one, enabling animals to distinguish between chance coincidence and a true causal relationship between two events. The function of conditioning, it is now argued, is to enable animals to discover the causal structure of their world.

Matching to sample
Some animals clearly can solve problems by reasoning and can discriminate between objects, patterns and colors. For example, pigeons can learn readily enough to peck at a red light rather than green when the "sample" light shown on that trial is red, and to peck green rather than red when the sample is green. But have they really learned the rule: choose the light that is the same as the sample? If they had, they should continue to respond correctly when new colors are shown, for instance, to choose yellow in preference to red when the sample is yellow. In fact their performance drops to chance, and they must laboriously learn the new problem from scratch – according to some investigators with essentially no signs of benefiting from their earlier experience at all. Not all birds are so foolish: crows, rooks, jays and jackdaws show excellent evidence of transfer to new stimuli. And monkeys, apes and dolphins have also displayed impressive transfer. An American psychologist, David Premack, has trained a female chimpanzee, Sarah, to discriminate between any pair of objects whether they are the same or different form each other (▶ page 31).

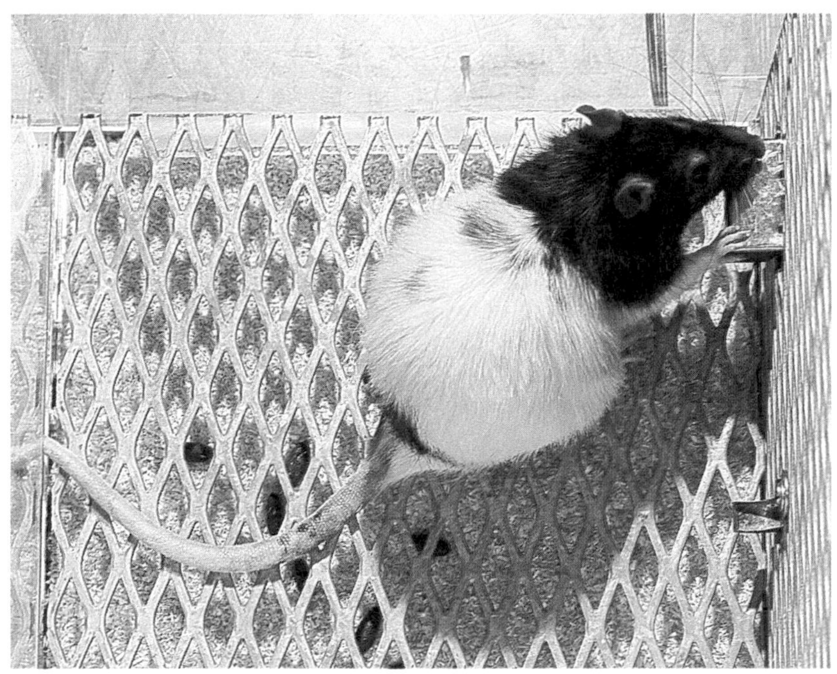

Knowing what will happen next

A rat is conditioned to press the lever in a Skinner box in order to obtain sugar pellets, which it then avidly consumes. The following day, in the rat's home cage, the experimenter gives it some sugar pellets and shortly thereafter an injection of a mild emetic that makes the rat feel ill. This experience conditions in the rat an aversion to sugar pellets, and it no longer eats them when they are offered. What will happen when the rat is now replaced in the Skinner box and given the opportunity to press the lever? As common sense suggests, the rat refuses to press the lever. But S-R theory predicts that if the rat has been conditioned to press the lever whenever placed in the box because a new S-R connection has been formed, the only way to weaken this habit would be not to reinforce it for doing so. Inducing an aversion to sugar pellets should have no effect on this tendency of the animal. The fact that it does shows that the rat is capable of putting together two pieces of knowledge, that lever presses produce sugar pellets and that eating sugar pellets makes it ill, and drawing the conclusion that there is no point in pressing the lever.

▲ An aerial view of a rat in a "Skinner box", pressing a lever that protrudes into the compartment, thereby operating some circuitry that causes the delivery of a small pellet of food to the container shown at the bottom of the picture. Soon the rat learns to press the lever repeatedly.

◄ B. F. Skinner (***♦ page 24******) created this type of problem box in which a pigeon can perform a single action to produce a result. The bird must peck several times at, say, the red key and not the blue key, to receive a reward, such as a seed. Edward Thorndike formed this instrumental learning.***

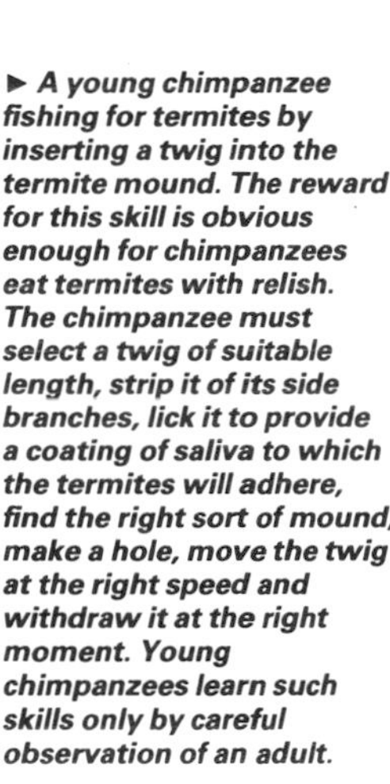

► A young chimpanzee fishing for termites by inserting a twig into the termite mound. The reward for this skill is obvious enough for chimpanzees eat termites with relish. The chimpanzee must select a twig of suitable length, strip it of its side branches, lick it to provide a coating of saliva to which the termites will adhere, find the right sort of mound, make a hole, move the twig at the right speed and withdraw it at the right moment. Young chimpanzees learn such skills only by careful observation of an adult.

Rhesus macaques, in learning set experiments where success is to get each discrimination right after only one trial, have scored 97% correct over more than 50 problems

Not only is conditioning a more intelligent process than was once thought, it is clear that there are other processes, besides those of conditioning, underlying intelligent behavior. Even such a favorite example as a rat learning to find its way through a maze turns out to reveal unsuspected complexity. According to S-R theory (◆ page 22) the rat is learning to make a sequence of turning responses because these are followed by food. In fact, however, rats appear capable of acquiring something akin to an internal map of the maze. They learn that the goal box for which they are aiming is located, say, to the right of a lamp, on the opposite side of the room from the door, and to the left of the table containing the rat cages.

Discrimination of concepts and abstract properties of stimuli

An animal's ability to move around in space, therefore, is based neither on simple S-R habits nor even on guidance towards or away from a simple stimulus (◆ page 21), but rather on a flexible knowledge of the spatial layout of its environment. Furthermore, simple conditioning experiments have equally underestimated animals' perceptual understanding of their world. Pavlov used ticking metronomes, tuning forks, flashing lights or a soft touch on the animal's skin as the signal for the delivery of food. He thought it important to be able to specify precisely the nature of the stimulus he was using, and to be able to vary the stimulus in a precise, quantifiable manner in order to study generalization. Psychologists readily followed his example. Yet the world of an animal consists of objects, not disembodied lights and sounds. Remarkable but perhaps not surprising, therefore, is the discovery that pigeons rapidly learn to discriminate between two series of pictures, one of which always contains a particular class of object (a tree, say, or a fish, or a person), while the other consists of pictures containing anything else except the target object. This suggests that the birds can categorize any of a presumably indefinitely large set of particular features.

Other discrimination experiments have shown that many animals can respond to relatively abstract features of stimuli, regardless of the nature of the physical stimuli concerned. In studies of counting, animals learn to choose between one array containing, say, five stimuli and another containing only three. The actual stimuli used will vary from one trial to the next in order to rule out the possibility that the animal is responding in terms of the overall area, brightness or color of the two arrays. Animals can also discriminate temporal duration, learning to perform one response when presented with a 5-second stimulus and another when presented with one lasting 10 seconds.

Problem solving and reasoning

If animals can respond to the relationships between objects and to abstract features of their world, they can perceive the world in more complex ways than simple reflex theory had imagined. According to Kohler, perception was the key to intelligent behavior (◆ page 26). Reasoning or solving a difficult problem was a matter of perceiving the elements of the problem in a new manner. The chimpanzee that learned to fit two bamboo poles together to rake in a banana lying out of reach outside the cage was perceiving the poles as a raking device. Yet the behavior of Kohler's chimpanzees seems less impressive to us now than it did to him, for later investigators established that the sudden solutions he saw in fact depended upon a long earlier history of playing with and using the implements he provided appropriately.

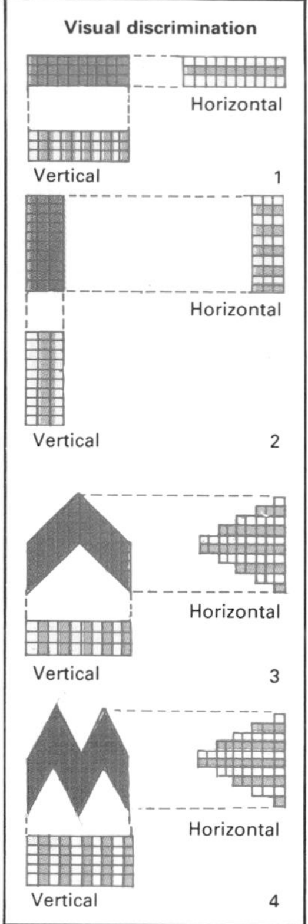

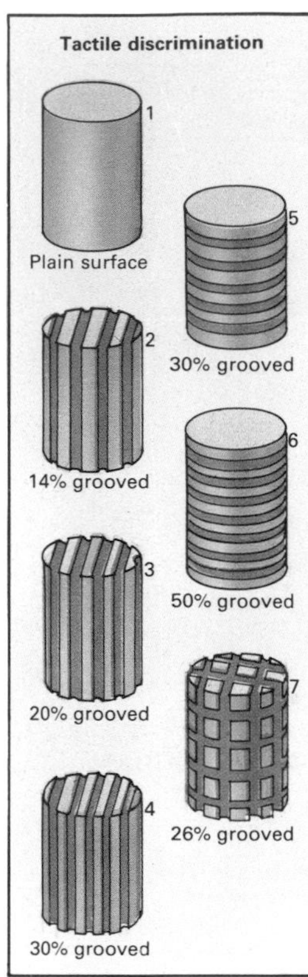

▲ Octopuses can be trained to discriminate between some objects by sight or touch. A visual discrimination between two rectangles is taught by rewarding the octopus with a small piece of fish whenever it grasps the vertical rectangle and punishing it with a weak shock whenever it touches the horizontal rectangle. But the octopus has great difficulty learning to discriminate between a V- and M-shaped block, because they have the same vertical-horizontal projection. Octopuses can also discriminate between objects solely by touch. Here they rely largely on local features, such as the proportion of grooved to smooth surface. Octopuses treat 4, 5 and 7 as the same.

Insight, learning sets and one-trial learning

Wolfgang Kohler regarded an ape's sudden, one-trial solution of a completely novel problem as a critical mark of "insight". Later investigations, therefore, by stressing the importance of the ape's prior experience, suggested that Kohler was wrong to attribute any insight to his animals, and the study of "learning sets" undertaken by H. F. Harlow in the 1950s confirmed these doubts. In Harlow's experiments, rhesus monkeys were trained on a long series of independent discrimination problems. For each problem the animal was required to choose between two objects, under one of which was concealed a peanut or raisin. Two new objects were introduced for each new problem. To begin with the monkeys learned each problem laboriously, but after 100 or so problems they were solving each new one in a single trial. On the first trial of a new problem they would necessarily score at chance since there would be no indication of which object was hiding the reward, but on the second trial they would repeat their first choice if that had been rewarded, or choose the other alternative if it had not. Apparently insightful solutions to novel

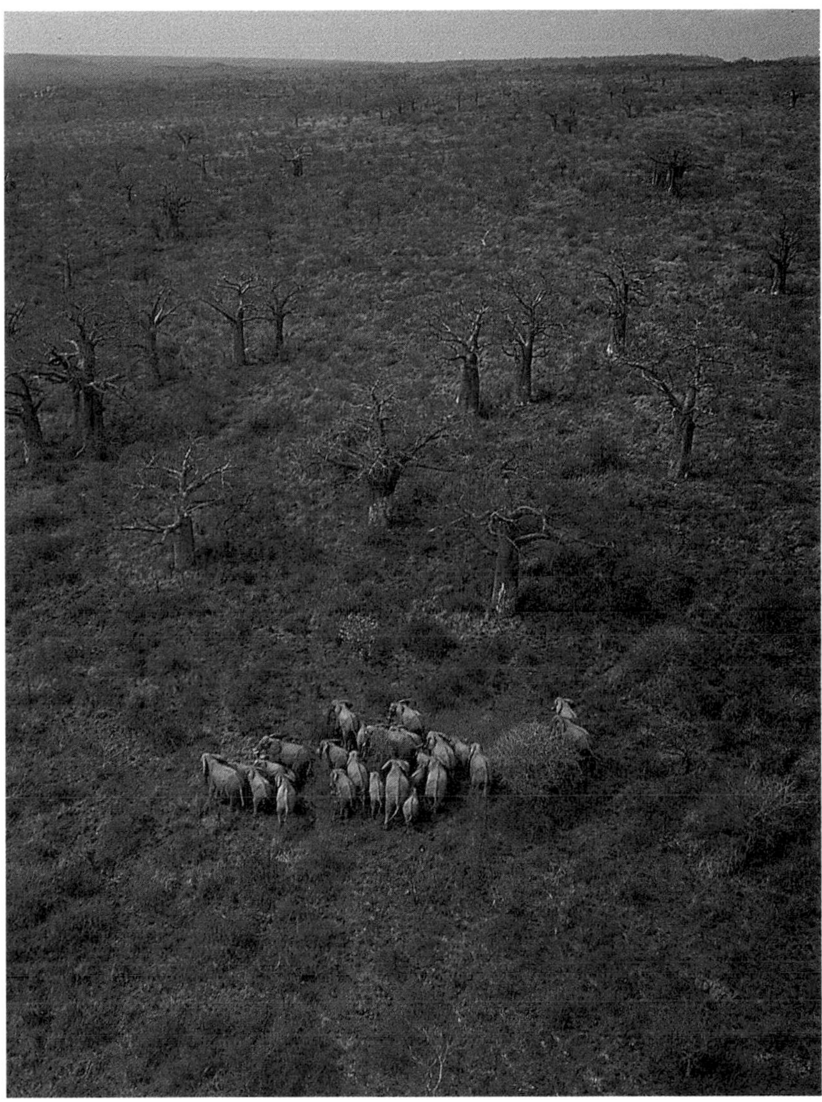

problems depended on a great deal of relevant past experience.

More recent experiments have suggested that a laborious training regime may be less important than the conditions under which animals are tested. American researcher Emil Menzel has observed one-trial learning in marmosets (a small South American primate, often thought to be relatively "primitive") with no prior training at all. His marmosets discriminate beween objects almost without error from the outset, and return to the one that contained food the day before, even though this was the first time they had seen it. The secret of Menzel's success is probably that he tests his animals as a group in a relatively natural setting. In their natural environment, marmosets look for fruit in social groups, and Menzel's testing procedures seek to mimic these features. His animals live in a communal room and are released together into a large cage furnished with trees, on some of which are placed various objects that marmosets naturally explore. They choose to explore novel objects in preference to familiar ones, and return to a novel object that contained hidden food on its first presentation in preference to one that did not.

▲ Even when there are no distinctive visual features in a landscape, animals can still find their way round by slowly building up a spatial map of their environment. This is partly a matter of responding to distinctive configurations of visual clues but may involve other sensory modalities.

► Rhesus monkeys can readily be shown to associate colors and shapes. First the monkey is conditioned to react to color by teaching it to push forward black objects when shown a black triangle (top left), and brown objects when shown a brown circle (top right). If the monkey is then presented with an uncolored triangle or circle (bottom part of diagram) it still pushes forward the same shapes. Such tests are simpler than learning-set experiments but still show that experience teaches.

Concept learning by pigeons

A common procedure for studying discriminative learning by animals is to train pigeons in a Skinner box to peck at a red light whenever it comes on (by rewarding them with the occasional delivery of food when they do so) and not to peck whenever the light is green (◆ page 26). Pigeons have excellent color vision and rapidly learn to peck only at the red light (the positive or rewarded stimulus), but they can equally learn to peck only at a circle rather than a triangle, or at a picture of one complex object or pattern rather than another. They can learn readily even if the actual stimuli shown vary somewhat from trial to trial, provided that all positive stimuli contain a constant critical feature (say a red circle in the bottom left corner) and all negative, or unrewarded, stimuli another (say a green triangle in the top right corner).

The Harvard University psychologist Richard Herrnstein has shown something much more striking: that pigeons can learn quite rapidly to solve a discrimination where not only do the stimuli (pictures) shown vary from trial to trial, but the only distinguishing feature is that all positive pictures contain, somewhere within them, a particular object, such as a tree, or part of a tree. The tree can be broadleaf or a conifer, covered in leaves or with bare branches, viewed from above, the side or below, in the foreground, plain in view, partially obscured by other objects in the picture, or with no more visible than a few branches in the top right corner of the picture. In order to solve this problem, Herrnstein has argued, pigeons must, in some sense, be credited with having the concept of a tree.

That birds which nest in trees should be able to categorize an object as a tree may not seem too surprising. But Herrnstein has gone on to show that the critical object need not be one which the pigeons' evolutionary history must have equipped them to deal with. They can solve a discrimination where all the stimuli are underwater photographs, with positive stimuli containing, somewhere within them, a fish and negative ones containing no fish. They have even learned to recognize an individual woman, responding correctly whenever the scene contains her (regardless of what she is wearing, whether she is in the foreground, facing the camera, looking away, etc.) and refraining from responding to pictures of other people. Evidently pigeons have evolved a brain that makes possible quite complex adaptive behavior.

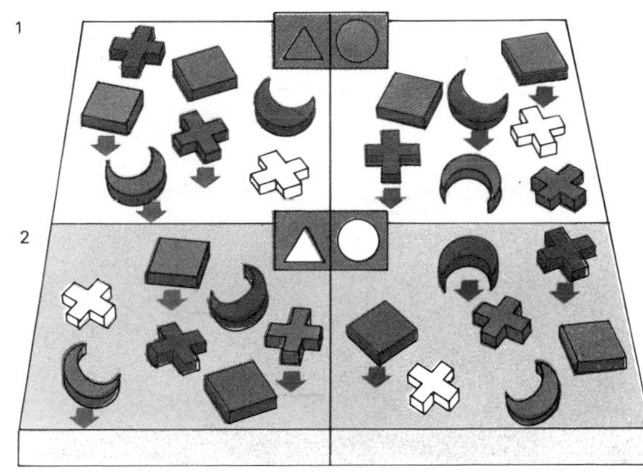

Computers can be programmed to solve differential equations, prove theorems in formal logic, recognize patterns, and understand language, but cannot effect any cognitive functions

▲ ***Learning sets have been studied in some animals not often seen in the laboratory. Among the more efficient mammals are raccoons – skilful and manipulative animals.***

► ***The orangutan is a rather solitary animal, and in zoos can look morose. But like the chimpanzee and, apparently, unlike monkeys, it does show great self-recognition.***

Perceiving relationships

Monkeys and chimps have been shown to be able to solve problems of the form: if A is heavier than B and B is heavier than C, then A must be heavier than C. Another form of reasoning task, familiar to anyone who has taken an IQ test, is the analogy: black is to white as night is to (?). One female chimpanzee, Sarah, has shown an impressive ability to solve such analogical reasoning problems (➧ page 31). The solution of an analogy requires one to see whether the relationship between one pair of items (black and white) is the same as the relationship between another pair (night and day). Unlike inferential reasoning, analogical reasoning does depend on perceptual processes, the perception and abstraction of relationships.

Many animals have solved discrimination tasks that appear to require their perception of relationships. One version of the task is known as "matching to sample"; the animal is required to choose one of two alternatives that matches (is the same as) the sample shown on that trial. Pigeons have no difficulty solving this problem, but it is questionable whether they are really responding to the relationship between correct alternative and sample, for, unlike monkeys or even crows, they show rather little evidence of transferring the solution to new stimuli (➧ page 26). Sarah herself has no difficulty with the simple version of the task, but can also abstract a particular feature or

Self-awareness in apes

We use the word consciousness and related terms such as awareness in more than one sense. But one quite specific use is in the phrase "self-awareness". We know who we are and are conscious of our own identity. Are animals? To test this, in the early 1970s the American psychologist George Gallup investigated whether an animal can recognize itself in a mirror. While his animals were asleep he painted a small red spot on one of their ears. When the animals woke up and looked in the mirror, they picked at the spot. Both chimpanzees and orangutans displayed such behavior (but not until they had seen themselves in the mirror, thus ruling out the possibility that they could feel the spot of paint). Similar experiments with monkeys have yielded uniformly negative results, thus providing one of the clearest instances yet shown of an evolutionary discontinuity in mental processes.

attribute of the sample and choose correctly between two alternatives, both physically different from the sample, but one displaying the appropriate attribute. Shown half of an apple as the sample, she will correctly choose a half-filled jug of water in preference to a full one. Shown a whole apple, she chooses the filled jug. It is this ability to detect the similarity between attributes of, or relationships between, objects that is required to solve analogies.

Consciousness and thought

The real reason why we are quite confident that we are not mere machines is because we are conscious. But are other animals conscious? How can we decide? We believe that our fellow human beings are conscious partly by analogy with ourselves. I know that I am conscious; you say and do things much as I do; therefore you are probably conscious also. Of course other animals behave, in some ways, rather as we do. They look excited in anticipation of food, contented having eaten it, disdainful if it was not what they were hoping for. The sheepdog watches the sheep intently and listens carefully for the shepherd's call. Is it not conscious of what it is doing and might be required to do next? But are animals really enough like us to count as conscious? Is the thrush catching the early worm conscious of what it is doing or why? Is the worm?

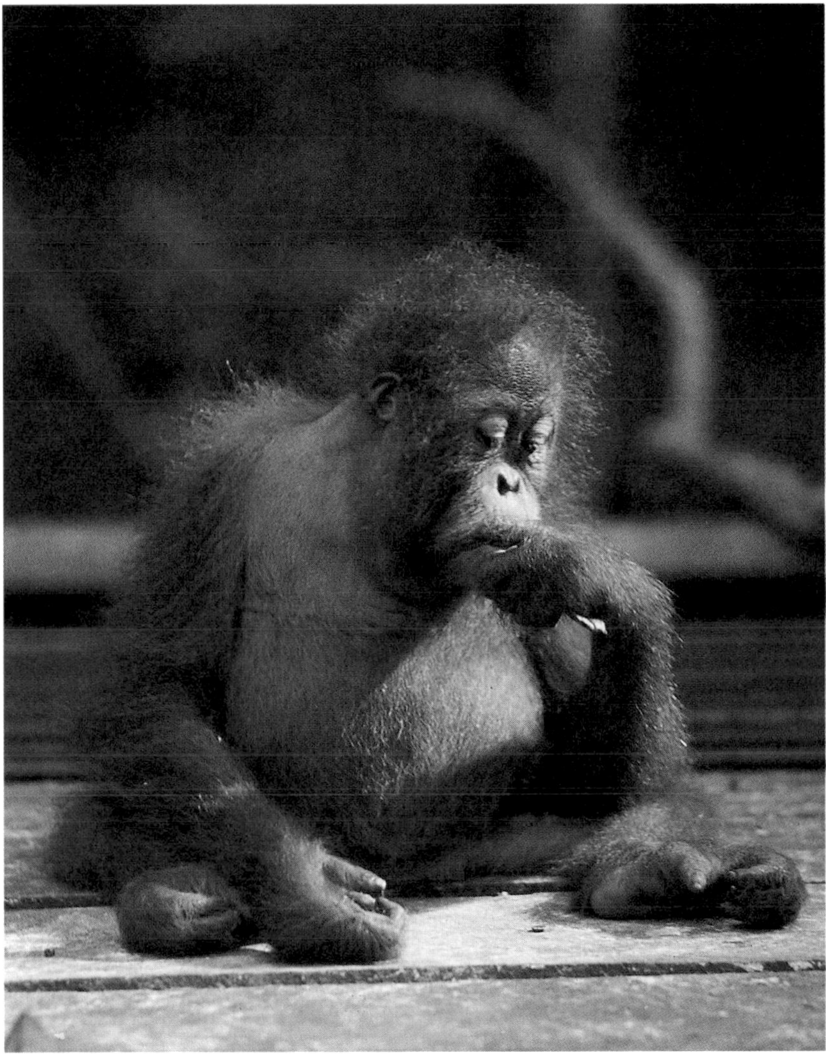

Reasoning by monkeys and chimpanzees
Squirrel monkeys can be trained to choose the heavier of two colored boxes by rewarding them with a peanut or raisin whenever they choose correctly. In this way they can learn a whole series of problems: to choose the heavier red box in preference to a green box; to choose green in preference to the lighter yellow; yellow in preference to blue; and blue in preference to black. The test for transitive inference is to give them new pairings of these boxes, say, green and blue. Young children can solve this problem by the age of 4 or 5, but squirrel monkeys are equally proficient. There remains some question as to how they solve the problem, but there is no doubt they are putting together information from different learning experiences in order to attain the solution. This is one definition of inferential reasoning.

The solution to an analogy problem involves the detection of a relationship between two terms and of a similar relationship between two further terms. David Premack's chimpanzee, Sarah (♦ page 26), has learned to perform one response whenever two objects that she is shown are the same (two apples, two biscuits, or two tin cans) and another when they are different (an apple and a banana, or a banana and a tin can). She can now be asked whether the relationship between one pair of objects is the same as the relationship between another pair. An example of a simple geometrical analogy is shown in the figure below: is the relationship between the two shapes on the left the same as that between the two on the right? Sarah correctly answers "same", since in both pairs the top figure is transformed into the bottom by changing its color from red to green and adding a central dot. Sarah also solves analogies between everyday objects with which she is familiar, such as between a tin can and a can opener and a padlock and a key.

▼ One of Sarah's geometrical analogies. Her task is to decide whether the relationship between the top and bottom shapes on the left is the same as that between the two shapes on the right. If satisfied it is the same, she places her "equal" sign between the two pairs. If convinced there is a difference, she must place her "not equal" sign between the two pairs.

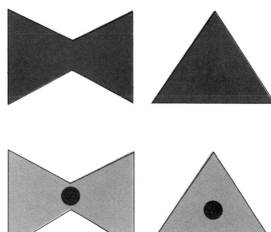

There are now several captive groups of apes that communicate among themselves using sign language but there is now evidence such communication occurs in the wild

When dealing with other animals we can only infer what is going on in their mind by observing their behavior. But we can ask other people. The importance of this distinction is suggested by observation of certain clinical patients. A patient with damage to the occipital lobe of the cortex (that part of the brain concerned with vision) may have a "blind spot", and report that he is quite unable to see a stimulus so located that it falls entirely within this blind area. But if the persistent experimenter asks him to guess the nature of the stimulus, or to press one button if it is a triangle and another if it is a circle, he will do so without error. He says that he cannot see the stimulus, but in some sense he clearly can. In other words he sees it, but is not conscious of seeing it.

How can we ever draw this distinction in the case of a dumb animal? The only way we have of asking animals what they can see is by training them to solve a discrimination problem. If they solve the discrimination, then they can clearly see the stimuli. But unless we can teach them to talk, we have no other way of asking them what they can see. Can we ever do that?

The relevance of language

Yet another of Descartes' grounds for distinguishing between ourselves and dumb animals (◀ page 19) is that we can speak and they cannot. And the distinction is not really blurred by talking parrots. It is not just the ability to produce sounds that happen to mimic human speech that counts. We understand what we are saying, the parrot does not (or so we believe). Here then has been one of the great challenges: can animals ever learn a language in the sense that the young child does? And if they can, would this open a new avenue for exploring their minds?

Dr Doolittle could talk to the animals because he learned their language. But ethologists studying the natural "languages" of various animals are now generally agreed that they are not remotely like any human language. Many species of animals do indeed communicate information to one another (▶ page 66). For example, the alarm calls of vervet monkeys warn other members of the group of the approach of a predator, and also indicate what kind of predator it is (lion, snake or bird of prey). By comparison with human language, however, these systems of communication are distinctly constrained. Their features are innately determined. The information they can provide is fixed and limited. They have no capacity for combining elements to generate an indefinite array of new "sentences". If we want to find out whether any animal is able to transcend these limits, we must resort to explicit training in some version of human language.

But what would count as evidence of true linguistic ability in, say, chimpanzees? That they could produce a string of sounds which we recognized as a meaningful English sentence? But parrots can do this and we credit them with nothing more than clever mimicking. Efforts to teach chimpanzees to talk have in fact been uniformly unsuccessful and it turns out that there is a critical anatomical difference between us and our closest animal relatives in the structure of the vocal system (▶ page 137). Nevertheless, speech is not a necessary ingredient of language. Written language, the sign language of the deaf, and the dance of the foraging honeybee, are perfectly good examples of unspoken language. The realization by several scientists in the 1960s that one might try to teach apes a non-spoken form of human language has revolutionized this whole area of understanding.

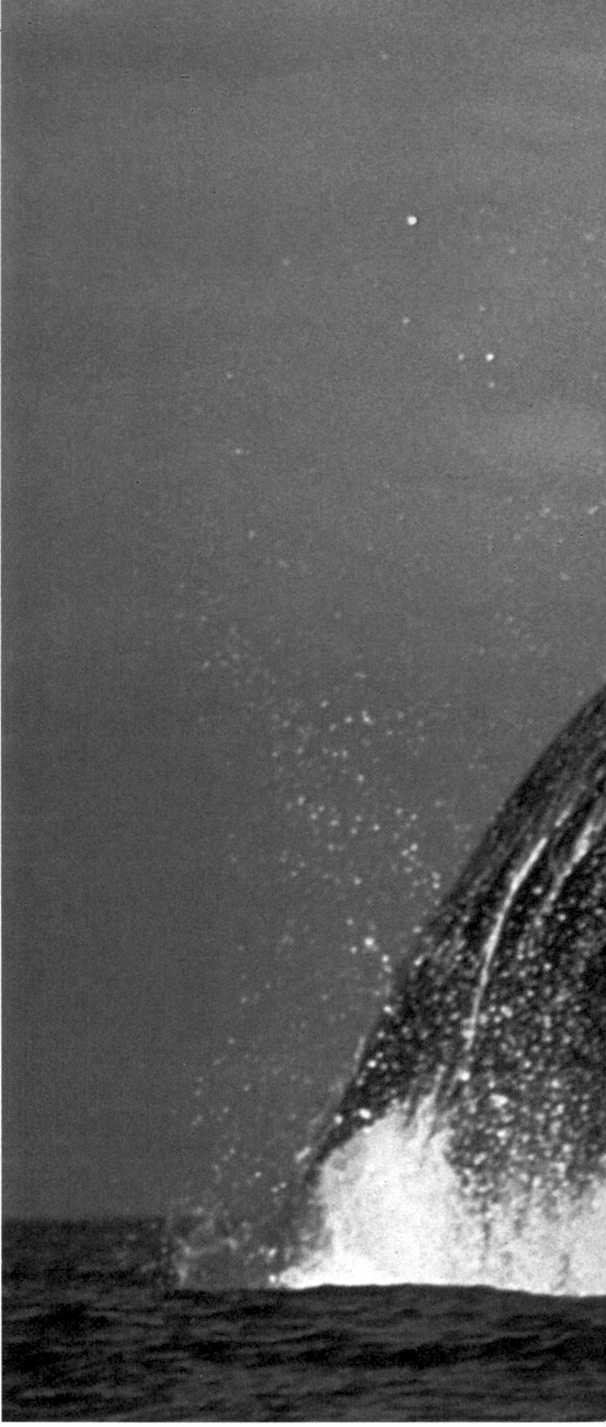

◄ The olive baboon's alarm calls convey a rich array of information, with distinct calls for such different predators as snakes, leopards and birds of prey. In response to the leopard call, baboons will hasten to climb the nearest tree. When they hear the eagle call, however, they look up at the sky. They can also respond appropriately to the alarm calls of other species living in the same region for terrestrial and airborne predators. But they are remarkably insensitive to the various visual clues of a predator's presence, showing no fear of the quite distinctive track of a python or of a freshly killed antelope hauled up into a tree by a leopard.

▲► The song of the humpback whale, which can last up to half an hour, is one of the most complex forms of display or communication seen in any animal. Different individuals develop their own variations of the song, and a given animal can repeat its own intricate variant almost note for note, as can be seen in the figure. Here, the two spectrographic tracings, labelled 1 and 2, represent two repetitions of its song performed by one whale near Bermuda. The song is very loud, being detectable over many kilometers, and sounds eerily beautiful to most human listeners. The notes are similar to those in our own music.

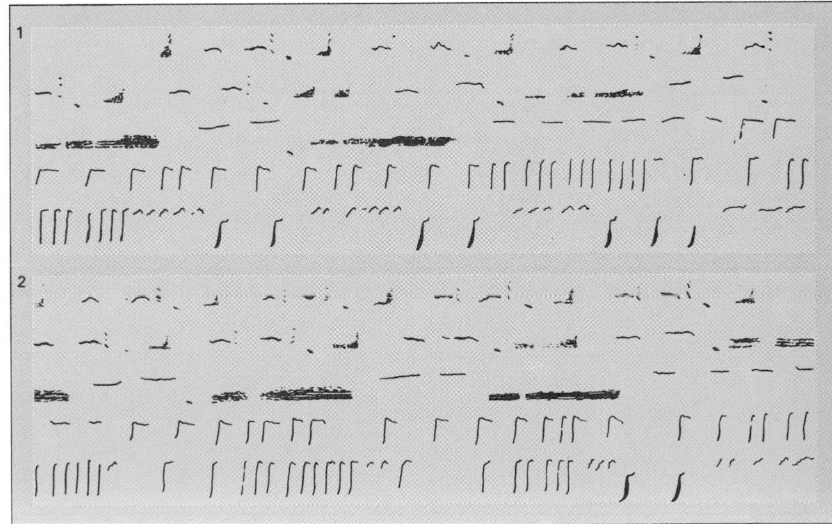

See also
Looking at Animals 9-18
Animals as Machines 19-24
Groups and Troops 47-62
Territory 63-70
An Ape Called Man 129-140

Descartes' theory exploded?

The first chimpanzee to be taught sign language was a young female, named Washoe, brought up by Beatrice and Allen Gardner at the University of Nevada. Since then a dozen or more apes have learned at least some rudiments of sign language. David Premack has taught Sarah (◆ page 26) to respond to and use plastic tokens, while American primatologists, Duane and Sue Rumbaugh, trained another female chimpanzee, Lana, to operate the keyboard of a computer terminal furnished with distinctive colored symbols.

Enthusiasts argued that the final barrier between the human and animal mind had at last been breached. However, sceptics soon began to argue that the only relation the behavior of Washoe, Sarah and Lana had to human language was in the minds of their trainers.

What the enthusiastic trainer sees as words are no more than arbitrary stimuli arbitrarily associated with certain objects or events, just as the sound of the metronome is associated with the delivery of food in Pavlov's conditioning experiments (◆ page 23).

Perhaps not all signing apes have been guilty of mere mimicry, however. A recent analysis of an orangutan named Chantek, trained by Lyn Miles, has shown impressive similarities between an ape and a typical young child. Like 2- or 3-year-old children, Chantek neither interrupts nor imitates her trainer, and about one third of her signs seem to be quite spontaneous.

Was Descartes right? Is the possession of language a uniquely human characteristic? The general consensus now is probably "yes". Yet if attempts to teach apes language have so far failed, they have revealed a striking difference between the way in which young children and apes use language or signs. With few exceptions, the majority of trained apes' spontaneous communication is in the form of requests. Most of a child's remarks are spontaneous descriptions of an object or event that they have observed, a prediction of what they are about to do, or an attempt to attract their companion's attention to something that interests them. Language is used for social interaction. It is not, perhaps, entirely surprising that other species should be less interested in social interaction with us than we are, but it may well be that this is the critical difference between us and them.

The achievements of Washoe, Sarah, Lana and Nim

When the Gardners announced that chimpanzee Washoe not only had a vocabulary of over 150 signs, but also used them in several dozen combinations of two- or three-unit strings, one eminent and initially sceptical psychologist said that "it was rather as if a seismometer left on the Moon had started to tap out 'S-O-S'".

Sarah could apparently use, and understand her trainers' use of, tokens that referred to objects (apple, banana, bucket), attributes of objects (red, round), actions (take, give, put) and even prepositions (in, under). She responded appropriately when instructed "place apple in bucket", and answered correctly that apples are red (the token for apple was in fact a blue triangle). Lana would busily type instructions on her keyboard, "Please machine give Lana drink".

Herbert Terrace of Columbia University trained a chimpanzee, Nim, to use sign language, initially, so he thought, with as much success as Washoe. But when Terrace analyzed complete videotapes of training sessions, it rapidly became clear just how little Nim's behavior resembled that of even a 2-year-old child. The interpretations of strings of signs as sentences was often imposed on the data by the human observer as forced by Nim's actual behavior. Where that behavior did suggest a grouping of signs into particular combinations, this was as often as not because Nim had been prompted. If Nim made a single sign and paused, he would be encouraged to continue. If he continued with further, unrelated signs after he had finished a "sentence", the trainer interpreted them as the start of a new sentence. A famous example of Washoe's supposedly inventive use of language was her signing "water-bird" on her first sight of a swan. But all Washoe did was to produce two signs in succession, for all we know because she saw both some water and a bird. We put the hyphen between the two signs to convert "water" into an adjective or modifier.

◀ At the Gorilla Foundation in the USA, captive gorillas are under tests to explore their intellectual processes, personality traits and social behavior. Some have been shown to need little or no training to perform simple tasks; are quite proficient at understanding spoken words; and can use symbols to request objects, describe them, or to direct their trainer's attention towards something they have noticed.

Sex and Species

4

Reproducing alone or in pairs...The evolutionary advantages of sex...Attracting the opposite sex... Courtship – antagonism between males...Whether to mate, fight or flee...PERSPECTIVE...Binary fission... Energy investment in mating and potential rewards ...Breeding cycles and seasons...One male, many females or vice versa...Using horns, tusks and teeth to secure a mate...Breeding territories and satellites

If organisms did not reproduce they would become extinct. However successful or dominant an individual animal or plant may be, it is bound to die, sooner or later, from accident or from disease (➧ page 198). All the genes it contains would die out with it if it had not already produced offspring that contain copies of those genes. Thus, however an animal or plant conducts its life, however its structure and physiology are organized, it has to incorporate mechanisms and techniques for reproduction.

The simplest way for an organism to reproduce is to divide to form two or more new organisms. This is asexual reproduction. There are many variants of this. One highly specialized but widespread form, particularly among lower organisms, is parthenogenesis. Here, the female produces an egg that develops into a new individual without fertilization. The second mode of reproduction is sexual. This involves two individuals each of which produces special sex cells, or gametes, that fuse to produce a zygote, which develops into the new organism (➧ page 242).

On the face of it, asexual reproduction is by far the more efficient. It involves only one organism, whereas sexual reproduction requires two, and the finding of a suitable mate is very chancy (➧ page 38). In addition, the aim of reproduction is to pass on the parental genes to the offspring. When an organism reproduces asexually it passes on copies of all its genes. An individual that reproduces sexually passes on only half of its genes, and its offspring acquires the other half of the genes it needs from the other parent.

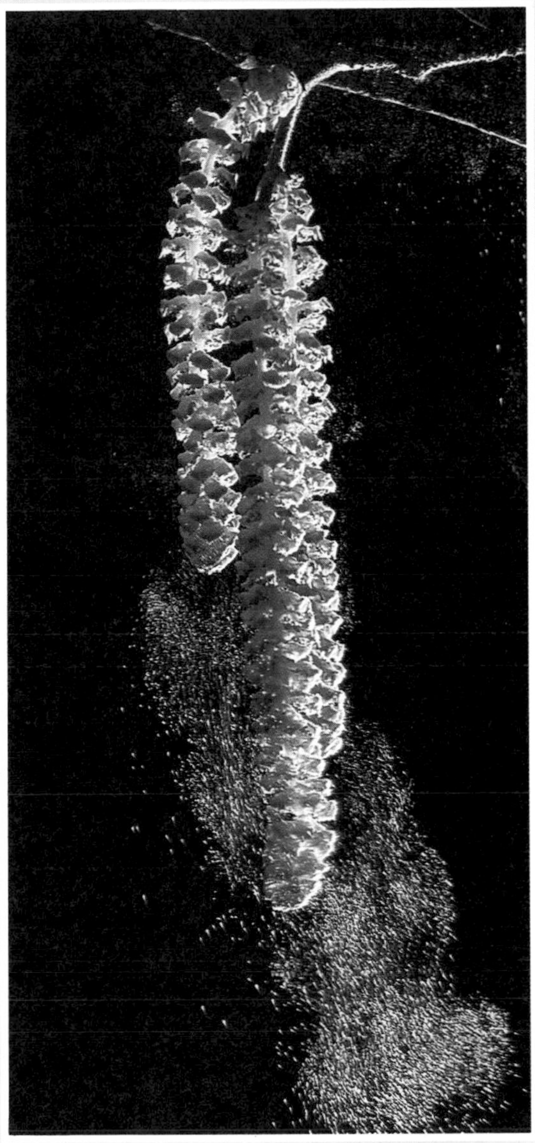

◀ Snails are hermaphrodite (bisexual). Courtship culminates in exchanging "darts" of sperm (clearly visible on the underside of the right-hand snail just below its head).

▲ Many plants, such as hazel bushes, rely on the wind to carry pollen from male flowers to fertilize female flowers. To ensure success, they must produce millions of pollen grains.

Methods of asexual reproduction

Single-celled organisms of different types, including bacteria and many protozoa, reproduce simply by splitting into two (➧ page 36). This is binary fission. Other organisms, both single-celled such as yeasts, animals such as hydra, and many plants, produce offshoots. This is budding. Among plants, dandelions and hawkweeds reproduce exclusively by parthenogenesis; ova develop into seeds without fertilization by pollen. Among insects, both greenfly and honey bees practise parthenogenesis at some time in their lives (➧ page 61). In bees, the unfertilized eggs laid by the queens develop parthenogenetically into males, while the fertilized eggs develop either into queens or workers, depending on how the larvae are fed.

The protozoan Paramecium *reproduces asexually by binary fission several times a day but eventually the clones age. It then reproduces sexually to rejuvenate its genetic material.*

The advantages of sexual reproduction

Despite the difficulties, sexual reproduction is extremely widespread. Even bacteria, which reproduce asexually at a phenomenal rate, also possess mechanisms for sharing genetic information in ways that are comparable to the sexual reproduction of higher organisms. Most plants reproduce sexually as well as asexually. Only a few have abandoned sex altogether. Most "higher" animals, including all but a few vertebrates, are exclusively sexual.

Presumably, then, sex has advantages that outweigh its obvious disadvantages or else the sexual breeders would by now have been swamped by the asexual. What exactly these advantages are is not yet certain, but biologists now agree that two kinds of influence are at work. The first is that organisms which reproduce sexually can evolve more quickly. Evolution implies the acquisition of new characteristics, which are brought about by changes in the genes known as mutations. But mutations occur only rarely, and most of them are not beneficial (➧ page 206). In an organism that reproduces only by asexual means, it would take millions of generations to accumulate enough new, beneficial mutations to produce significant evolutionary change. In populations of organisms that reproduce sexually, however, mutations arising in different individuals can be combined in the offspring of the next generation, and all the mutations arising within all the individuals in a population could be combined in any one individual within a few generations.

Nonetheless, more rapid evolution alone cannot account for the dominance of sexual reproduction. In theory, a sexually reproducing organism that suddenly acquired the ability to reproduce asexually could swamp all its sexually reproducing competitors within a short time simply by breeding twice as quickly. It seems, then, that sexual reproduction must also bring short-term benefits, though it is not certain what these are.

All forms of reproduction exact a price. Organisms have to invest some of their energy to form the bud or the gametes. It is important to invest the correct amount. For example, codfish lay around two million eggs of which, on average, two reach maturity. If they laid only a few hundred eggs, the chances are that none would survive. If the female tried to produce ten million eggs, she would use so much of her body resources that she would die in the attempt. In contrast, animals such as the horse produce one large offspring at a time, which they look after. In general, the larger the offspring at birth, the greater the chance of surviving the first few days, when it is extremely vulnerable. Yet if a mare tried to produce a fetus that was too big, she would probably be unable to carry it to term. But organisms, and particularly animals, that reproduce sexually have an additional price to pay. They must devote a part of their lives to finding suitable sexual partners.

A short-term benefit of sex

One of the leading theories for the success of sexual reproduction comes from Professor William Hamilton, at Oxford University. He points out that in sexual reproduction the offspring are different physiologically from their parents. Thus, he suggests, they are not liable to be prone to exactly the same diseases as their parents, or not, at least, to the same degree. In short, it pays organisms to alter slightly from generation to generation just to stay one step ahead of the parasites that seek to invade them. This need encourages sex to remain the prime means of reproduction.

▼ Laboratory experiments on fruit flies* Drosophila *have helped in understanding the effects of genetic mutations on a male fly's ability to court and mate with females. Here, males of a wingless (apterous) mutant form (with grey bodies and white eyes) attempt to court females (with red eyes and yellow bodies).

► Spawning large numbers of eggs to be fertilized by the male clasped to her back, a female tree frog leaves her offspring to fate. She relies on producing so many eggs that at least some are bound to survive by chance alone. Frog spawn is a much favored food of many fish and birds. Some frogs produce several thousand eggs.

▼ In the protozoan* Paramecium*, sexual reproduction occurs following the partial fusion of two individuals (1) during which they exchange genetic material (2-3). Then they separate (4), and each one's genetic nuclei divide into four identical sets (5) that provide the basis for four new individuals produced by fission (6).

► While frogs count on luck to ensure the survival of their offspring, birds opt for the alternative strategy of having fewer offspring and devoting more time and energy to caring for them. The robin lays between four and six eggs. The female incubates the eggs for 14 days. Both parents feed the nestlings which are able to fly at 15 days old.

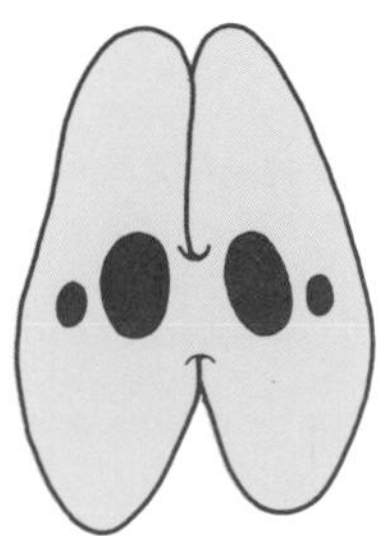

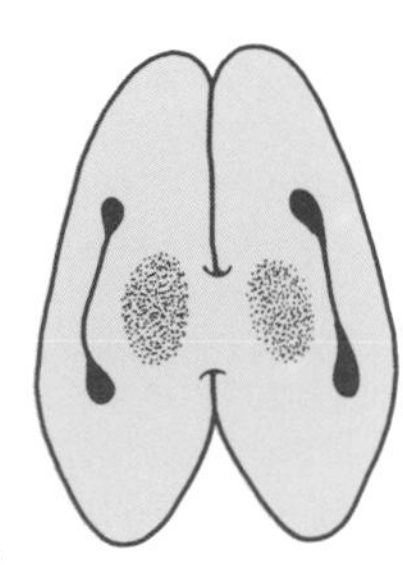

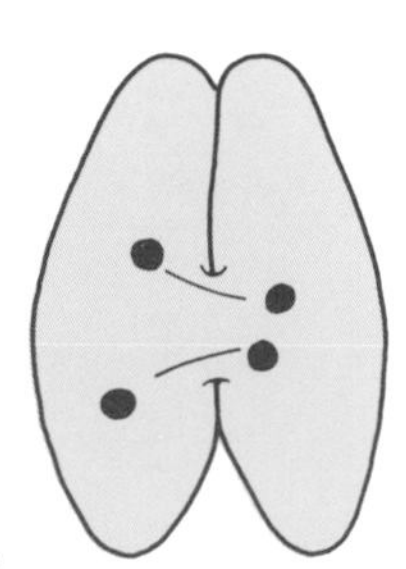

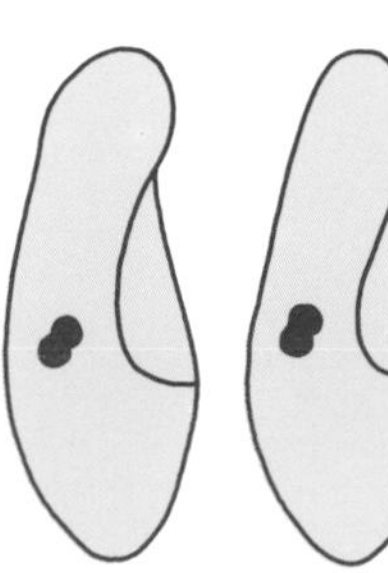

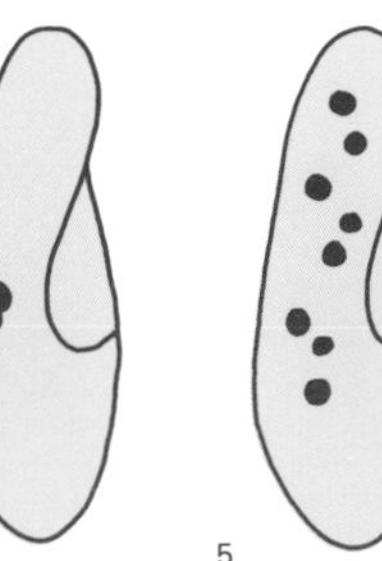

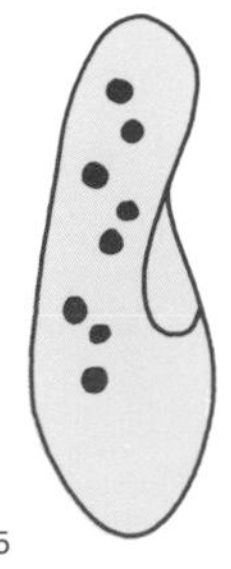

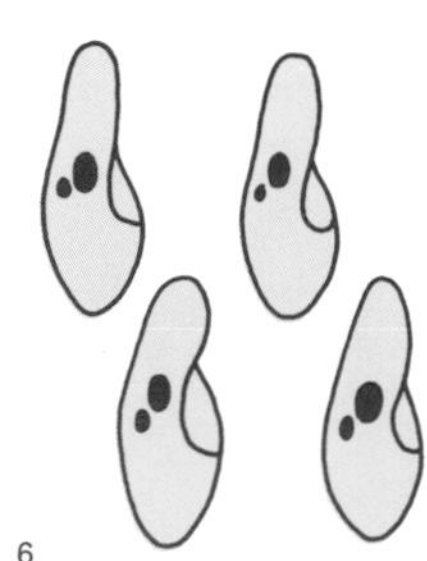

Mate now, pay later

An animal has a limited energy budget available to apportion between growth and reproduction. How it does this determines its life history. If it opts to invest at an early stage in reproductive activity, it will not have much energy to spare for growth. While there may be advantages to engaging in reproduction early on, the resulting loss of growth potential may place an early reproducer at a disadvantage in later years when it has to compete with larger contemporaries that preferred continued growth over early reproduction. This is well illustrated by the bluegill sunfish, a small fish that occurs widely in North American streams and lakes. Males normally reproduce when they are about 5 years old. Some males, however, begin to reproduce a year earlier by "stealing" matings from territorial males. These males, known as sneaks, grow more slowly than their nonreproductive contemporaries. As a result, they never get large enough to hold their own territories. Because of their small size, they suffer high frequencies of wounds and have shorter lifespans.

▲ *Frond-like antennae on male luna moths can detect the female's scent from many kilometers away.*

▼ *The female gelada advertizes her sexual receptivity with fluid-filled vesicles on her chest.*

Breeding seasons and courtship rituals

A species that reproduces by budding or self-fertilization can produce offspring wherever and whenever it chooses. But an animal that reproduces sexually has to find a mate in order to do so. This raises a number of difficulties that have to be resolved if the species is to survive. First, only a small proportion of the members of the opposite sex that the animal meets will be ready to mate at that moment. Second, no individual wants to waste its genes on a mate that is not suitable (➧ page 202). Thus each individual will want to ensure that a prospective mate is the best available. This is usually achieved by a period of courtship during which the partners assess each other.

Often, animals cannot reproduce until complex hormonal and tissue changes have occurred, such as maturation of testes and ovaries. Many species, from plankton to antelope, breed seasonally so as to ensure that all members of the species come into reproductive condition at the same time. Breeding seasons can be as short as a few hours or as long as several weeks, but rarely exceed a month in length and are normally confined to one particular time of the year.

Breeding seasons are often timed to ensure that the young are born when food is abundant. In various species of antelope and in gelada baboons, the seasonal onset of vegetation growth stimulates hormonal

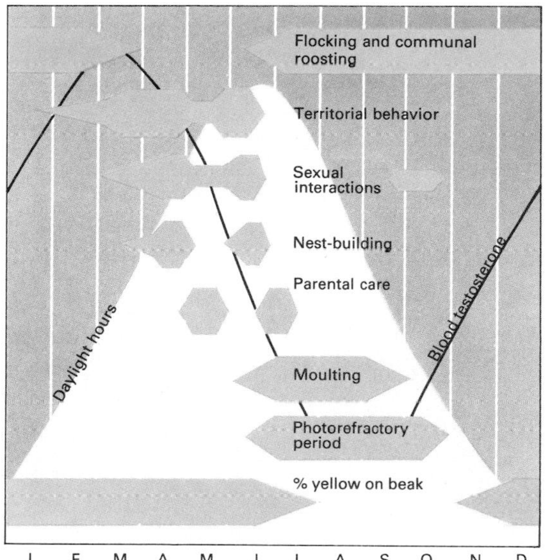

▲ **All mammals suckle their young for a period after birth. These lion cubs will not be weaned until they are eight months old.**

◄ **The starling's annual reproductive cycle consists of a sequence of phases set in train by a late winter peak in testosterone levels. In spring the flocks disperse and the males establish their territories. Pair-bonding and nest-building follow, then the eggs are laid and incubated before the final arduous task of feeding the hatched young. With a possible second brood before the summer molt, the cycle ends with the winter flocks reforming. But already the hormonal events that will trigger the next cycle are underway.**

changes in the animal that trigger the reproductive cycle. In northern and southern latitudes, where there is a marked change in temperatures and in the length of daylight between seasons, many species such as the goat use the changing day-length as a cue to start.

Some species use special signals to notify potential partners of their readiness to mate. Most female mammals undergo a period known as estrus during which they are willing to mate. These periods are always tied into the ovulatory cycle so that ovulation (the release of eggs from the ovaries) occurs at the same time that mating is likely to take place. This ensures that fertilization occurs while the eggs are still fresh.

In species such as macaque monkeys, baboons and chimpanzees, the females have evolved external signs to show males they are receptive. In these species, the female has an area of sensitive skin on her bottom that swells or changes color under the influence of the reproductive hormones. The sexual swelling reaches its peak just about the time of ovulation. Smell is another important cue of sexual condition, especially among insects. At night female moths waft special sex scents (called pheromones) that the males detect via their antennae. The male Atlas moth of Asia has such large sensitive antennae that it can detect a female's scent at concentrations as low as one molecule per cubic meter of air.

Rearing patterns

In evolutionary terms, an animal's problem is to contribute as many genes as possible to future generations. In order to be able to do this, it must ensure that its own offspring reach maturity so that they too can breed.

*Some species simply trust to luck, hoping that by producing very large numbers of offspring, some at least will survive to reach adulthood. Ocean living species like jellyfish and molluscs have free-swimming larvae that float in the sea, where many millions of them fall prey to fish and other predators (**▶ page 121**).*

Because losses are very high and much energy is wasted in producing offspring that have little or no chance of survival, some species try to improve their offspring's chances of reaching adulthood. Parental care has, in fact, evolved in many different groups of animals. Few, however, have been so heavily committed to this way as the mammals. These feed their infants on a special food (milk) produced by the mother's body.

The drawback with parental care is that an animal's time and energy is limited. Consequently, it faces the choice between having many young, each of which receives only a limited amount of parental care, or having only a few young at a time and lavishing a great deal more care on each.

The ways in which the different species look after their young vary considerably. Some crabs carry the developing eggs cemented onto their shells. The male midwife toad carries the string of eggs wound around his legs until the tadpoles hatch out. The male of one species of South American frog swallows the eggs after they have been spawned. The young frogs develop safe from predators inside his stomach until they are old enough to crawl back out of his mouth.

Many species of small fish spawn in nests on the stream bed. The male is responsible for guarding the nest and looking after the eggs. In order to prevent the water in the nest becoming fetid and the eggs being killed by fungus, he uses his tail as a fan to force clean water through the nest. Most birds lay their eggs in nests and then brood them until they hatch, after which they have to feed the voracious chicks. But a few species such as the American cowbird leave that onerous task to other species in whose nests they lay their eggs.

Induced and spontaneous ovulators

In some species, the females ovulate only after being courted. Courtship may involve ritual displays. In pigeons, for example, the male's bowing and cooing stimulates the female's reproductive system to prepare itself for ovulation. Such species are known as induced ovulators. Other induced ovulators, such as rats and mice, can also induce spontaneous abortions of the developing fetuses. Even the smell of another male is sufficient to cause a pregnant female rat to abort a litter and come back into reproductive condition in order to mate with the new male. These species contrast with those such as the primates (which include human beings) that are spontaneous ovulators in which the females ovulate on a regular cyclical basis whether or not mating takes place.

During the breeding season of giraffes a succession of bulls of increasing dominance court a receptive cow, stimulating her to be ready to mate with the dominant bull of the area

Choosing the best mate

Even if an animal is ready to mate, it will not necessarily be willing to mate with any individual that comes along. In many species of bird, the male's help is essential for both incubating the eggs and feeding the chicks that subsequently hatch out. A female needs to be convinced that a prospective mate will not desert her immediately after mating. If he does, she risks being unable to rear all the young on her own. Courtship has evolved as a means of allowing the female to assess the honesty and the quality of a prospective mate.

The diversity of courtship rituals in the animal kingdom defies summary. Gulls present their mates with morsels of food, a behavior known as "courtship feeding" (➧ page 42). Dance flies are only about 1 cm long, but the male will carry an insect he has caught and offer it to a female while rising and falling in the air in a graceful dance. When the female accepts his offering and begins to feed on it, the male mates with her. In many fishes, such as the three-spined stickleback, the male tries to attract passing females down to a nest that he has prepared on the stream bed or seafloor by swimming in a characteristically elaborate way. Male birds of paradise hang upside-down to display their delicate filamentous feathers.

Even in those species where the male's help is not necessary for rearing the young, the female must choose a mate that will give her healthy young. In species such as the sage grouse of North America and the birds of paradise from New Guinea, the females choose a suitable mate from among males that display together on an arena called a lek.

Optimal mate choice

An animal has to ensure that a prospective mate is sufficiently closely related to it that it will produce viable offspring, but not so closely related that inbreeding will be a serious problem. Inbreeding (mating with very close relatives) increases the chances that rare deleterious genes will come together and result in abnormalities in the offspring (➧ page 204). The best known example of this is the human blood disease hemophilia. This condition became especially common among the crowned heads of Europe at the end of the last century because they were all closely related, yet continued to marry mainly among themselves. Others are the blood disorders sickle cell anemia and thalassemia.

Studies of Japanese quail and other species have shown that birds that have been reared alone mate with cousins in preference to individuals that are more or less closely related. This preference for individuals that are genetically similar, but not too similar, has been termed optimal inbreeding.

Biologists do not yet know how animals discriminate degrees of genetic relatedness, though it is known that species as different as monkeys and mice can reliably distinguish between relatives and nonrelatives when meeting them for the first time. In many birds, it appears that during the juvenile stage a form of imprinting (➧ page 12) occurs so that individuals come to recognize their siblings' adult appearance and use that information when they mate.

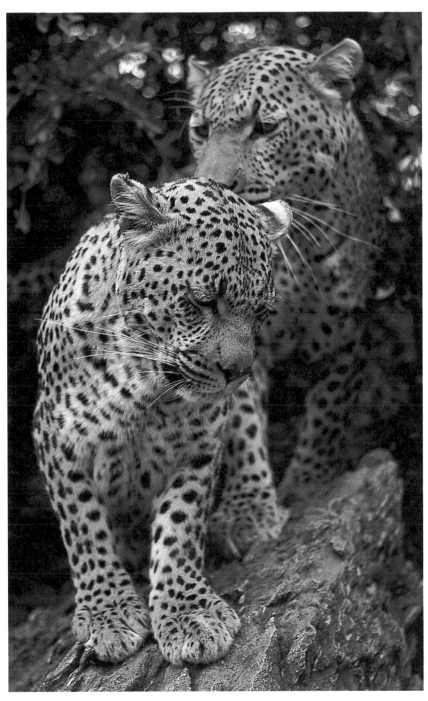

▲ Since only large male red deer stags can produce loud roars, an individual whose territorial status has been challenged will opt first for a roaring contest. Only if the two stags can match roar for roar (itself an exhausting activity) will they resort to physical fighting.

◄ In some insects, males that mate with a female can flush out the sperm of a previous mating. So that its sperm is not wasted in this way a male blue damselfly uses special pincers at the end of its abdomen to hold the female by the head while she lays her eggs.

► In many species, courtship involves elements of submissive behavior in order to reduce the female's tendency to flee from the more aggressive males. A male leopard nuzzles against the female's nape, both to calm her and to stimulate her willingness to mate.

▼ The courtship displays of water birds represent one of the high-points in the evolution of ritualized behavior. A male and female great crested grebe are shown at successive stages in the intricate ballet of their courtship ritual, which maintains a lifelong bond between them.

In elephant seals, the males can grow to 6m in length and weigh up to 4 tons while females rarely exceed 4m and 1 ton

▲► The common tern is a monogamous breeder. During courtship the male brings food to the female, who sits on the selected nest site (1). Terns generally pair for life. Moloch gibbons (2) perform a mutual duet that not only helps in forming the pair bond, but also to maintain and develop it. They too are monogamous. Northern elephant seals (3), howler monkeys (4) and capercaillies (5) are polygamous and display sexual dimorphism. The male capercaillie calls to the drab female on the courting ground, who adopts a mating invitation position.

The female does not always have to do the choosing herself. She can achieve the same effect by allowing males to fight over her. By defeating all the other males in the neighborhood, the winner demonstrates his fitness and prowess. Wild sheep and goats, for example, engage in ferocious fights in which the males crash horn against horn in bone-shattering contests.

Courtship may also be important to the male in allowing him to assess the female's quality. In species where the male's help is essential for rearing the offspring, a female who has already been mated can easily deceive another male into rearing her offspring if her mate deserts her. Courtship rituals give a male the opportunity to assess the female's honesty. A male pigeon, for example, will stop courting a female that responds too quickly and will chase her away. Female pigeons require the male's courtship to stimulate their reproductive physiology into readiness for egg-laying (◀ page 39). A female that responds too quickly is likely to have been courted (and therefore already mated) by another male.

Competition and the evolution of weapons

One of the consequences of competition for access to reproductive females has been the evolution of large body size among males. The relative sizes of males and females reflect the species' mating system. In species that mate monogamously, males and females are normally of the same size. In species that mate polyandrously, that is where one female mates with several males, the female is usually slightly larger and more brightly colored than the male. In polygynous species, where each male mates with many females, males are larger and more brightly colored than females. Among polygynous species, the difference in size between males and females is directly related to the ratio of breeding females to breeding males (the breeding sex ratio). This reflects the extent to which males have to fight in order to acquire breeding females. As the males become relatively bigger, they can monopolize an increasingly large share of the females. As a result, there is a steady pressure on the males to grow larger in order to be able to acquire females to mate. Competition for control over females has also led to the evolution of weapons (▶ page 44).

Mating Systems

A species' mating system describes the way in which the animals are arranged in time and space for the purposes of reproduction.

Promiscuous mating systems are those in which males and females mate repeatedly and more or less at random with each other. They are common among insects, which often form mating swarms that contain millions of individuals.

Polygamous systems are those in which a single member of one sex mates with several members of the other sex, each of who mates only with that one individual. Polygamy can take on several forms depending on the time that individuals associate together. Bighorn sheep males defend individual females in estrus as they locate them, moving from one female to the next as each becomes sexually receptive. Red deer stags hold harems that constantly change composition over time as females come and go across territorial boundaries. In many Old World monkeys, on the other hand, the females form stable social groups and individual males may retain control over the whole group for many years (▶ page 50).

Polygamy normally involves many females mating with a single male. This form is known as polygyny (many wives) to distinguish it from the much less common polyandry (many husbands) in which one female mates with many males.

In monogamous mating systems, a single male mates with a single female. Monogamous pairs may be either temporary or lifetime arrangements. Pairbonds that last a single breeding season to be replaced by a new mate the following year are described as serial polygamy. Monogamy usually occurs where the male's help is needed for the female to rear the offspring produced by a mating. About 90 percent of birds mate monogamously, but only about 5 percent of mammals do so. This is because male birds can help incubate eggs and feed hatchlings just as easily as the female. Young mammals are fed on their mother's milk and the males can rarely be of much help.

Polyandry
Mating systems that consist of a single female and several males are rather rare in nature. They do, however, occur in a number of birds (notably the Arctic sandpiper, jacanas, phalaropes, wild turkeys and Tasmanian native hens), sea horses and some species of frog, as well as in a few human cultures, for example the Tibetans. The reasons why polyandry has evolved remain obscure. It may be that differences in resource quality make it more worth a male's while to join another male on his territory than to set up a territory of his own where no females are likely to come (➧ page 68).

Animal Weapons

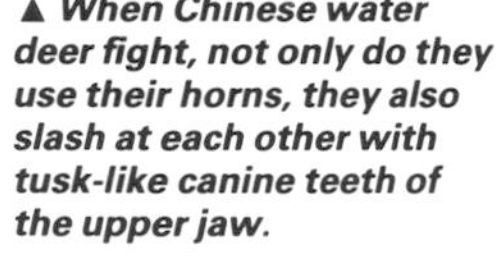

▲ When Chinese water deer fight, not only do they use their horns, they also slash at each other with tusk-like canine teeth of the upper jaw.

▲ For giraffes, contests involve prolonged neck-wrestling. Each animal tries to hook its head round that of its opponent and pull it to the ground.

The fight for supremacy

Weapons used by males of a species to gain control of a female for mating include, among mammals, canine teeth, tusks, horns and antlers. Some of these are of formidable proportions. The tusks of bull African elephants, for example, can weigh as much as 50kg each.

Most species' weapons are designed for use with a characteristic style of fighting. The heavily ridged horns of sheep and goats are designed to sustain the impact of battering-ram charges. The ridges on the horns help to prevent the heavy horns sliding off each other, thereby resulting in a damaging blow to the skull itself. Reedbuck have horns that curve forwards at the tip. These are designed to prevent the horns sliding past each other while the protagonists try to twist each other's heads over on to the floor. Deer such as the wapiti, red deer, caribou and fallow deer have large branching antlers with many tines. These provide a firm surface against which to push when the antlers interlock during contests. The short straight horns of small antelope like dikdik and klipspringer, on the other hand, are designed for jabbing – mainly at the rears of retreating opponents.

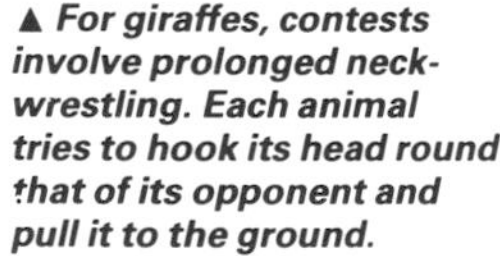

▼ Hippos use their formidable tusks to bite each other about the jaws and head. Although injury is often sustained, this is rarely serious.

▲ Large bovids such as bison engage in pushing matches in which they try to force each other backwards. Eventually one contestant concedes defeat.

► When white-tailed deer fight, they use their branching antlers as firm surfaces on which to push. They try to force each other off balance.

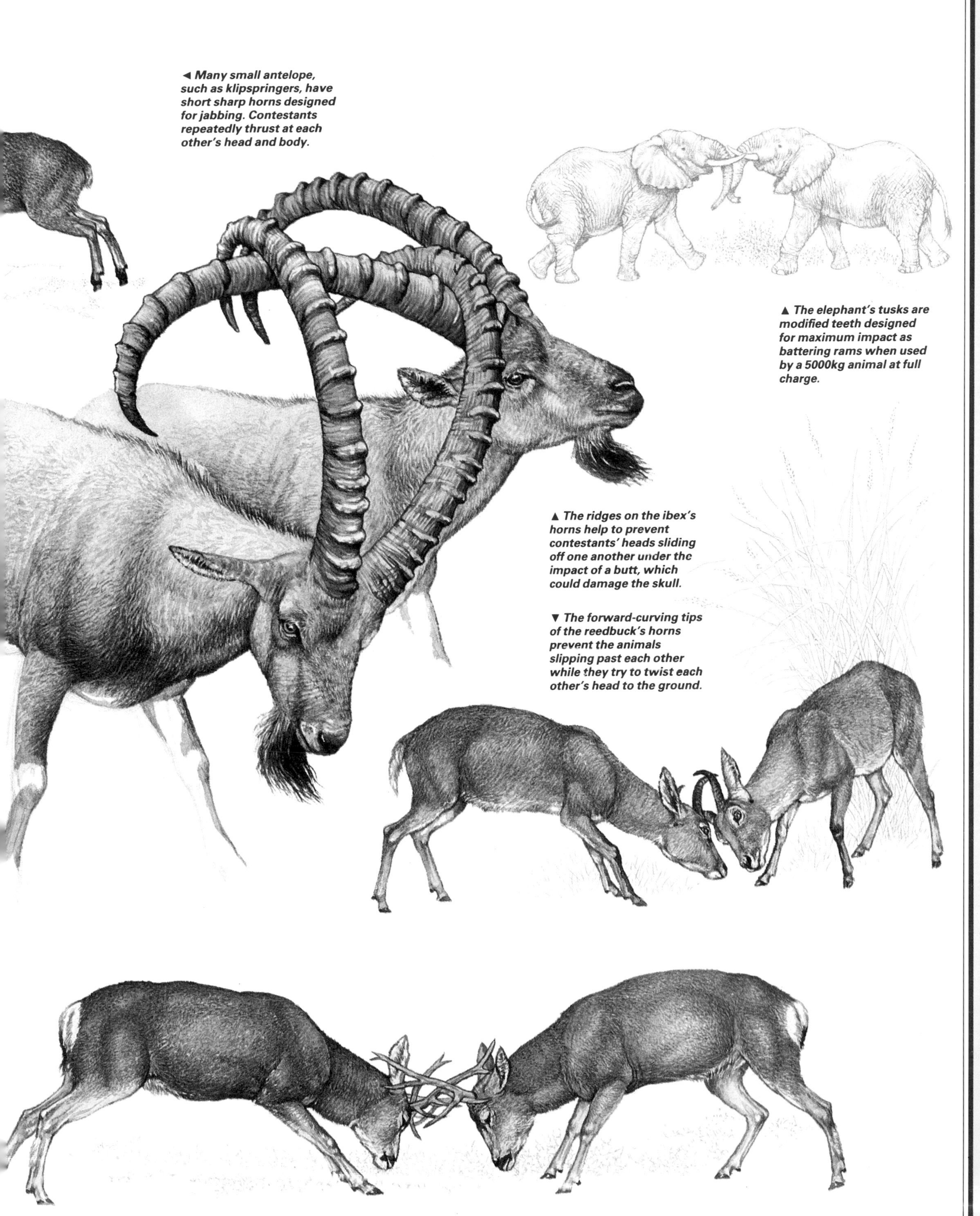

◀ *Many small antelope, such as klipspringers, have short sharp horns designed for jabbing. Contestants repeatedly thrust at each other's head and body.*

▲ *The elephant's tusks are modified teeth designed for maximum impact as battering rams when used by a 5000kg animal at full charge.*

▲ *The ridges on the ibex's horns help to prevent contestants' heads sliding off one another under the impact of a butt, which could damage the skull.*

▼ *The forward-curving tips of the reedbuck's horns prevent the animals slipping past each other while they try to twist each other's head to the ground.*

See also
Groups and Troops 47-62
Territory 63-70
Animal Migration 71-80
The Shadow of Extinction 195-208
Saving Wildlife 235-248

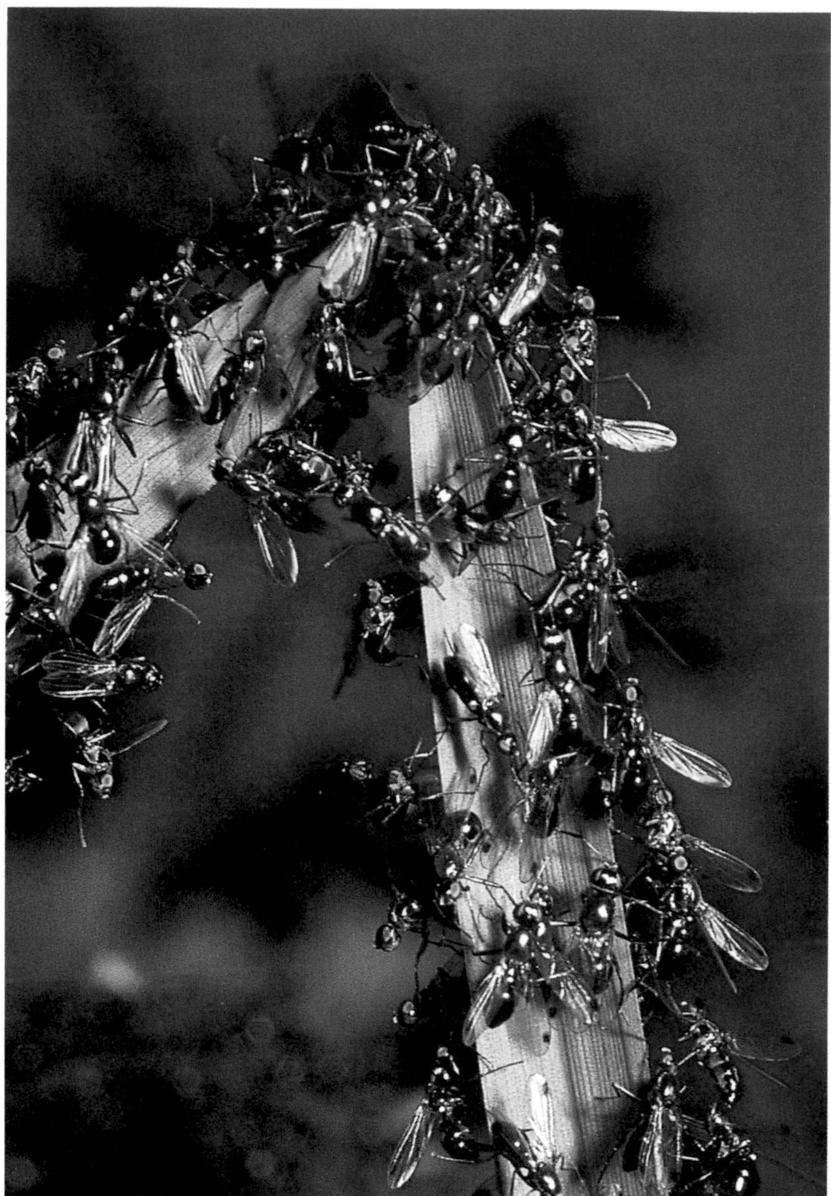

◄ *In many species of insects such as* Sepsis *flies, females visit the mating site only briefly to mate with the males congregated there. Since the males far outnumber females at any given moment, competition among the males for access to females is often very severe. Thus it is not uncommon to find several males trying to copulate with the same female at the same time.*

Because an animal's weapons are capable of delivering fatal wounds, physical contact usually occurs only as a last resort during fights. Initially, most species use displays to frighten off opponents. Cats erect their fur to make themselves look larger, red deer roar at each other and Siamese fighting fish flap their fins while swimming side by side. Such behavior gives a rough guide to the performer's prowess because strong individuals will tire less quickly or can produce deeper-sounding notes. Weaker opponents will usually retreat rather than risk injury in a fight. Since injuries are likely to occur when evenly matched opponents fight, many species have also evolved ways of protecting their more vulnerable areas. Lions (▶ page 62) and gelada baboons, for example, have massive capes that protect their shoulders from damaging during fierce jaw-fencing fights. Bison have similar capes to protect their shoulders from goring, while elephant seals (◀ page 43) have thickened pads of skin on their shoulders for the same purpose. The fighting can become so dangerous, however, that some males prefer to avoid it altogether and pursue less risky alternative strategies for acquiring mates.

Alternative mating strategies

Males usually compete either for control over reproductive females or for territories that give them exclusive access to the resources that females need for breeding.

The costs of fighting, measured in terms of the likelihood of being killed (◀ page 14), can become so high that many males gain no offspring because they die before they have the chance to breed. It may then pay some males to opt out of the fighting altogether. Providing such a male can obtain a few matings now and again, his longer lifespan will allow him to sire as many offspring as a male who fights to control a large harem of females but dies early as a result.

The ruff is a medium-sized bird of the wader family that occurs widely on the coastal mudflats of northwest Europe. Breeding males sport a dark cape or ruff around the neck which is erected during courtship displays. Each male holds a small mating territory from which he tries to attract females to come and mate with him. Territorial males are very aggressive towards other males in breeding plumage. In some cases, an additional male, called a satellite, shares the territory with the territorial male. This individual lacks the conspicuous ruff and always behaves submissively – indeed behaves like a female – whenever the territory owner approaches him. As a result, territorial males normally tolerate satellites. However, when the territory holder is preoccupied mating with a female, the satellite will mate with any extra females that come onto the territory. Although the satellite male does not gain as many matings per day as the territorial male, he does not have to bear the cost of defending the territory against intruders.

Other kinds of alternative mating strategies have also been observed in the wild. In many species of fish, for example, males in breeding condition develop conspicuous colour markings and defend territories. A territorial male courts passing females, some of whom spawn in the nest he has prepared on the streambed. During spawning, small dark-colored nonterritorial males that have been swimming in shoals near the surface dart down and try to lie between the spawning pair, releasing milt (semen) to fertilize a proportion of the eggs as they do so. Such behavior has been observed in sticklebacks, coral reef fish, darters, sunfish and salmon.

In many species of solitary bees and wasps, males defend flowers to which virgin females come to feed after they have emerged from their underground nests. When the competition for flowerheads becomes too fierce, some males stop trying to defend territories and instead cruise widely in search of females that are in the process of emerging in order to mate with them before they set out in search of food plants.

Groups and Troops

5

Types of groups...Group life and reproduction...Social life....Predator evasion and cooperative hunting... Lineages of related individuals...Changing from one group to another...PERSPECTIVE...Friendly relationships in monkeys...Sharing the workload...Vertebrate sterile workers...Altruistic behavior

The fact that there are two sexes means that individuals have to come together in order to reproduce. Although many species do this only for the brief purpose of mating, in the more social species animals live together on a seasonal or even a permanent basis.

A diversity of societies

Although they may form large swarms, many insects are asocial. Individuals come together only to exploit a local abundance of some resource, like flies round a piece of fish. That others happen to be present is, in many ways, accidental and the swarm members do not relate to each other as individuals. The simplest form of true social life is represented by semi-solitary species of mammal like otters, muntjac deer and orangutans. Although individuals of these species spend most of their time alone, they are not asocial. Their knowledge of other individuals in their local area is usually considerable and they have an extensive network of relationships with them.

In marked contrast to these simple social systems are the complex multilevel societies of birds like the white-fronted bee eater.

Clans, territories and home ranges

The white-fronted bee eater is a small, brightly colored bird that lives in colonies of up to 450 individuals in the sandy banks of river gullies throughout the plains of eastern Africa. The birds form monogamous pairs for breeding, but within the colony the pairs are organized into groups of up to 11 individuals called clans. Most of the birds in a clan are genetically related to each other, but unrelated individuals can join by becoming the mate of a clan member. Each clan occupies a specific nest chamber which the members defend aggressively against other individuals from the colony. Each clan also has its own separate feeding territory up to 7 km away from the roost. Within the clan's territory, each pair has its own special home range, though the ranges of individual pairs overlap by up to one third. The territories of different clans, in contrast, never overlap.

▲ Muntjac inhabit grassland and woodland in southeast Asia. Like many small antelope and deer, they are solitary animals, with each individual having its own ranging area. Males occupy larger ranges than females, so each male can access several females during the mating season.

◄ The clawless otter of Asia is a solitary creature yet lives in extended family groups. Breeding pairs form strong bonds, and the male otter helps to raise the young. In the closely related giant otters and smooth-coated otters the female is the dominant partner in the pair.

▲ A blue-footed booby comes in to land at its nest on the desolate cliffs of the Galapagos Islands to relieve its mate. In many seabirds, the male and female share responsibility for brooding the eggs, so allowing each of them to spend time at sea fishing.

◄ For the solitary tiger, the sheer size of its range (60-75 sq km) makes it difficult for a male to find a female when she is in estrus. But by advertizing their whereabouts by roaring, they can keep track of each other's movements.

► Grazing species such as Cape buffalo often form large herds as a means of defense against predators. Within these herds, the males try to establish moving territories in order to monopolize clusters of females.

The hamadryas baboon lives along the desert margins in the Horn of Africa. The basic social unit is the one male group consisting of a single mature male and one to three reproductive females, together with their dependent offspring. These form groups of two to five units called clans. All the adult males of a clan are related to each other. Two or three clans constitute a band which has a common home range (➧ page 50). Bands themselves may group together into troops that share sleeping cliffs, though only certain bands will tolerate each other's presence.

Between the two extremes of the semi-solitary species and those with highly structured multilevel societies there exists a wide variety of social systems. In many forest monkeys, pairs or small groups consisting of a male with half a dozen females occupy territories that are vigorously defended against other members of the species. At the opposite extreme, some species such as the caribou, bison and many of the larger antelope live in vast, loosely structured herds that roam widely over the plains.

The roots of society

It is the network of positive and negative relationships between individual animals that gives a society its distinctive structure. In many mammals, individuals form strong bonds of friendship, often expressed through close physical contact. Among monkeys, social grooming is used to reinforce these bonds. In the hamadryas baboon, however, the males physically enforce the cohesiveness of the one-male units by punishing females that stray too far from them – usually by biting them on the neck or shoulders. Male deer and antelope try to prevent females from wandering off their mating territories by herding them back into the center (➧ page 54).

In many species, the central core of the society consists of groups of closely related females. Among red deer and wild goats, for example, daughters take up residence with or near their mothers, so that groups of closely related females tend to range together and to associate together in loose groups.

Within many monkey societies females spend all their lives in the group into which they are born

Social grooming

Grooming is a very important activity for many species of animal. Picking through the fur helps to keep it clean and healthy. In many cases, animals not only groom themselves but also devote a great deal of time to grooming each other.

While hygiene has been the key factor favoring the evolution of self-grooming, it is not at all clear why social grooming should have evolved. One theory is that animals groom each other to clean those areas that an individual cannot reach itself. This is supported by the observation that most social grooming in monkeys concentrates on exactly these parts of the body – the top of the head, the neck and the back.

However, this cannot explain the fact that highly social species such as horses, monkeys and chimpanzees spend far more time in social grooming than is necessary to keep the fur clean and the skin healthy. Whereas a colobus monkey spends only about 5 percent of its time in social grooming, a baboon of approximately the same size spends up to 15 percent.

In the highly social species, grooming performs an important additional function in reinforcing social relationships. Animals invariably form coalitions more easily with those individuals with whom they groom most often.

Grooming may also be used to pacify another animal after a squabble or to establish a friendly relationship between two strangers. Being groomed is a very relaxing experience. Animals sometimes even fall asleep while they are being groomed. This helps to reduce tension and defuse any aggression that is present.

Defensive strategy

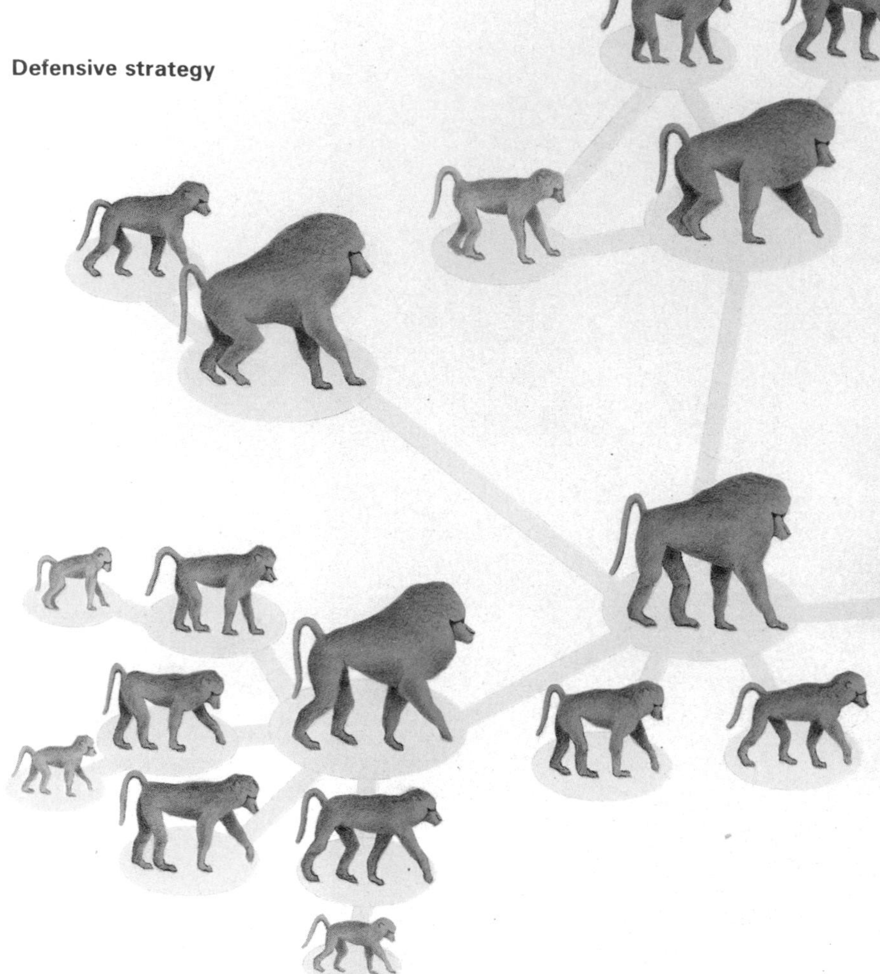

Who grooms who?

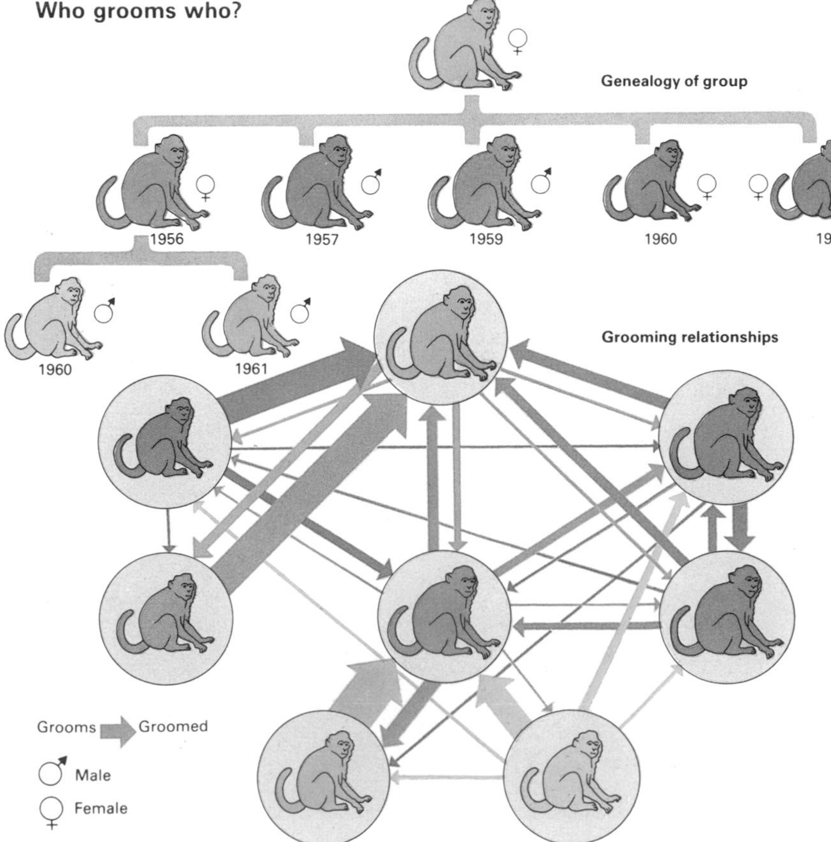

Pairbonds

Successful reproduction often depends on individuals establishing a well-coordinated relationship. In many birds, for example, this relationship is necessary to ensure successful incubation of the eggs as well as the subsequent feeding of the hatchlings. Among monogamous antelope and primates, the relationship literally lasts a lifetime (➧ page 42).

For such a relationship to survive, the behavior of the two animals has to be closely meshed. In some birds, the male and female have to alternate incubation roles. While one member of the pair sits on the nest, the other goes off to feed. For a seabird, this may mean flying a considerable distance out to sea and the partner may be away for a long time. The most extreme example of this is the Emperor penguin which breeds on the icy wastes of the Antarctic during the southern winter (➧ page 53).

In all these cases, the bird on the nest has a problem if its partner does not return at the right time. It has to choose between starving in order to continue incubating the eggs or abandoning the eggs to certain death so that it can go and feed.

◀ In a typical monkey family, as above, a female grooms most with her own mother or daughters (lower diagram). However, a son, once independent of the mother, tends to spend more time interacting with other juvenile males in play groups. The direction and extent of grooming between individuals shows the hierarchy in the family.

▼ ***When a troop of baboons moves across the open grassland, the females often cluster in the center near to particular males with whom they have special relationships. Juvenile males stay at the front and rear.***

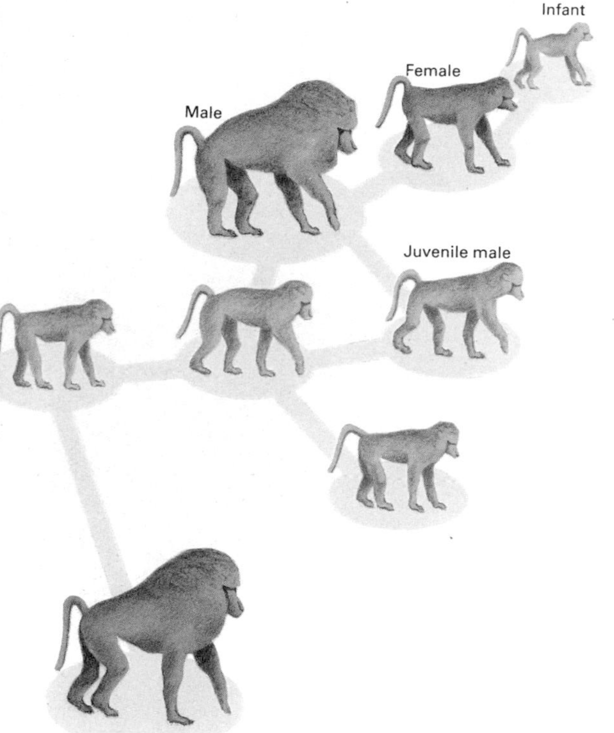

Even in species that seem to lack sophisticated social systems there may still be a considerable structuring of the local population. Individuals do not wander randomly through the habitat, interacting with whomever they happen to encounter. Rather, they restrict their activities to a specific area whose boundaries are often defined by the interactions that the animal has with its neighbors. Knowledge of these neighbors as individuals is often considerable and a subtle structuring of relationships may be found to exist on closer inspection.

A good example of this is the white-crowned sparrow that lives along the Pacific coast of North America. The birds live in monogamous pairs, each on its own territory. There is, however, a marked clustering of pairs into distinct geographical units. These units share a common song dialect that differs from those used by neighboring clusters. The clusters are also genetically distinct. Young birds tend to mate with members of their own dialect group.

A mate must be reliable and return punctually to relieve its partner or it risks losing the offspring that it is trying to rear. If this happens, then both partners have wasted considerable energy.

Klipspringer are small African antelope. Males establish territories in the breeding season. Like many small mammals, predation poses a particularly serious problem for them. One way in which they reduce the risk of being caught by a predator is for the pair to take it in turns to feed and keep watch. While one feeds, the other stands on a prominent rock where it has a good view of the surroundings. This meshing of roles does not come naturally. It has to be learned. New pairs are much less well coordinated than long-established ones.

The poor meshing of newly formed pairs is often reflected in a poor reproductive performance. In kittiwakes, pairs that breed for the second year in succession fledge 12 percent more chicks than they did when breeding for the first time. Failure to develop a compatible sharing of incubation duties is one of the main causes of a failure to breed successfully in this species. Unsuccessful pairs often desert each other and find new mates in the following breeding season.

► ***The males of many species of monkey use infants to reduce the risk of attack from more poweful individuals. In this sequence, a male macaque who is being threatened by another male (in front of it, just out of the picture) persuades a nearby infant to come to him so that he can gain protection from its proximity.***

Within a group of elephants, "baby-sitting" by adolescent females and mutual suckling of calves by related mothers is common

Family units and bond groups

Vervet monkeys are common throughout the woodlands of Africa. They live in closed social groups of 10-15 individuals with relatively little movement of individuals between groups. Yet the monkeys respond with peering and surprise when the recorded calls of individual members of another group are played back through loudspeakers placed in the wrong territory. They clearly know most of the individuals in the groups around them and associate those individuals with particular territories (➧ page 64).

The African elephant is another species with a complex, multitiered society. African elephant society is matrilineal: it is built around the mother and her offspring, in this case two or three dependent offspring. These family units associate closely with one or two other similar units in a bond group. Bond groups in turn associate in clans. Although individual mother-offspring units wander freely over vast areas, they associate most frequently with other mother-offspring units of their own family unit and least frequently with those of different clans even though all of them use the same ranging area.

The pattern of the relationships is reflected in the way the animals greet each other when they meet up during foraging. Greetings between members of the same family unit are very intense and excited. The animals run alongside each other with their heads and ears raised. They back into and rub against each other, and entwine trunks to the accompaniment of deep rumblings, trumpeting and ear-flapping. Animals of different clans do no more than quietly insert their trunks into each other's mouths.

The polygyny threshold

In many species, a female's success in rearing offspring depends on help from the male (➧ page 42). Many species consequently form monogamous pairs and the male helps the female to rear their offspring. If the male dies or deserts the female, she is usually able to raise only a small proportion of that brood by her own efforts.

A great deal, however, depends on the richness of the habitat. In a rich habitat containing many insects, a bird will have to work less hard to obtain all the food its chicks require. Consequently, it may pay a female to join an already mated male on a rich territory rather than an unmated male on a poor territory, even though she may have to feed her nestlings without any help from the male. Dunnocks and red-winged blackbirds of North America do this. If the territory is rich enough, she will just be able to do this on her own.

The difference in richness between the two territories at which a female would do better to join an already paired male is called the polygyny threshold.

A key factor favoring the evolution of polygyny in these cases is that territories vary considerably in their quality. The majority of North American songbirds that mate polygamously nest in marsh or grassland habitats, both of which are much more variable in their food productivity than forests, where monogamy is the predominant mating pattern among songbirds.

◀ An elephant's relationships extend far beyond the circle of its immediate family to include most individuals in the local population. Groups of elephants numbering between about 30 and 60 individuals are common and comprise three or four families. To have reached this size, the groups' families must have been in close association and harmony for more than 100 years.

► The emperor penguin boasts the most enduring pairbond in the animal world. As the long Antarctic winter draws in, the females migrate northward leaving their mates to incubate the eggs. Standing motionless with their single egg tucked beneath their "brood skirt", the males lose up to half their body weight during the four months of intense cold before their mates return to relieve them in the spring.

▼ Bee eaters nest in sandy banks where they can excavate nest chambers easily. Scarcity of suitable sites obliges these carmine bee eaters to live colonially, each colony defending its sandbank against rival groups. Under such conditions, competition for nest holes can become intense, predisposing offspring to remain with their parents as helpers-at-the-nest rather than seeking holes of their own.

Helpers-at-the-nest

In many monogamous species, offspring from one litter often stay on with their parents to help rear the next litter. This behavior has been observed in more than 140 species of bird including scrub jays, woodpeckers, bee-eaters and kingfishers. It also occurs among such mammals as mongooses, jackals and foxes.

Helpers play an important part in feeding the young while these are still totally dependent on the parents for food. In a study of Florida scrub jays, the number of fledglings produced by an adult pair doubled when they had helpers. Among jackals in Africa, pairs on their own managed to rear an average of 1.3 pups to the age of 14 weeks, whereas pairs with helpers reared 3.6 pups.

Helpers can also reduce the risks run by the parents. Repeated trips to find food for the young expose the parents to predation. Adult scrub jays suffer a 20 percent mortality rate each year under normal conditions, but the presence of helpers reduces this to about 13 percent. If one parent does die, the survivor is normally unable to raise the litter on its own. In such cases, helpers can offset the loss of a parent (or even both parents).

In evolutionary terms, staying on to help means that the helper has to forgo its own chances of reproduction. However, if the helper is related to the litter it is helping to rear, then whatever genes it shares with its younger brothers and sisters will get passed into the next generation as a result of its helping behavior. In this way, an animal can make an indirect contribution to the next generation without actually breeding itself (▶ page [illegible])

A queen termite may reach a size of 14cm long and 3.5cm across and produce 30,000 eggs a day. Her king, constantly at her side, remains his normal size

***►* Naked mole rats have only a few scattered hairs and thus have the poorest ability of any mammal to maintain body temperature. Huddling together helps them to keep warm.**

***►►* The silvery flashes of many fish moving rapidly in different directions probably serves to confuse a predator, thus providing one reason for forming schools.**

***◄* The females of most large antelope such as impala form groups in order to benefit from the increased chances of detecting an approaching predator.**

***▼* For termites, successful reproduction depends on a cooperative effort between workers, soldiers, king and queen, both for the construction of their near impregnable mud mounds and for tending the fungus gardens that provide them with their food.**

Why form groups?

Grouping patterns usually reflect important environmental problems that affect the animals' ability to survive and reproduce in a given habitat. Biologists think that group-living evolved for at least three reasons, though in any given case only one of these is likely to be important. Groups allow animals: to avoid predation, to rear their offspring more effectively, or to forage more efficiently.

Bunching together is a common response to the threat of attack by a predator. This allows animals with horns, for example, to face the predator with a formidable wall of dangerous weapons. Generally speaking, predators are reluctant to attack large groups, preferring to select their victims from among the sick and disabled. Forming groups can help to reduce the risk of predation in other ways too. An animal in a group can rely on its group-mates to warn it of danger, for example. It can therefore devote more of its time to feeding without having to spend so much time scanning the surroundings for approaching predators (➧ page 111).

In many species of birds, the male and female form a temporary pair in order to be able to feed their voracious young more easily than either could do alone. In general, an individual male is able to contribute most descendants to future generations by mating with as many females as possible and leaving the females to rear the resulting offspring. However, if the female cannot rear as many offspring on her own as she could with the male's help, then it may be to the male's advantage to stay with her after mating in order to help out. It is for this reason that monogamy is so common among birds. Here, incubation of the eggs and the feeding of the hatchlings can be done equally well by either parent. Although there are exceptions (➧ page 52), the energy and time costs involved in rearing the young are so great that a pair working together can raise many more young than either of them can working alone.

Termite mound architecture

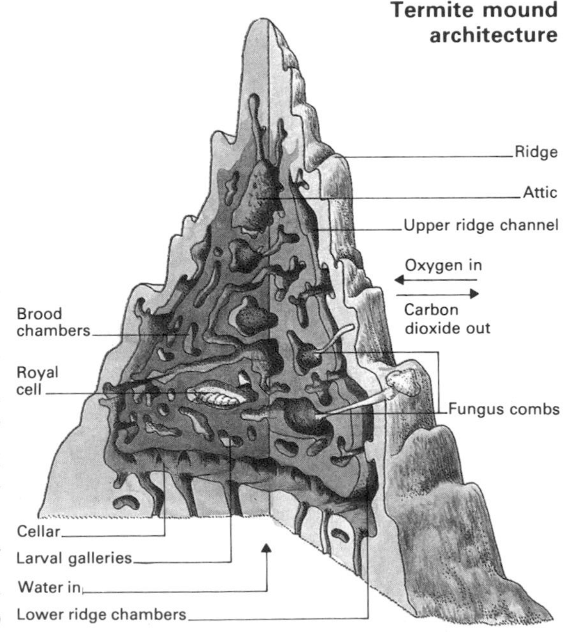

Division of labor

Older offspring sometimes stay on with their parents as helpers-at-the-nest in order to assist with the rearing of litters of younger brothers and sisters. This process has been carried to an extreme in the social insects where whole cohorts of offspring become sterile workers that gather food, clean the home, look after the larvae and guard the hive or nest against predators in order to ensure that their reproductive brothers and sisters are reared successfully (➧ page 61). So far as is known, physiologically sterile helpers occur in only one species of vertebrate – the naked mole rat.

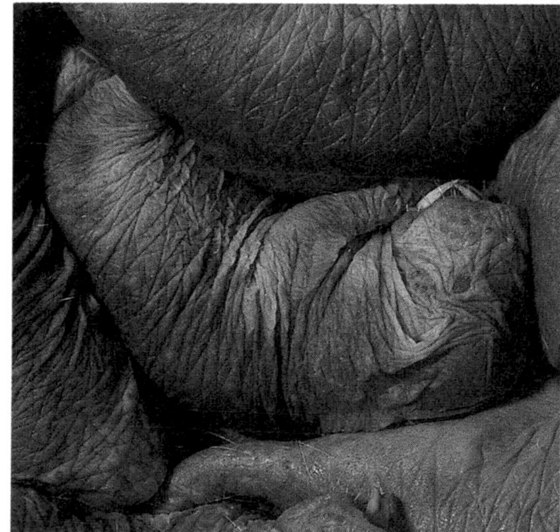

The caste system of naked mole rats

Naked mole rats inhabit the drier areas of eastern Africa where they live in complex systems of burrows. From a deep central nest chamber, an extensive 3-4km long network of tunnels extends out over an area of about 12ha. These tunnels lie quite near the surface and the plant roots and tubers that protrude into them provide the colony with all its food requirements. Except when washed out during the rainy season, mole rats spend their entire lives underground.

Each tunnel system is owned by a colony of 40 to 50 mole rats. Only one female in the colony is reproductively active and she produces up to four litters of 12 pups each every year. All the members of a colony are thus the offspring of the reproductive female. Successive litters form castes of workers that are responsible for all the work that has to be done in the colony.

The youngest and smallest do most of the work digging tunnels, transporting soil and bringing food back to the nest. Each worker has its own small set of tunnels for which it is responsible, so that its activities are spatially quite restricted. Many animals remain as workers throughout their lives, but some continue to grow and eventually join a caste of larger workers. These are involved in maintenance tasks less often, and spend more of their time patrolling widely throughout the entire tunnel system. Their exact function is unclear at present. However, if the tunnel system is broken open, it is these larger workers that are usually caught nearest the opening, suggesting that they may be active in the defense of the colony.

Finally, the largest animals become a caste of nonworkers. Most of these are reproductive males, but there are a few nonreproductive females as well. These females appear to be postreproductive and may act as "aunts" to look after the newborn pups. In contrast, while the females in the other two castes of workers are also nonreproductive, they are far from being sterile. Their ovaries are quiescent. If the colony's reproductive female dies or is removed, one of these females will suddenly start to grow and become reproductively active and will eventually take her place.

The ability of sheepdogs to herd sheep is a direct behavioral carry-over from the cooperative hunting techniques of their ancestral wolves

Groups may also allow their members to obtain their food more efficiently. In general, this explanation is likely to apply only to carnivores that hunt large prey or to species that live in unpredictable environments. Lions form groups because doing so enables them to capture large prey such as zebra and buffalo. Weaver birds are grain-eating finches that occur widely through the savanna grasslands of Africa (➧ page 108). They form nesting colonies that contain upwards of a million birds. Although these giant colonies do provide protection against predators, they also act as information centers where birds can find out about the availability of food sources over a very wide area. Flocks of birds forage out in all directions from the central roost during the day. Those birds that found little food on the previous day hang back and wait to follow birds that confidently head off for good feeding areas that they know about.

► **A single tall tree can provide space for thousands of weaver birds to build their nests, each hung from a terminal twig where it is safe from predators.**

Cooperative hunters

Several species of large carnivore live in groups. Wolves live in families that consist of a single adult pair and their offspring, some of whom may be mature but nonreproductive. African hunting dogs and hyena, on the other hand, live in large packs that contain up to 80 individuals. By hunting in groups, these species are able to bring down prey that are significantly larger than themselves. Wolves hunting in packs, for example, can bring down moose even though these massive animals are many times their own size. (An adult male moose can grow to 230cm high at the shoulder and weigh up to 800kg.)

Hunting success with all prey increases steadily as group size increases. About 15 percent of hunts result in a successful kill when lions hunt alone, but this rises to a maximum of 33 percent with groups of six or more. This is most marked when hunting large prey such as zebra. In this case, the success rate for groups of six or more lions hunting together is 43 percent. (By contrast, another member of the cat family, the tiger, is a solitary hunter and its success rate with prey rarely exceeds 8 percent of hunts.)

Groups are more successful in hunting because the members of the group can coordinate their behavior to prevent the prey animal escaping. When a predator hunts alone, it has to follow the twists and turns of the prey as it tries to escape. This often results in the predator overshooting when the prey reverses its direction suddenly or even losing it altogether when it goes into cover. When predators hunt in groups, however, some individuals can station themselves so that their companions drive the prey towards them.

Predators are also more likely to be successful if they do not have to chase a prey animal continuously but can take turns at doing so with other group members. Wolves and hunting dogs, for example, select a victim and then run it down, but only one member of the pack actually worries the victim at any one time. The others lope along behind at a more relaxed pace or save energy by anticipating the prey's movements in order to cut corners. When the lead animal tires, it drops out to rest and another fresher animal takes its place. Only the healthiest prey animals can succeed in outrunning predators when they hunt in this way.

▲ **Revealing their wolf ancestry, two Australian cattle dogs round up a herd of cattle. By coordinating their actions with each other and with the herdsman, they behave like a wolf pack during a hunt.**

► **Nose down to the scent, African hunting dogs run their prey down until it collapses with exhaustion. Though not especially fast, the dogs' endurance is remarkable and, by taking it in turns to rest and to harry the quarry, a pack can wear down animals as large as zebra. Each pack comprises up to 10 adults.**

Wolf and sheepdog similarities

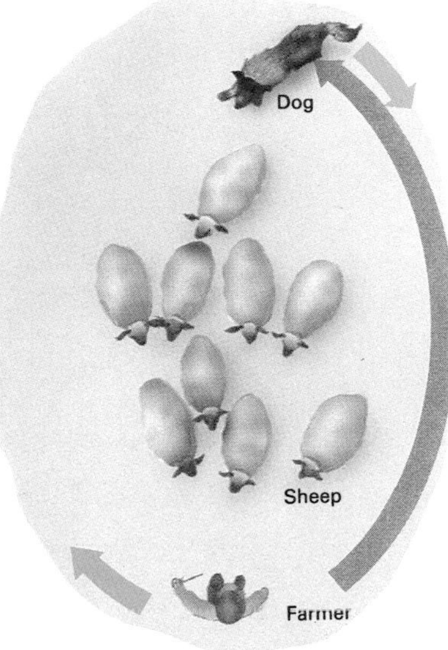

◀ ***When a wolf pack hunts caribou, it may force its quarry into a confined space such as a clump of trees. Then an individual may work its way around to the front of the prey while the other wolves drive the animals into the trap (left). Another strategy involves separating a young caribou from defending adults. The youngster is then easy prey for just one wolf. A sheepdog (right) uses the same techniques, responding to the shepherd's instructions by going beyond the flock so that between them the dog and the shepherd can drive the sheep to the left towards the sheepfold. Domestic dogs, like wolves, are very social animals. When allowed to run wild, they will sometimes form stable packs – a legacy from their wolf ancestors.***

Among African huntings dogs only the dominant female in a group breeds. The other adult females may breed occasionally but their success rate is very poor

► Within the "pecking order" in domestic fowl, low-ranking birds never peck higher-ranking ones. In a newly formed flock of roosters, individuals beneath the most dominant bird compete with one another until a stable, linear hierarchy is established (week 36). Encounters involve repeated threats and fights (♦ page 14), but after the issue has been settled, each individual gives way to its superiors with the minimum of hostilities. In large flocks, some low-ranking individuals form alliances which en masse are able to intimidate higher-ranking individuals. The simple hierarchy then breaks down and is constantly in a state of flux.

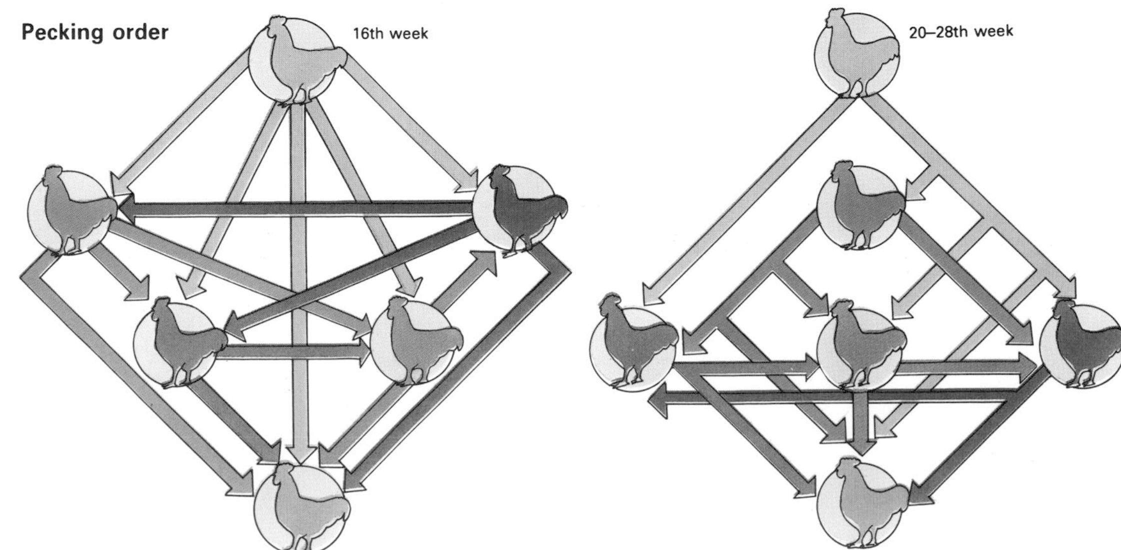

Group-life has disadvantages

While there are important advantages to living in groups, group-life also has consequences which are less desirable. Because animals are crowded together into a small area when they live in groups, tension inevitably arises over social space and access to the best feeding places and sleeping sites. One effect of this competition is the formation of dominance hierarchies. The more powerful animals within a group are able to prevent others getting access to the best resources. As a result, individuals at the bottom of the hierarchy are forced to occupy peripheral positions at a roost where they are more likely to be caught by predators and they have to make do with food of poorer quality. Low-ranking animals are also harassed by individuals above them in the hierarchy. The stress caused by this often results in reduced fertility through disruption of reproductive physiology. In some societies, individuals seem to defer more to related individuals than to non-related ones; family loyalties seem to operate.

Faced with these problems, animals usually try to find ways of reducing the disadvantages of low status in a group. One possibility is to leave the group and join another. Among gelada baboons, for instance (◆ page 50), low-ranking females who experience reduced fertility as a result of harassment by dominant females become increasingly willing to desert in favor of another male as the group size increases. Eventually, the group will undergo fission and some of the females will split off to form a new unit with another male. In doing so, each reduces the number of females dominant to her even though her own competitive abilities remain unchanged.

◄◄ Two male arrow poison frogs compete over territorial rights.

◄ A female house mouse can produce as many as 50 offspring in a season. However, when food resources are limited or overcrowding becomes a problem this figure may fall to single figures. Lemmings, close relatives of mice, respond to overcrowding by mass migrations.

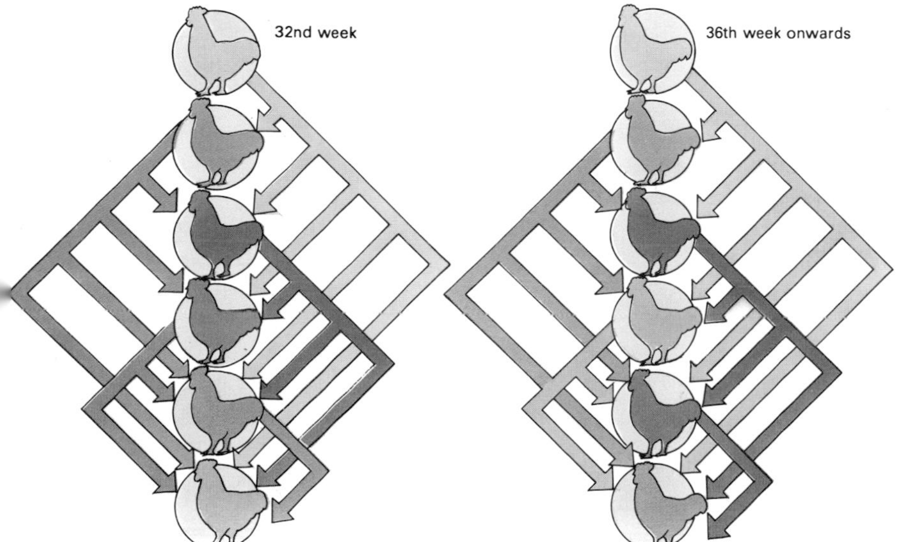

Reproductive suppression

One of the most serious side-effects of social life is that stress can lead to infertility. Both physical injury and the psychologial stress resulting from harassment causes large quantities of endogenous opiates (the brain's own painkillers) to be released into the bloodstream. These are intended to buffer the body against the pain of injury or stress. However, the opiates also interfere with the animal's reproductive physiology. In consequence, production of the hormones that control the reproductive cycle is blocked, and thus results in reproductive suppression. Studies of women athletes under severe physical training regimes have shown that menstrual cycles are often seriously disrupted because the pain caused by hard exercise causes the release of endogenous opiates. There is also evidence to suggest that males may suffer from reduced fertility under similar circumstances.

This phenomenon has been noted in a wide variety of species living in many different kinds of social systems. In many strictly monogamous species (◆ page 42), such as marmoset monkeys and klipspringer antelope, an offspring that remains on its parent's territory will not go through puberty. As soon as it leaves or is removed, however, the animal undergoes puberty at once and becomes sexually active.

Reproductive suppression need not always be total. Among many species of monkeys females experience increasing amounts of harassment from higher ranking females as their own rank in the hierarchy declines. As a result, they suffer from increasing frequencies of infertile estrus cycles. This does not mean that they never conceive, but it does mean that it usually takes them longer to become pregnant than it does a higher ranking female. Consequently, they have fewer offspring over the course of a lifetime than dominant females. Among gelada baboons, for example, the top-ranking female in a unit gives birth to an infant once every 21 months on average. The fifth ranking female, on the other hand, gives birth once every 33 months, while the tenth ranking female does so only once every 70 months on average.

Even though females suffer from reproductive suppression by remaining in a group, they may prefer to do so rather than to face the uncertainties of leaving the group to live on their own. Female African hunting dogs suffer very much higher rates of predation while migrating between groups than they do as members of a group (◆ page 74).

A female has to balance the relative costs and benefits of group-living before deciding whether or not to leave a group. If the risk of being caught by a predator is high, it will be to her advantage to remain in a group providing she can expect to breed eventually. Staying in the group at least guarantees a certain minimum reproductive output, whereas leaving offers a prospect of better conditions (if all goes well) that is at the same time offset by a risk of total failure if the animal is caught by a predator before it has had time to join another group. Within the group there is also the option of helping close relatives to raise young and thereby perpetuate some of one's own genes (◆ page 61).

▲ *A meerkat acts as baby-sitter for its young sisters and brothers while their mother is away.*

▼ *Among Bedouin, if a man dies his widow is taken over by his brother to perpetuate some of his genes.*

Forming coalitions

Some animals may not be able to change groups. Many species are reluctant to accept strangers and will chase them away from their groups (➧ page 68). Even in those cases where tolerance is displayed, moving from one group to another invariably means spending time alone and this increases the risk of predation. So many animals prefer to remain in their groups and to buffer themselves against the consequences of low dominance rank by forming coalitions with other individuals. By forming an alliance, two low-ranking animals may be able to defeat a more dominant animal.

For a coalition to work, one individual has to be prepared to risk injury in order to further the interests of its ally. For such behavior to evolve, an animal must be paid back in some way for the losses it incurs. This can be solved in one of two ways. First, coalitions formed with relatives are less demanding in this sense than those formed with nonrelatives because relatives have a proportion of their genes in common. An animal that supports a close relative against a more dominant individual helps that relative to reproduce more successfully. It will be paid back through the extra copies of the genes it shares with the relative that are contributed to the next generation. This is known as kin selection.

Kin selection

In most sexually reproducing species, each individual receives half its genes from one parent and half from the other. In turn, it passes on half its genes to each of its offspring. Relatives thus share a proportion of their genes by virtue of their descent from a common ancestor.

The probability of sharing a given gene can be calculated very precisely, since at each successive generation the probability of that gene being passed on from parent to child is exactly one half. Hence, parents and offspring are related to each other by a half, as are full siblings (brothers and sisters born of the same two parents). Two half-siblings (siblings with only one parent in common) or an aunt and her niece are related to each other by one quarter, as are grandparents to their grandchildren. Two cousins are related to each other by one eighth, and so are great-grandparents to their great-grandchildren. The probability of two individuals of the same species having any particular gene in common is given the mathematical symbol "r".

*In 1964, the English biologist William Hamilton (b. 1936) pointed out that a given gene could be passed on from one generation to the next in two quite different ways. One is the conventional mode of reproduction in which the gene passes directly from a parent to its own offspring. The other is by helping a relative that also has that gene to reproduce more successfully (**➧ page 53**).*

The problem from an individual's point of view is that in order to help its relative reproduce, it invariably has to use time and/or energy that it could otherwise have devoted to its own reproduction. Helping another may also place the helper at some risk of injury or death (as when it defends the relative against attack by a predator or even another member of their group). Biologists refer to such behavior as altruistic because an individual risks giving up some (or all) of its own reproduction in order to further the reproduction of another individual. However, providing the number of extra genes passed on by the relative as a direct result of the altruist's assistance exceeds the number of its own genes that the altruist fails to pass on, then helping behavior of this kind will evolve.

This relationship can be expressed formally in the equation: $C < rB$, where C is the number of offspring that the altruist loses as a result of its behavior, B the number of extra offspring that the recipient gains as a result of that behavior and r is the coefficient of relationship between them. The variable r does not literally mean "degree of relatedness" but that is what it refers to. In practice it varies from 0 (meaning no relatedness) to 1 (the value for genetically identical individuals).

The coefficient of relationship (or degree of relatedness) converts the recipient's extra offspring into offspring-equivalents from the altruists's point of view. It takes into account the fact that the probability of both altruist and recipient sharing any given gene decreases as they become less closely related. One consequence of this is that the recipient will have to produce proportionately many more offspring to make helping worth the altruist's while as they become less closely related. From an evolutionary point of view, one's own offspring are worth very much more than anyone else's offspring, except the offspring of an identical twin. Identical twins have exactly the same set of genes because they are produced from a single fertilized egg. The offspring of one is related to the other by a half, exactly the same value as a parent to its own offspring.

*The principle of kin selection has helped to explain such behaviors as alarm calling in birds and small mammals, sterile workers in bees and termites (**➧ page 54**), coalition formation in monkeys and tolerance of others at feeding sites in almost all species (**➧ page 63**).*

▼ Genetic relatedness (below left). In diploid species that possess two sets of chromosomes (i.e. most vertebrates), each parent contributes half its own genes to each offspring (upper diagram). In litters sired by the same male, siblings also share half their genes in common. In the honeybee, on the other hand, males possess only a single set of genes (haploid) even though females possess the normal two sets (lower diagram). Since one male does most of the breeding in a hive, sisters normally have the same set of genes inherited through the father, together with the usual half set from the mother. Hence, sisters share three-quarters of their genes in common with each other, but share only the usual half set with their own offspring. Being more closely related to their sisters, it pays them to help their mother produce more sisters than to breed themselves. All vespid wasps (below) are highly social and form colonies of hundreds or thousands of individuals that occupy elaborately constructed paper nests.

Genetic relatedness

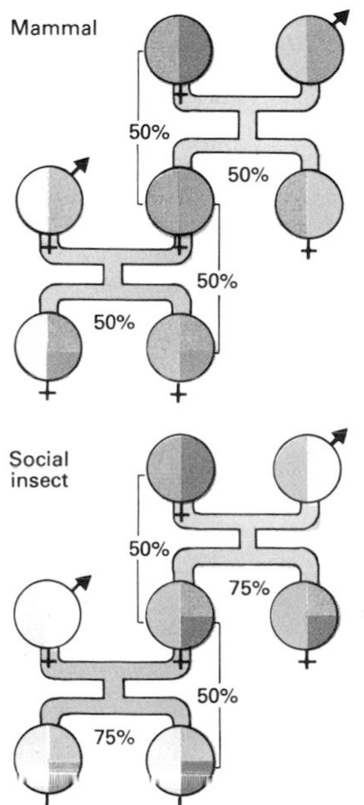

See also
The Limits of Animals 25-34
Sex and Species 35-46
Territory 63-70
Communities of the Land 97-116
The Shadow of Extinction 195-208
Saving Wildlife 235-248

▲ Male lions often form an alliance to take over a pride of females. Depending on competition from rival males, they will control the pride for between 2 and 10 years.

► An alarm call from a single sentinel is enough to warn an entire colony of ground-dwelling and burrowing long-tailed marmots of an approaching predator.

Unrelated individuals will be less willing to give away support since they cannot expect to gain through kin selection. Coalitions between unrelated individuals therefore tend to be reciprocal in nature. Support is given on the assumption that it will be returned in kind in the near future. A male baboon will support another against a third male that is trying to take an estrus female away from his ally. Although the first male supports the second, he is not himself able to mate with the female. Instead, when he later finds himself in a similar position, he can count on his ally coming to his aid.

Both male lions and male wild turkeys form coalitions in order to gain control over groups of females. These coalitions sometimes consist of brothers or half-brothers, but coalitions of unrelated males are also common. Large coalitions are more successful both in gaining control over prides of females and in retaining control in the face of competition from other males. Each male in a large coalition consequently fathers more cubs than males in small coalitions.

Territory

6

What constitutes a territory?...Territory size and species' body size relationships...Defending food and breeding rights from competitors...Breakdown in the territorial system...PERSPECTIVE...Fight or play a waiting game?...Marking boundaries...Resident always wins...The best way of exploiting resources

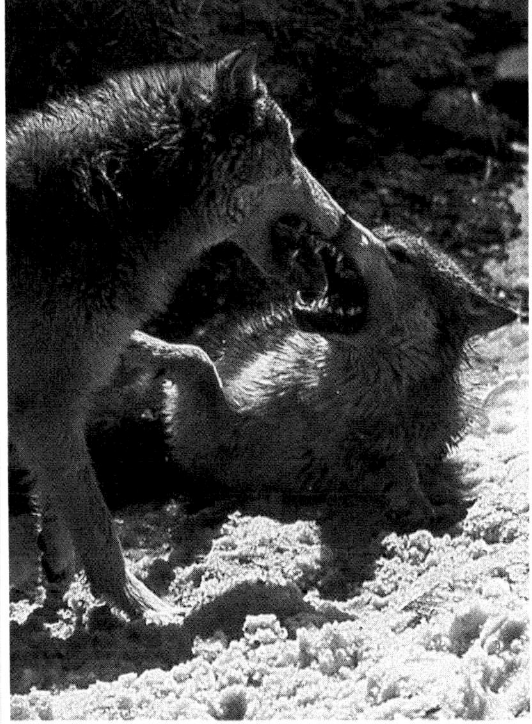

Animals need to avoid overcrowding because that both reduces the available food resources and disrupts the animals' reproductive physiology (◆ page 59). Most animals therefore maintain a space around them within which they will not tolerate unrelated individuals. When this defense of space is extended to cover a clearly defined area within which all the animal's activities are confined, biologists consider that animal to be defending a territory.

What is territorial behavior?

A territory is an area defended against intruders of the same species by an individual or group. Animals defend territories not only by chasing intruders out, but also by various indirect means of signaling territorial occupancy such as songs, displays and the marking of boundaries (◆ page 66). Songbirds like the American song sparrows defend territories in this way. On the other hand, animals such as the bison and the caribou do not defend territories, but range over a wide area from which they make no attempt to exclude other individuals. Between these two extremes lies a variety of behavior patterns. Many colonially nesting birds such as seagulls, herons and flamingos defend a small area immediately around their nests in large colonies that contain upwards of a million pairs. Is this territorial behavior or defense of personal space?

▼ A pair of Cape gannets (foreground right) defend their territory against an intruder (on left). In the vast colonies of many seabirds, nesting pairs are often so densely packed that squabbles with neighbors create a cacophony that acts as a constant background to breeding. Colonies of 100,000 are common.

▲ Each family of wolves occupies a territory that is too large to patrol effectively. Instead, they rely on strongly scented urine deposited around the territory to notify intruders that the territory is occupied. Within the territory, dominance interactions are common, with males often fighting one another.

During the course of a year, a baboon will travel about 2,500km within an area of 35km², constantly moving between feeding areas, waterholes and sleeping sites

***►* Relative territory size (circles) and group size in various primate species. Territory size depends on: (1) the species' diet (leaf-eating howler monkeys have smaller ranges than fruit-eating langurs), (2) where the animals feed (arboreal howlers and langurs have smaller ranges than terrestrial baboons and gorillas), (3) the quality of the habitat (desert-living bushmen have larger ranges than forest-living gorillas or woodland baboons) and (4) the size of the group (arboreal fruit-eating gibbons have smaller groups than the ecologically similar langurs). Within populations, territories overlap by varying degrees.**

African bushman 708km² for group of 20

Baboon 24km² for group of 40

Mountain gorilla 16km² for group of 17

North Indian langur 4.8km² for group of 25

Howler monkey 0.8km² for group of 17

Gibbon 0.16km² for group of 4

Territories provide their owners with exclusive access to one or more key resources. In most cases, these are either females with whom to mate or the food resources and/or nest sites that the female needs in order to rear young. Male bluetits in Europe and indigo buntings in North America acquire territories early in the spring from which they sing to attract mates. Once a male has acquired a mate, the pair set about building a nest and then rearing the young birds that eventually hatch out. The territory provides the hardworking parents with an exclusive supply of insects to feed to their voracious young.

Territories that function purely for the purposes of mating are held by male hartebeest and kob, both large antelopes that are widely distributed on the African plains. During the breeding season, the males defend territories that are too small to provide even a single adult with enough food to survive (➧ page 68). These territories are strictly mating territories which give the owner exclusive rights to mate with any females that happen to come onto the territory. Male impala – another African antelope – move their territories around so as to follow the seasonal movements of the female groups as these search for the best grazing conditions.

Determinants of territory size

The size of a territory depends on the purpose it serves for the territory owner. Those that are used exclusively for mating are usually very much smaller than those used mainly for feeding. Sage grouse, for example, defend small mating territories that are only 5-10m in diameter on communal display areas that are called leks. Though a very much smaller bird, the nuthatch, defends a feeding territory that is well over 200 m in diameter.

The size of a feeding territory is also determined by the species' diet, by the number and size of the animals that have to live on it, and by the quality of the habitat. Fruiting trees tend to be rather patchy in their distribution and animals have to move long distances from one to the next. Fruit-eaters therefore tend to have larger territories than leaf-eaters. Thus, in tropical rainforests (➧ page 100), the fruit-eating common gibbon, a small Asian ape, defends a territory of about 55ha, whereas the siamang, a leaf-eating gibbon, can manage quite adequately on only 25ha despite being a much larger animal.

Since large animals eat more than small ones, territory size increases steadily as the species' body size increases, a relationship that has been widely found in groups of animals as different as birds and monkeys. Similarly, the more animals that have to live on the territory, the larger is the area needed. Finally, for any given type of diet, the richer the habitat in terms of the amount of food it produces, the smaller the area an animal will require in order to obtain the food it needs to sustain itself over a given period. The dunlin is a small wading bird. In southern Alaska it defends a territory of about 300 m². Further to the north, where the summers are shorter and the food supply less predictable, territories average 1500 m².

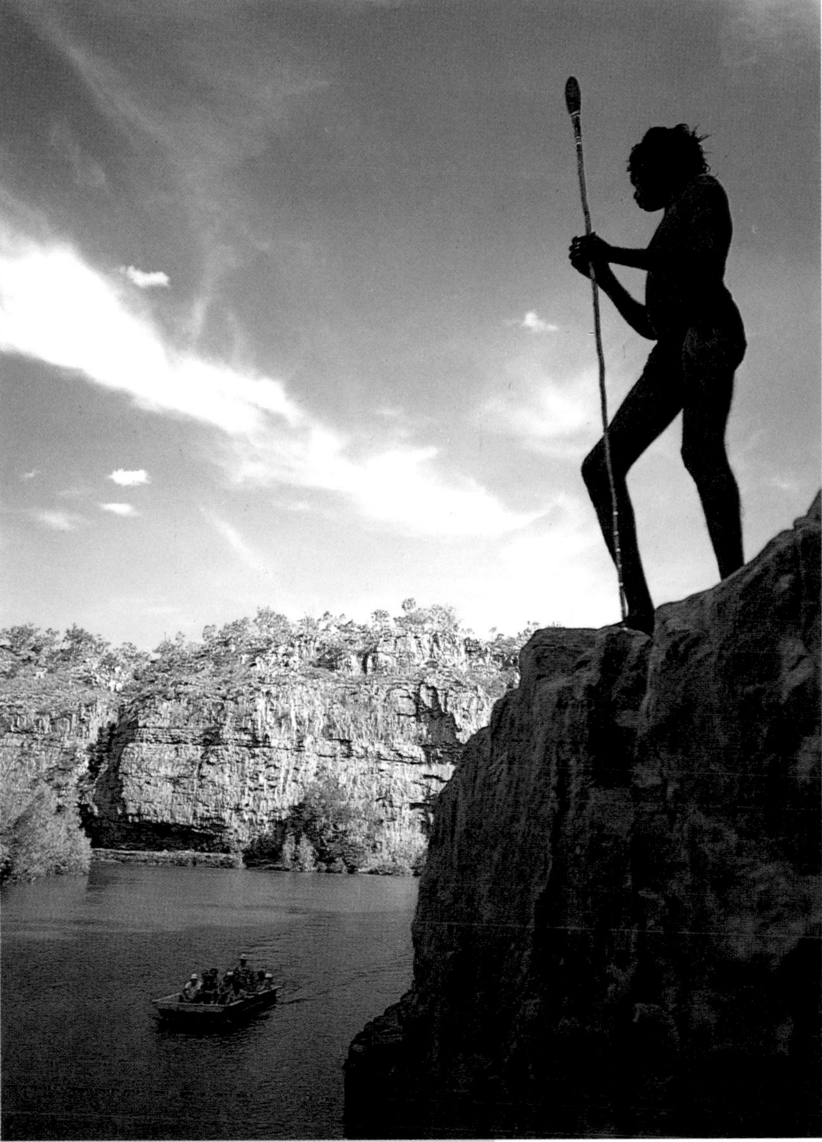

How to acquire a territory

In species that hold breeding territories during only a small part of the year, the first individuals to arrive at the breeding site each year obviously acquire empty territories. Latecomers, however, are obliged to fight males who are already in possession of a territory. In such cases, a male's fighting qualities are of paramount importance.

Generally, territorial males are larger and stronger than nonterritorial males. But, territory-holders lose weight rapidly because they cannot feed properly while defending their territories against intruders. They are easily displaced by nonterritorial males in better condition. This weight-loss has been observed in species as different as green treefrogs and hartebeest antelope. Hunger may even force them to abandon their territories before they are displaced.

Since the costs of fighting for territories often become considerable, males sometimes pursue alternative methods of acquiring them. One of these is to join a territorial male as a satellite (♦ page 46). A subordinate male waterbuck, for example, makes no attempt to challenge the territory-owner in whose territory he lives. He invariably behaves submissively when challenged and makes no attempt to compete directly for the females. Instead, he waits until the territory-holder is forced to leave the territory in order to feed. Then, he simply takes it over.

In many species of treefrog, males defend positions around the edges of ponds from which they attract females ready to spawn by giving loud mating calls. Many of these males have silent satellites sitting near them. These males are waiting for the owner to leave his calling station in order to spawn in the pond with a female.

▲ In the arid deserts of Australia, survival for the aborigines depends on securing access to essential resources. Permanent rivers provide not only water but key food plants on which a group must rely during the dry season. Territories extend over several hundred square kilometers.

◄ Male frogs and toads, such as the Costa Rican puddle frog, defend calling sites in places where females come to spawn. Croaking serves both to advise other males that the site is occupied and to attract the females to come and mate (♦ page 37). The sexual cell is species specific.

► Successful reproduction for social-living prairie dogs depends on having a burrow located in a safe area, such as the center of a large colony (♦ page 111). A family group (or coterie) will vigorously defend its burrow against strange intruders with both loud calls and chases.

Territorial Signals

◄ Howler monkeys from South America give very loud and distinctive roars that are taken up by neighboring groups so that waves of these calls can be heard spreading through the forest around dawn. They establish feeding areas that change each day.

▼ Songbirds such as yellowhammers advertise their territorial status by singing from conspicuous perches. Nonterritorial males soon take over any territories in which males can no longer be heard singing. Robins behave in a similar fashion.

SONGS

SIGHTS

▲ Coyotes use both calls and urine marking to define their territory. Their howl is unique and consists of a series of high-pitched staccato yelps followed by a long siren wail. Other vocalizations include barks, bark-howls, group yip-howls and group howls.

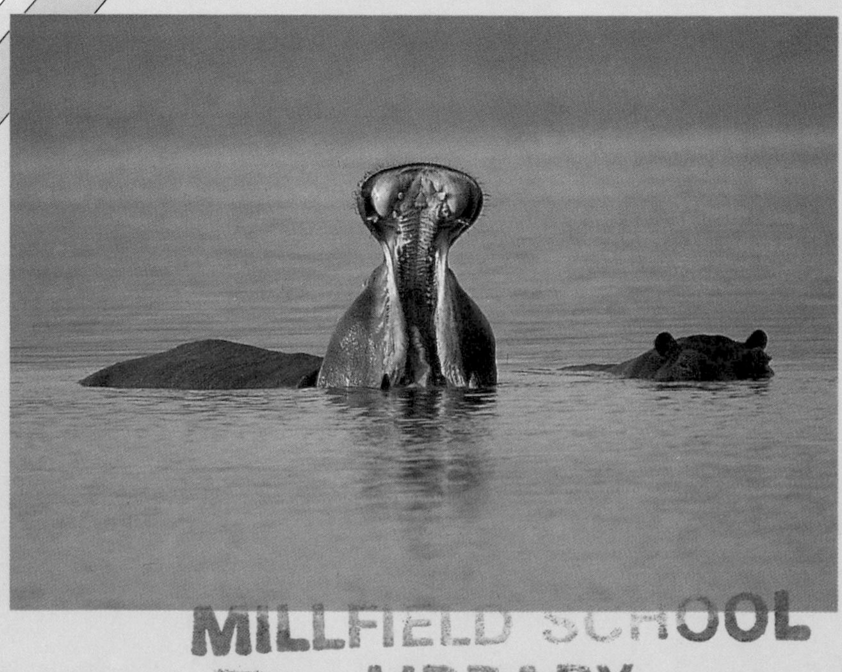

► Hippopotamuses live either in nursery groups, bachelor groups or alone (usually solitary males). Group males maintain territories containing the females and young. Non-territorial males do not breed. Territorial displays include mouth opening.

► *Male bison, during the mating season, compete for territories holding females. Contestants may threaten each other by rolling in dust (occasionally urinating in the dust before rolling in it) and by standing broadside to one another rather than settling the dispute by physical attack, which costs both time and energy (◆ page 14).*

Sights, songs and scents

If animals were only able to prevent other individuals from intruding onto their territory by being physically present themselves, they would be so overworked patroling their boundaries that they would have no time left for the other important activities of feeding, mating and rearing offspring. Instead, they advertise their ownership of a territory in various indirect ways that save time and energy.

These involve one, or a combination, of visual, aural and chemical signals. Human beings have mimicked many of these tactics, although territorial behavior of this kind does not have the same evolutionary connotations (◆ page 14).

SIGHTS

SCENTS

▲ *Cheetahs and other cats advertise land occupancy largely by scent and visual signals. They warn off intruders by depositing urine, feces and anal gland-secretions, and making scrapes in the ground. Vocalizations play little part.*

► *Territorial klipspringer mark the boundaries of their territory by placing secretions from a gland just below the eye onto twigs and grass stems. The same stems are used over and over again so that they eventually acquire small balls of dried secretion.*

◄ *Most female mongooses remain in their natal groups for a long period but eventually emigrate. They may meet other emigrants of the opposite sex to form a new pack on unoccupied territory. Pack members mark their territory with scent from anal glands.*

Nikolas Tinbergen noticed that a territorial male stickleback adopted a territorial posture on noticing the distant passage of a bright red mail van

The competition for territories determines how easily a territory-owner can keep other animals off its area. Animals often take larger territories than they need. Later, as the pressure from conspecifics increases, the size of the territory is compressed. Colobus monkeys are leaf-eaters of African forests. In habitats where colobus occur at low densities, each group's territory is typically 15ha in area, often with considerable overlap. In areas of high density, territories are as small as 2ha with virtually no overlap. This minimum value corresponds to the most heavily used core area in the largest territories and seems to be the smallest area that the animals need to live in.

Economics of territory defense

Animals defend territories only to ensure access to a limited resource such as food or breeding females. Defending a territory uses up time and energy that could be spent mating females and rearing offspring. Consequently, an animal is wasting effort defending a resource that other individuals can obtain elsewhere without fighting.

Grazers such as cattle, sheep and many of the larger antelope and deer usually roam their ranging areas in large herds because grass is too evenly distributed to be worth defending (▶ page 108). Individual trees in fruit and waterholes, on the other hand, are worth defending against other individuals because there are fewer of them. Whoever gains exclusive access to one has an advantage over its neighbors. Throughout the wooded areas of Africa, groups of vervet monkeys live in territories that are defended by the males. Among the many limited resources are waterholes. During drought years, groups with waterholes that do not dry up completely during dry periods suffer less mortality than those with less productive waterholes.

Defendability is also an important consideration. If the animal cannot patrol its territorial boundary on a regular basis it will be too easy for other animals to exploit the area's valuable resources. Among monkeys, territories are usually limited to a diameter which is less than the distance the group can travel in one day (◀ page 64). This ensures that all parts of the territory are visited regularly so that intruders can be evicted before they have a chance to settle.

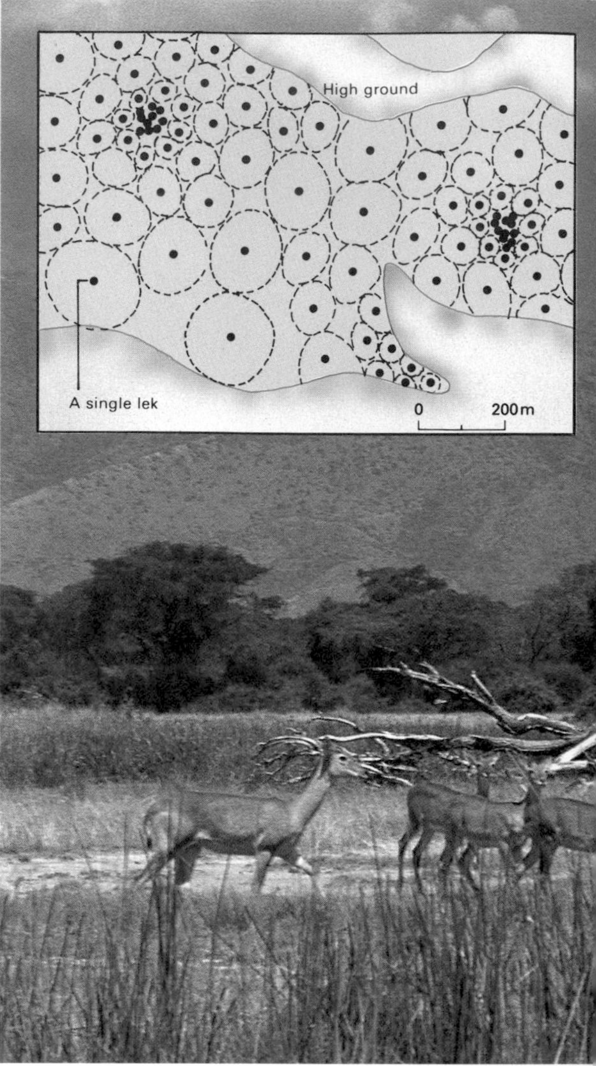

▲ *Against a backdrop of Uganda's Ruwenzori Mountains, a group of female kob move across the floodplain of a river from one feeding area to another. During the breeding season, the males establish mating territories on leks (map inset) located in the areas most frequented by the female herds, such as the lowland grasslands. The male approaches a female with head held high.*

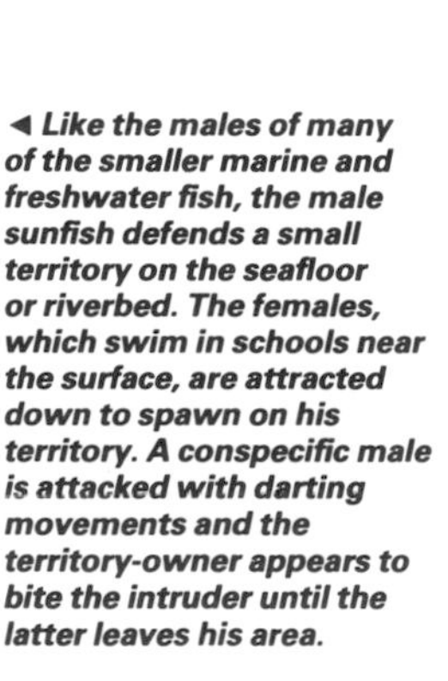

◀ *Like the males of many of the smaller marine and freshwater fish, the male sunfish defends a small territory on the seafloor or riverbed. The females, which swim in schools near the surface, are attracted down to spawn on his territory. A conspecific male is attacked with darting movements and the territory-owner appears to bite the intruder until the latter leaves his area.*

Rules for deciding disputes

When two European robins confront each other on a territory, the intruder will invariably retreat after a short scuffle. The territory-owner will chase the intruder, but it will not chase it very far beyond its own territorial boundary. At that point, the roles suddenly reverse and the former intruder will chase the first male off his territory. The usual rule is: Resident always wins.

Severe fights occur only when intruders are deliberately trying to challenge territory-holders for the ownership of the territory. So strong is the respect for territorial boundaries that it is usually very difficult to persuade intruders and owners to fight at all seriously. However, it is sometimes possible to deceive two males into thinking that both of them own the same territory (◆ page 17).

Where there are plenty of territories available in the habitat, there is little point in an animal contesting an owner's control over his territory. The intruder can easily obtain an empty territory elsewhere without having to fight for one. Iguana lizards live along the sandy banks of rivers and lakes in Central America, where the females dig 2m-long nest burrows in which to lay their eggs. Because the effort of digging in the hot sun is so great, the females keep having to interrupt their digging to go and rest in the shade. While they are away, other females often take over their holes. When the owner returns, her willingness to fight the usurper depends on how deep the burrow is. If it is very shallow, it is less trouble for her to abandon it and start a new hole elsewhere. But if she has already dug to a considerable depth, she will fight the intruder in order to get it back.

◀ The marine iguana of the Galapagos Islands defends its territory with a vigorous bobbing display. If its opponent declines to give way, a head-to-head pushing match may result.

▼ If a wandering male large white butterfly comes across a territorial male on a flowerhead, it is likely to retreat rather than compete for the food resource.

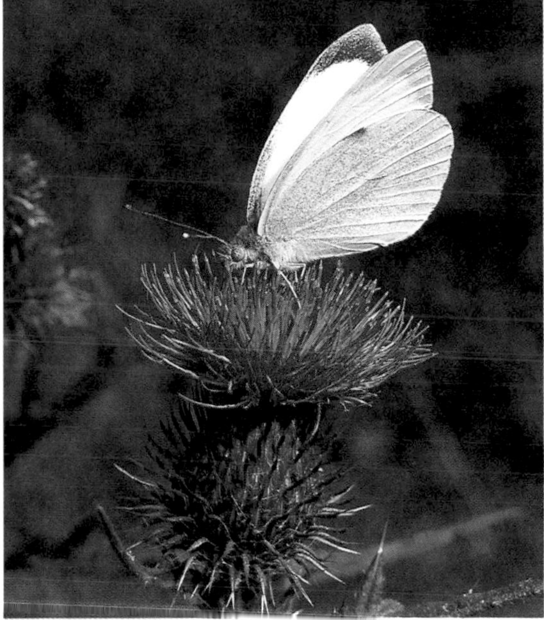

See also
The Limits of Animals 25-34
Sex and Species 35-46
Groups and Troops 47-62
Animal Migration 71-80
Communities of the Land 97-116

In some species, the territorial system may break down altogether when the pressure from competitors becomes too great. Bull elephant seals defend mating territories on the beaches where the females haul out to give birth or mate (◆ page 43). The males defend these territories against other nonterritorial males who try to steal copulations with the females. If the number of intruding males rises too much, the territory-holder is unable to prevent other males mating with his females while he is distracted trying to repel intruders elsewhere. At this point, most males stop defending their territories and instead concentrate on defending individual females as they come into estrus. A similar collapse of territorial systems has been noted in dragonflies, mason bees, starlings and house mice.

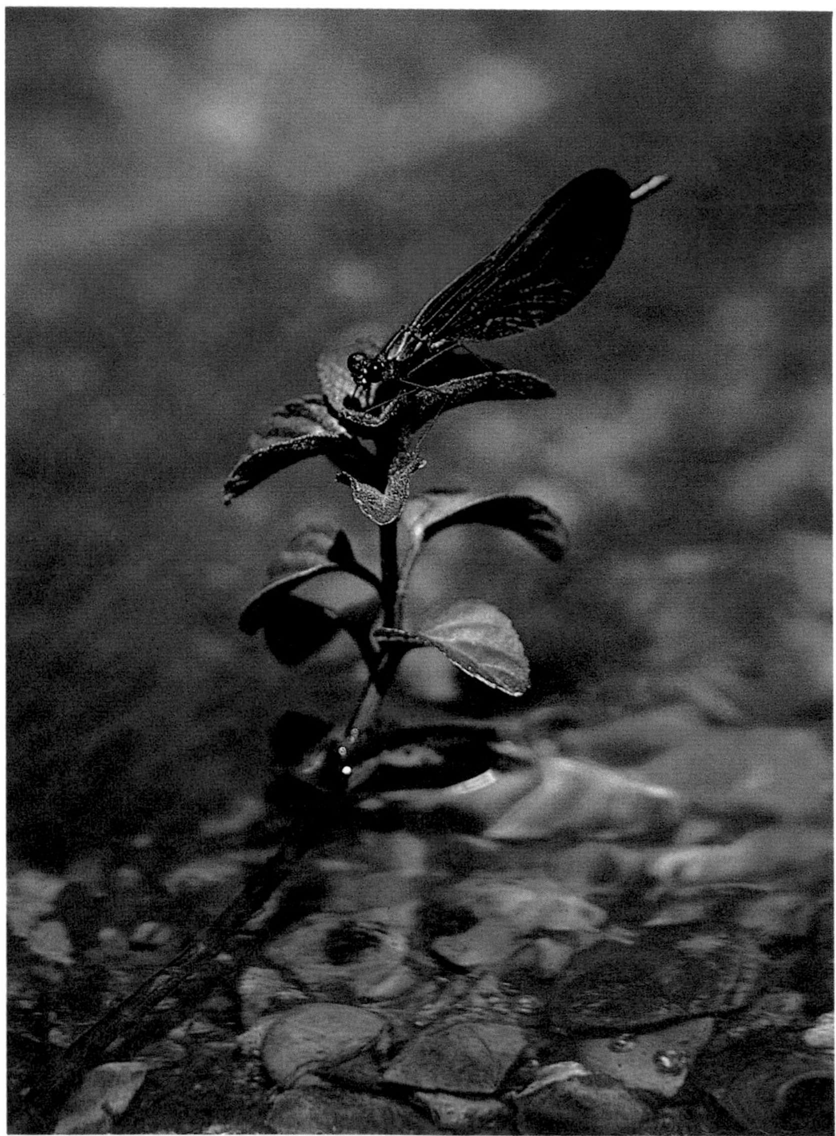

▲ ***A sprig of wild mint provides a male damselfly with a perch on which to rest while waiting for the arrival of females in search of suitable sites in which to lay their eggs. In some damselfly, the eggs are laid while the male holds the female's neck.***

Optimal foraging

Each day an animal has to find enough food to meet its nutritional requirements while avoiding being caught by a predator. The risk of predation invariably means that animals try to minimize the time they spend foraging in exposed positions. This can usually be done most efficiently by maximizing the rate at which key nutrients are ingested.

Based on this principle, biologists predict that animals should exploit the resources nearest to their home base first. Most nesting birds and other "central-place" foragers do just this. Similarly, animals that use up a food source should not visit it again until it has had time to renew itself. Nectar-gathering birds like South American hummingbirds and African sunbirds have been shown to avoid revisiting a flower before it has renewed its nectar resource.

Much finer predictions can be made about how an animal should evaluate the individual prey items it comes across when foraging. Great tits and bluegill sunfish, for example, have been shown to prefer the larger (and hence more profitable) of two prey items whenever large prey are encountered sufficiently often to make it worth ignoring small prey even though these may be more abundant.

Many insectivorous birds have been shown to shift from one feeding site to another depending on which provides the best return in terms of food acquired for time spent searching. Studies of juncos have also demonstrated that birds can be sensitive to the risk they run of not getting enough food to survive. When offered a choice of two sites, they prefer the one that offers the more constant amount of food when not too hungry. But they shift to the more variable site when starving because this at least sometimes produces enough food to satisfy their needs.

Blackbird territories

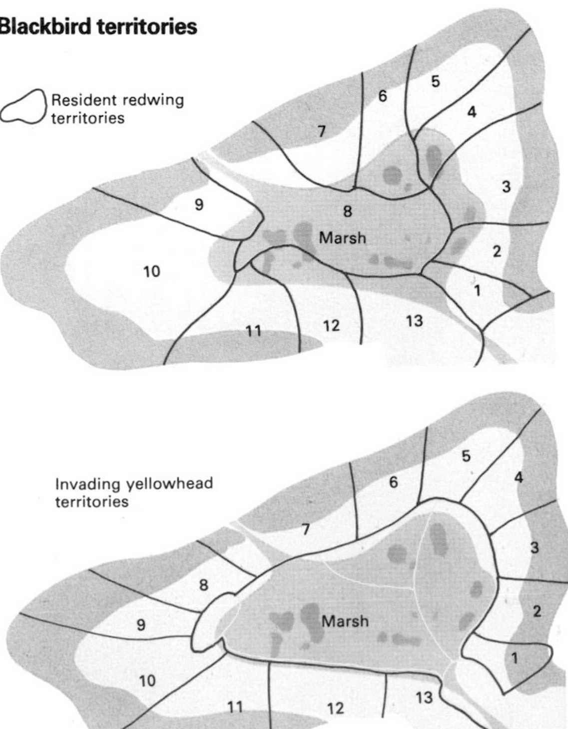

► ***Redwing blackbirds prefer to hold breeding territories in and around a marsh (top). But when the larger yellowheads arrive, they can monopolize the prime habitat in the marsh and force the redwings to the less desirable areas around the edge (below).***

Animal Migration

7

Why do animals migrate?...Strategies of migration... Migrants and residents...Magnetic, celestial and solar cues...Threats to migration...PERSPECTIVE...Green turtles find their way...Migratory routes of birds... Investigation and analysis of migratory behavior

Few animals remain stationary for any length of time: if they did, they would soon exhaust the food they had available. Instead, most animals travel round their habitat in search of fresh pastures, moving on as the food sources in each area are depleted. In some cases, these movements are localized. Shrews, for example, may spend their entire lives in areas measured in tens of square meters. Other species have taken the nomadic lifestyle to an extreme and regularly commute each year over large distances in order to take advantage of differences in the annual cycles in the food supplies of widely separated areas. In eastern Africa, millions of wildebeest and other antelope undertake a migration every year that takes them in a great circular route round the grasslands of the Serengeti plains (➧ page 108). They follow in the wake of the passing summer rains that bring flushes of green growth to the countryside as they sweep across the continent. In Canada, the great herds of caribou range up into the Arctic tundra during the summer and then travel more than 800km southwards as the winter snows begin to cover up the grazing (➧ page 112).

Few species, however, are as ambitious as the Arctic tern. It spends its summers above the Arctic circle, then flies south to spend its winters 6,000km away on the coasts of the Antarctic, thereby exploiting the local summer abundance of each polar region in turn.

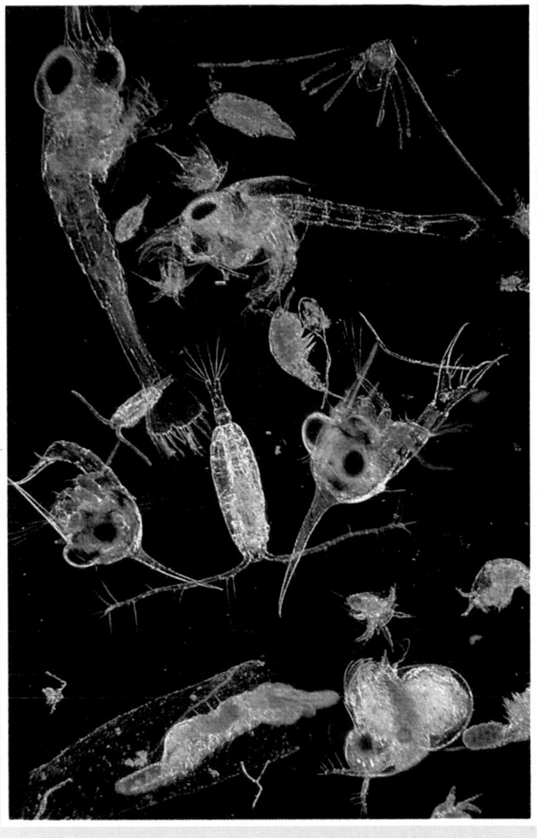

▲ *The myriad larvae, copepods and other organisms that make up the plankton swarms of oceanic waters migrate towards the surface at night and sink to lower depths in the day.*

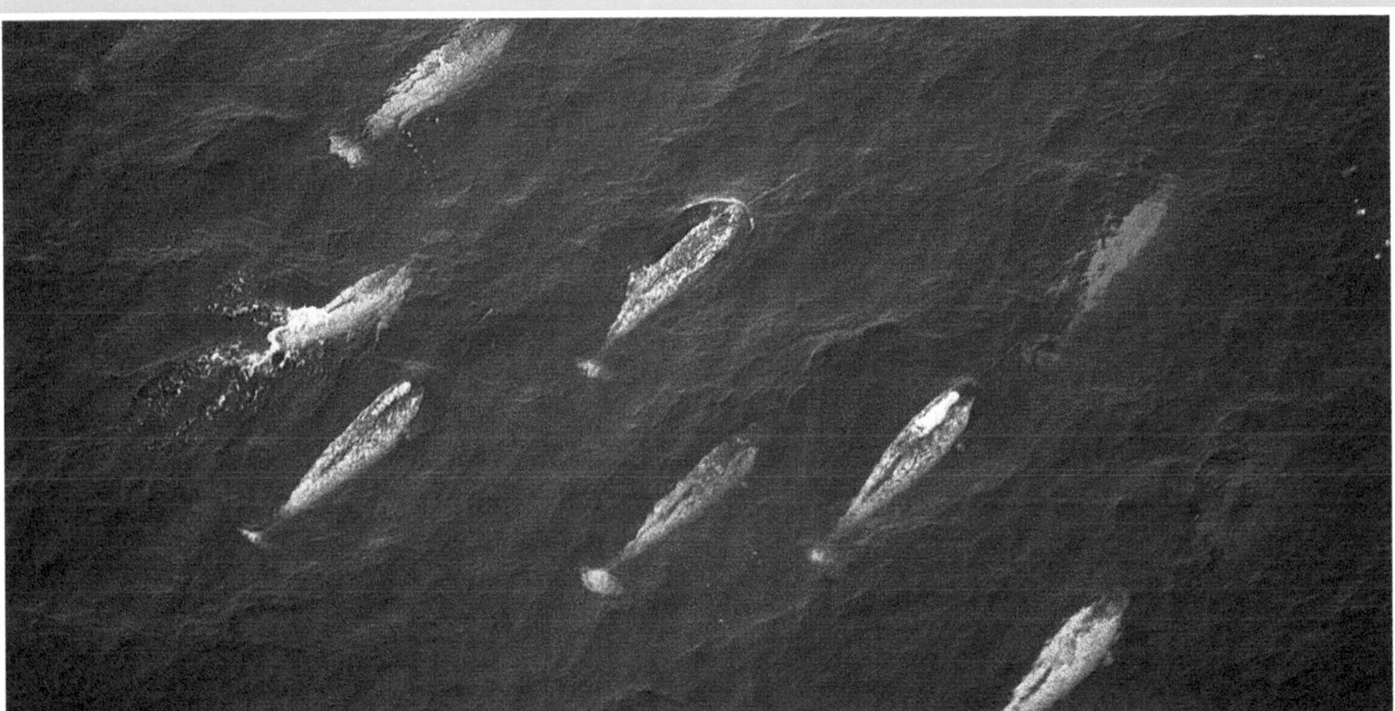

▼ *In the north Atlantic, a school of narwhals moves in close formation. Narwhals spend most of their time within the Arctic Circle, but some migrate as far south as Britain.*

In North Africa, locusts follow a circular migratory route in search of food in swarms of upward of 100 million individuals

The vanishing turtles mystery

One of the most persistent mysteries of animal migration concerns a population of Atlantic green turtles that breed on the isolated mid-Atlantic island of Ascension, but spend the time between breeding seasons off the Brazilian coast more than 2,000km to the west.

It is easy to see how the turtles get from Ascension to Brazil because the South Equatorial current flows westwards from Africa via Ascension to the east coast of South America (see map, right). The turtles have only to drift on the ocean current to find their feeding grounds off Brazil. The problem is how they manage to get back to Ascension, for the South Equatorial current flows at nearly 2km per hour. Since this is the maximum speed that a turtle can maintain for any distance, a turtle swimming flat out against the current would only just manage to remain stationary.

Among the explanations that have been suggested as to how the turtles might find their way back to Ascension is that they use the deep counter-currents that flow eastwards across the Atlantic 200m or more below the surface. Another is that they float northwards into the Caribbean with the South Equatorial current to catch the Gulf Stream northeastwards across to Europe. From there, they can catch the North Equatorial current down to the west coast of Africa where, remembering not to get swept back across to the Caribbean by it, they can once more pick up the South Equatorial current.

A third possibility is that, on reaching Brazil, they drift southwards on the Brazilian current until they can pick up the West Wind Drift just above Antarctic waters. This would take them across to Africa where they could catch the Benguela current back up the west coast to meet the South Equatorial once more off the Bight of Benin.

Unfortunately, all these routes take the turtles into waters whose temperatures are far below their normal active temperature of 20°C. This means that the turtles would have to go into hibernation in the water to survive, and so might miss crucial migration points where they need to change currents. This ought to mean that turtles were washed up occasionally on north European coasts, but there are no records of this happening.

The real problem is that although studies of tagged individuals confirm that the turtles spend time off both Brazil and Ascension Island, we have no idea where they go to in between.

► When the vast herds of wildebeest migrate across the Serengeti plains in search of fresh grazing, they are quite undeterred by the rivers that bar their way. When the rivers are in spate, many are carried away by the torrent and drown. The migratory urge drives the herd on.

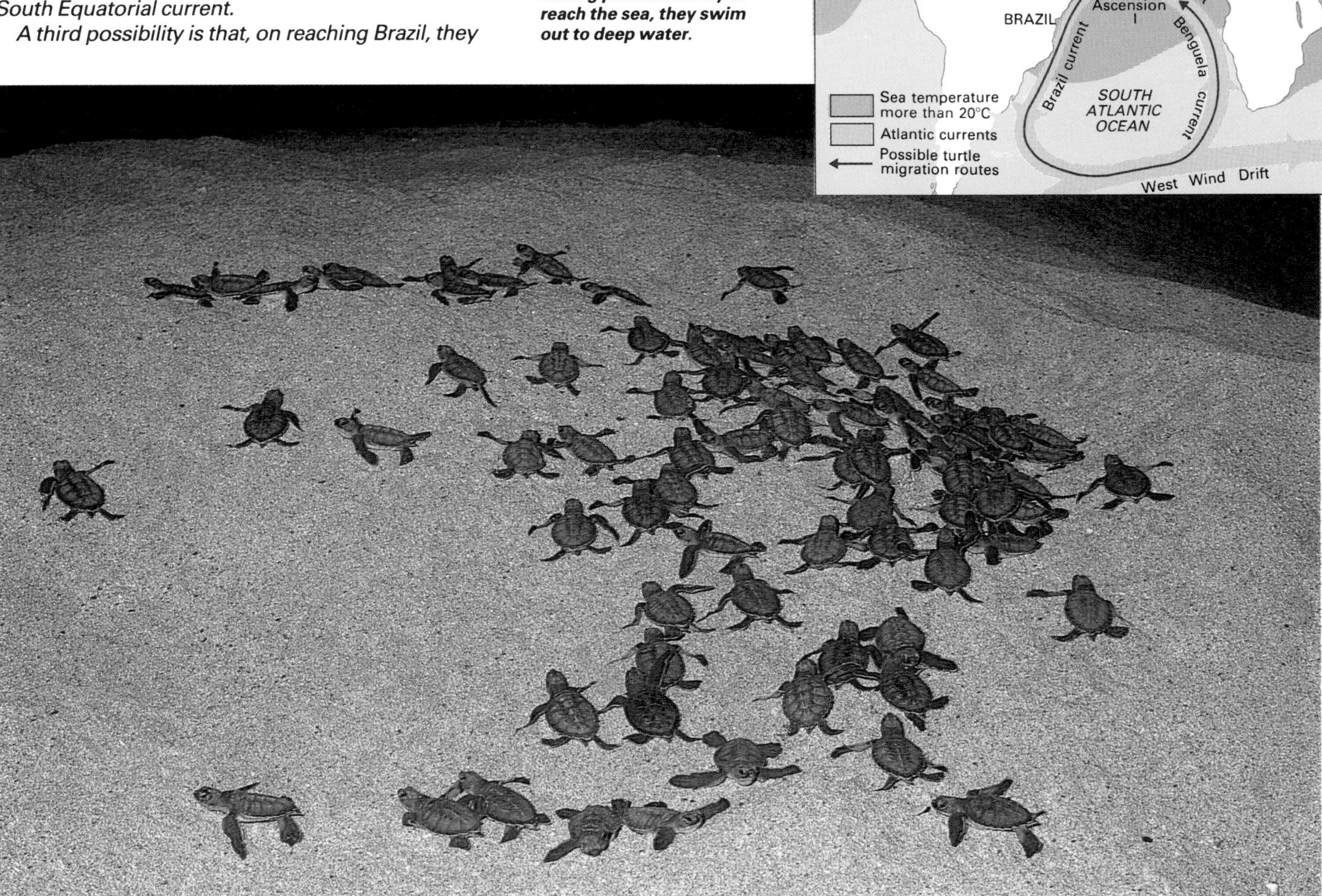

▼ The westward-flowing currents of the mid-Atlantic allow the green turtles that breed on Ascension Island to drift to their feeding grounds off the coast of Brazil. Biologists do not know how they get back to breed on Ascension. To navigate they probably use the angle of the Sun.

▼ When green turtles hatch on the sandy beaches of Ascension Island, their first challenge is to find their way down to the sea. As they struggle across the moonlit sand, many fall prey to predators. They know instinctively which way to head, reacting to the brighter light over the sea, which reflects more light than the land behind them. Having paddled wildly to reach the sea, they swim out to deep water.

Salmon are able to trace and identify the taste of the river in which they hatched over distances of several hundred kilometers

Many whales and seals make long-distance migrations in order to exploit the seasonal richness of the polar habitats. They spend their summers in Antarctic or Arctic waters, then head towards the equator during the polar winters. Other animals make once-in-a-lifetime journeys. Salmon are spawned in the headwaters of rivers along the coasts of Europe and North America. As young fish, they head downriver to spend the next 4 years or so out in mid-ocean. Then they return to spawn in the very same streams in which they began life. Biologists now know that salmon become imprinted on the characteristic taste of the streams in which they grow up (◆ page 12). Adult eels find their way back from the European rivers in which they spent most of their adult life to spawning grounds in the Sargasso Sea, which comprise a floating mass of seaweed and flotsam in mid-Atlantic. When they made the initial outward journey to Europe, they did so as leaf-like larvae that drift on the surface currents flowing northeast towards the European continental shelf from the Caribbean. Since the larvae pass through several metamorphoses before becoming ocean-going adults, they can have no direct recollection of their spawning grounds to guide them.

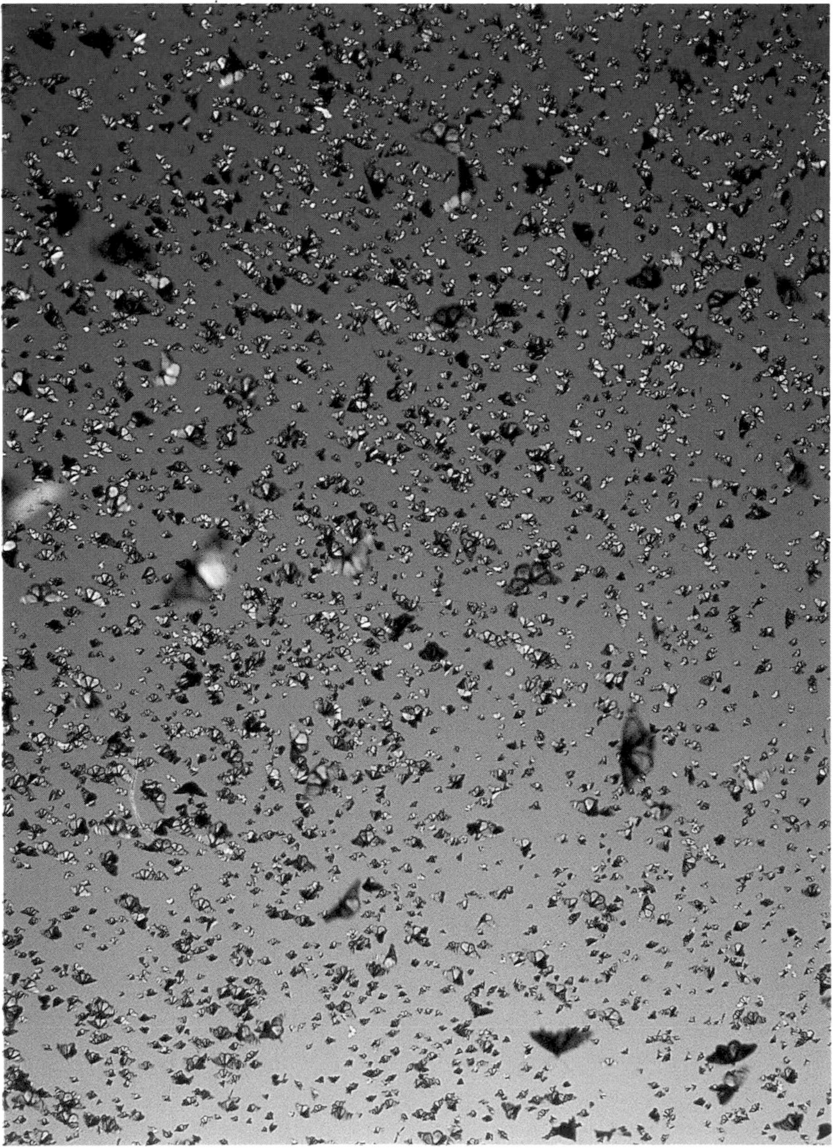

Movements of fruit-, seed- and insect-eating birds

The African violet-backed starling follows the fruiting cycles of fig trees, throughout the year moving from tree to tree through the forest as each comes into fruit. Vast flocks of quelea weaver birds move great distances across the African savanna in search of grain and other seeds (◆ page 56).

The northernmost latitudes provide habitats that are especially productive during their short summer seasons once the winter snows have thawed. Millions of ducks and geese are attracted by the tundra grasslands (◆ page 112). A host of swifts, warblers, wagtails, chats and flycatchers feast on the billions of insects that swarm out of the waterlogged marshes, and in turn attract falcons, hobbies, harriers and other birds of prey.

◄ The great annual migrations of the monarch butterfly of North America are a remarkable feat for an insect. An individual that emerges during the summer in the Great Lakes region on the Canadian border will fly as far as 3,000 km southward ahead of the advancing snows to winter in Florida or Mexico. Here it congregates on trees in its millions. The following spring, these individuals fly north. Many die on the way, but not before they have produced offspring, which continue on to the Great Lakes. A small proportion of the monarch population remain in the north throughout winter. They overwinter as pupae.

▲ *Penguins gather in Antarctica in vast numbers to exploit the annual summer superabundance of krill and the fish that feed on these crustacea. But as the southern winter sets in, the krill stocks begin to fall and species such as the Adelie penguin must head north to its breeding grounds in warmer waters. Only the male emperor penguin remains behind to sit out the rigors of the long Antarctic winter night (◆ page 53). Behavioral rhythms such as migration are synchronized with the seasons by changing daylength.*

◄ *Salmon, having spent some 6-10 years out at sea as adults, return to their birthplace to breed. They follow the gradient of taste of their home river back into the headwaters of the streams in which they hatched. These newly hatched salmon are known as fry or alevins.*

The majority of birds migrate flying at altitudes of between 100 and 1,500m but some geese have been seen crossing the Himalayas at 9,500m

Choosing whether to go or stay

Some species, known as partial migrants, contain both individuals that migrate and others that remain in the same area all year round (residents) (➧ page 91). Among birds, European examples of these include the blackbird, song thrush and lapwing. In each case, part of the population migrates from central Europe up into Scandinavia to breed during the spring, while the rest remain to breed within the species' normal winter range.

One possible explanation is that migrants and non-migrants are genetically different and that these two types differ in their success in breeding under different environmental conditions. In good years migrants do better because they can exploit the superabundance of food along the Arctic circle. In poor years, non-migrants are at an advantage because they thrive on the more stable food supplies that are found on the winter range during spring. Since, on average, good and bad years alternate, both types are held in balance in the total population.

An alternative suggestion is that all individuals in the population are really migrants, but that each animal's willingness to migrate differ slightly. The distinction appears because some individuals' "migration threshold" causes them to set off for the summer northern breeding grounds much earlier, whereas others need to be more convinced that it is worth their while to embark on the long and risky business of migration.

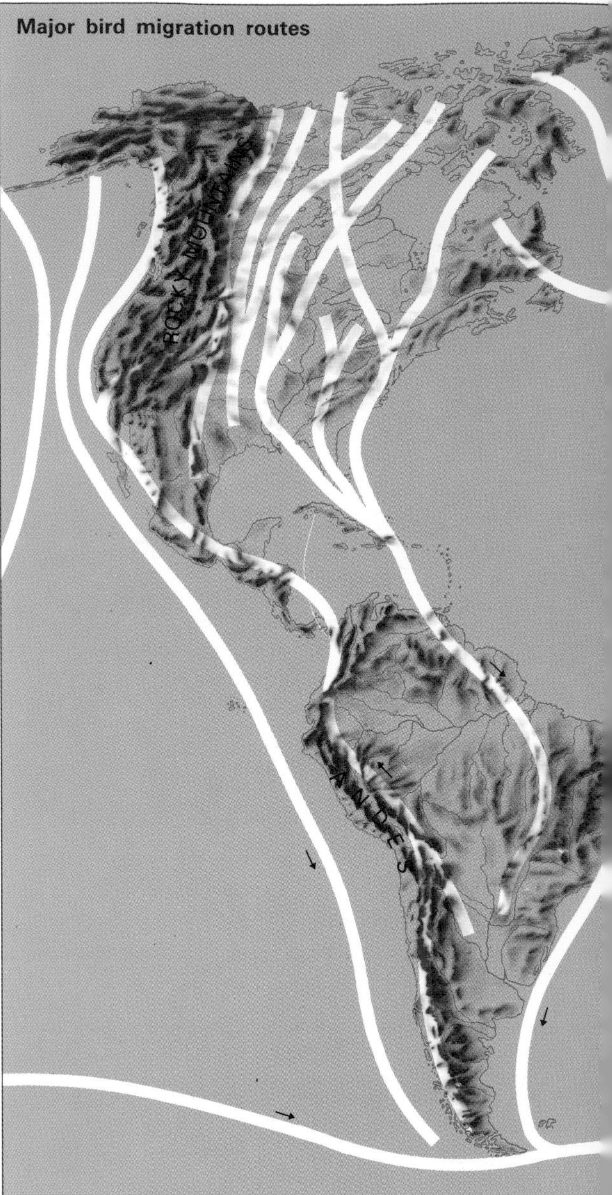

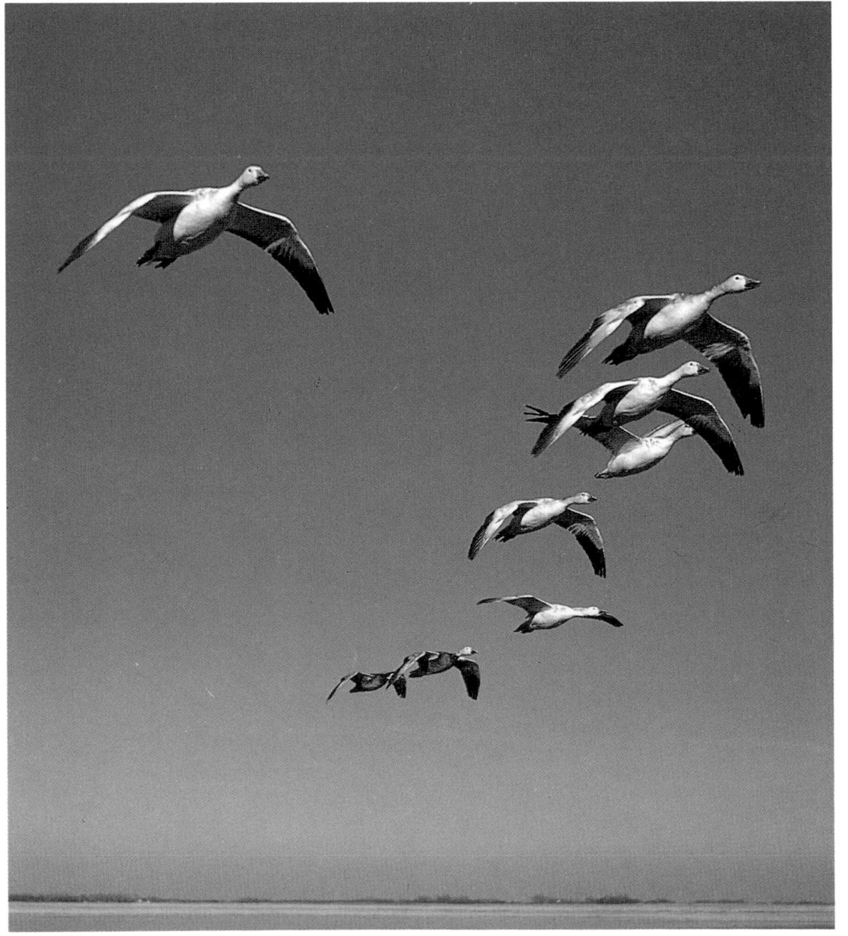

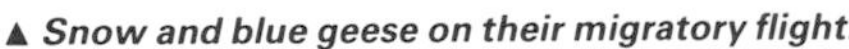
▲ ***Snow and blue geese on their migratory flight.***

Exploratory migrations

In North America young Canada geese accompany their parents on their first migration but in the following year they retrace the route all over again mid-year before accompanying their parents once more the following autumn. It seems as though their mid-year migration serves to remind them of the route while allowing them time to make exploratory flights along the way.

Studies of European lesser black-backed gulls suggest that parents accompany the young on their first southward migration in early autumn, but then return north to Britain before winter sets in. Cross-fostering experiments in which the eggs of lesser black-backed gulls have been raised by the non-migratory herring gulls have shown that juveniles acquire the migration habits of their foster-parents rather than those of their biological parents. The exploration theory of migration may explain why migrant birds are seen in unusual places.

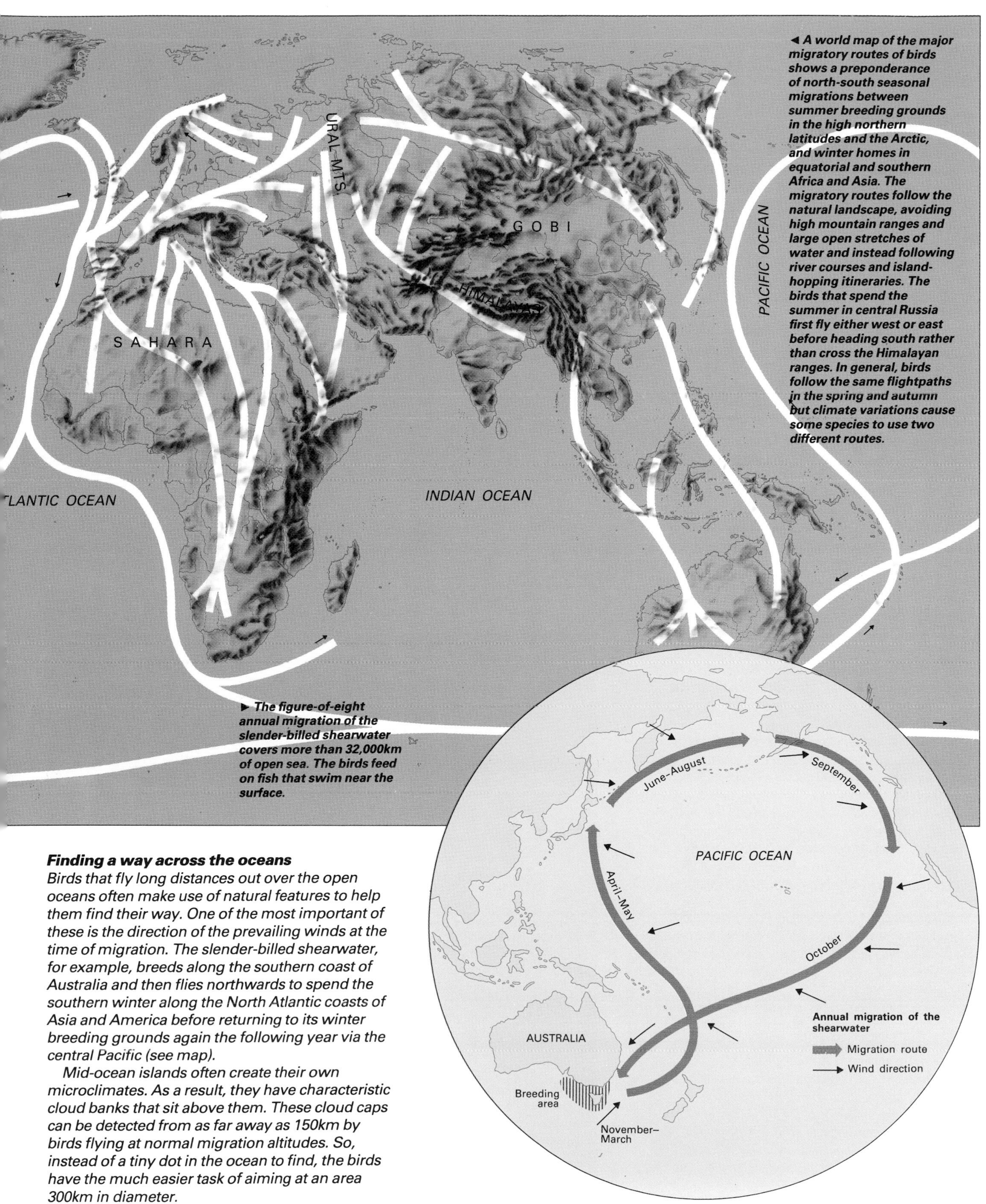

◄ *A world map of the major migratory routes of birds shows a preponderance of north-south seasonal migrations between summer breeding grounds in the high northern latitudes and the Arctic, and winter homes in equatorial and southern Africa and Asia. The migratory routes follow the natural landscape, avoiding high mountain ranges and large open stretches of water and instead following river courses and island-hopping itineraries. The birds that spend the summer in central Russia first fly either west or east before heading south rather than cross the Himalayan ranges. In general, birds follow the same flightpaths in the spring and autumn but climate variations cause some species to use two different routes.*

► *The figure-of-eight annual migration of the slender-billed shearwater covers more than 32,000km of open sea. The birds feed on fish that swim near the surface.*

Finding a way across the oceans

Birds that fly long distances out over the open oceans often make use of natural features to help them find their way. One of the most important of these is the direction of the prevailing winds at the time of migration. The slender-billed shearwater, for example, breeds along the southern coast of Australia and then flies northwards to spend the southern winter along the North Atlantic coasts of Asia and America before returning to its winter breeding grounds again the following year via the central Pacific (see map).

Mid-ocean islands often create their own microclimates. As a result, they have characteristic cloud banks that sit above them. These cloud caps can be detected from as far away as 150km by birds flying at normal migration altitudes. So, instead of a tiny dot in the ocean to find, the birds have the much easier task of aiming at an area 300km in diameter.

Animals' Maps and Compasses

Piloting, orienteering, navigating

How animals manage to find their way over the immense distances they cover during migrations has remained one of the most enduring mysteries of the natural world. During the past 30 years, however, biologists have learned that animals use one, or a combination of, three quite different processes: piloting, orienteering and navigating.

Many of the migrant birds of Europe and North America travel down traditional flightways (♦ page 77), following routes whose main features they have learned and remembered during earlier flights with more experienced individuals. They thus find their way by piloting, or flying from one easily identified landmark to the next. From high up, features such as mountain peaks or large rivers like the Nile or the Mississippi provide conspicuous features that can be used to guide migrating birds. The European migrants do not fly directly across the trackless Mediterranean Sea to Africa, but converge on peninsulas like Spain, Italy or the Balkans where they can follow coastlines or island-hop without losing sight of land.

Not all migrants rely on piloting. Those birds that fly across the world's oceans to breed on tiny isolated islands like Ascension or Tristan da Cunha have no cues to use while out of sight of land. Instead, they rely on orienteering to keep them heading in the correct direction. Long-distance migrants can call on a number of mechanisms to do this. Many nocturnal migrants, for example, use the stars as a point of reference just as humans do. Experiments have shown that birds in the Northern Hemisphere learn to focus on one of the stars in the cluster around the Pole Star that remain in a fixed position in the sky because they lie above the Earth's axis of rotation. Diurnal migrants may use the position of the Sun, though this requires a mechanism for compensating for its movement across the sky each day.

Recently, it has been shown that bees as well as species ranging from pigeons to human beings can detect variations in the Earth's magnetic field, thus providing them with a built-in magnetic compass. Pigeons are also known to use smell, while some species may be able to identify from thousands of kilometers away the characteristic sound signatures in the very low "infra-sound" frequency range that are given off by prominent geographical features as winds pass over them.

Orienteering and piloting are very different, however, from true navigation. Navigating requires not only a mechanism for steering a course but also a map from which the compass bearing connecting any two points can be calculated. Although a number of attempts have been made to show that homing pigeons have internal maps which they can use in navigating, no unequivocal evidence has so far been found to prove the point.

A more likely explanation for the remarkable abilities of animals like the homing pigeon is that they have a very much wider knowledge of the surrounding landscape than we suppose (♦ page 13) and that they use a variety of cues (including smell and the stars, perhaps) to orienteer until they come within the area of their detailed knowledge where they can finally pilot their way home.

***►** Birds use the Sun's position to estimate compass directions. Proof of this comes from experiments on pigeons that were kept under artificial lighting so that their day started earlier or later than it should do. Birds that had their clocks advanced by six hours presumed that the Sun was further to the west than it really was and therefore headed to the northeast when released to the north of their home loft, whereas control birds raised under normal daylight headed due south. Birds that had their clocks delayed by six hours thought that the Sun was further to the east (near diagram) and so headed northwest.*

Day

Six hours early

Six hours late

Shift in bird's perception of direction

Direction of Sun's movement

Direction of flight

Control group of birds

Release point

***▼** Radar-tracking of flocks of migrating birds proved to be an invaluable source of information on their nighttime movements. On a radar screen, individual birds show up as dots.*

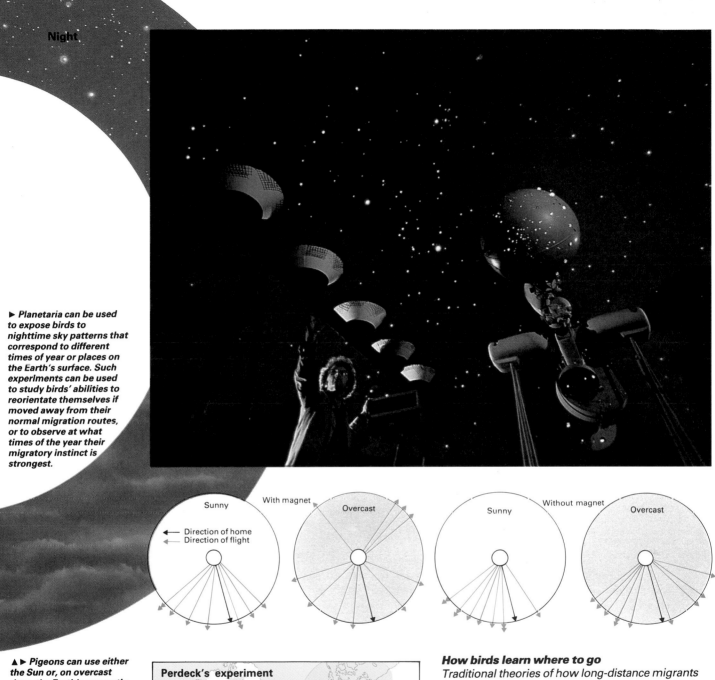

Night

► *Planetaria can be used to expose birds to nighttime sky patterns that correspond to different times of year or places on the Earth's surface. Such experiments can be used to study birds' abilities to reorientate themselves if moved away from their normal migration routes, or to observe at what times of the year their migratory instinct is strongest.*

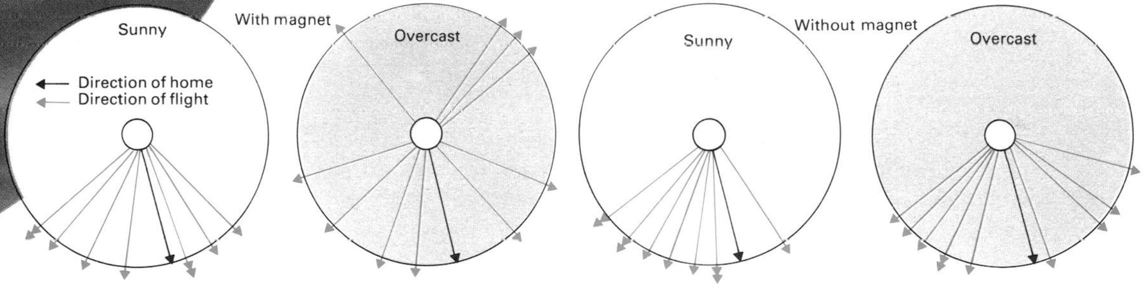

With magnet

Sunny

Overcast

Direction of home

Direction of flight

Without magnet

Sunny

Overcast

▲► *Pigeons can use either the Sun or, on overcast days, the Earth's magnetic field to estimate the compass direction of their home loft (on right). When fitted with strong magnets, their magnetic sense is confused and, with no Sun to cue them in, they fly off at random on release.*

► *When starlings that breed in the Netherlands and normally winter in western France were transported to Switzerland, the adults were recaptured in their normal wintering areas. They adjusted their migration to compensate. Juveniles were recaptured in Spain, relying on traditional flyways rather than the geography.*

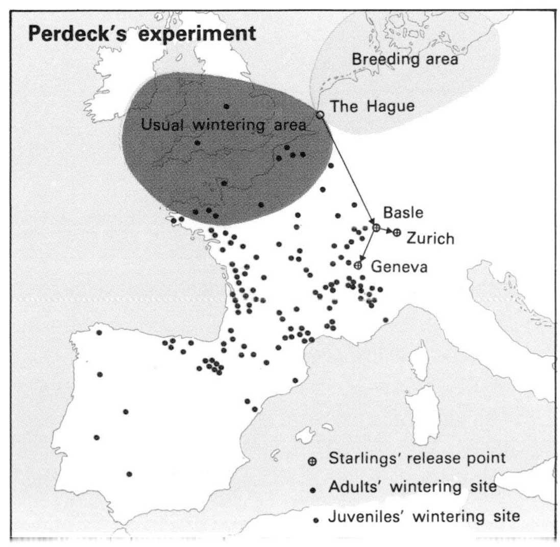

How birds learn where to go

*Traditional theories of how long-distance migrants such as birds know where to go have emphasized genetic inheritance of the correct behavior. These have tended to focus on the inheritance either of a tendency to fly in a predetermined direction for a specific length of time or of a preference for being at a certain latitude at a particular time of year. Either way, these theories place excessive demands on the genetic processes of inheritance by requiring them to pass on from one generation to the next information about complex environmental factors. An alternative hypothesis is that animals as intelligent as birds learn much of what they need to know through experience as they grow (**◆ page 18**). Information about migration routes and summer and winter ranges can come either from accompanying adults or from undertaking exploratory flights of increasing length on their own.*

See also
Territory 63-70
Communities of the Land 97-116
Communities of the Water 117-128
The Last Great Hunt 181-190
The Shadow of Extinction 195-208

Man and migration

Migration routes often follow traditional paths that have been used since time immemorial. The ancient Egyptians and Greeks, for example, recorded the regular passage of migrant birds down through the eastern Mediterranean on their way to and from Africa. In Canada, the passage of countless caribou over hundreds of thousands of years has worn down narrow paths to depths of 60cm or more in the rocky surface of the landscape.

Because many of these migration routes have become fixed traditions, problems often arise when human economic development results in migration routes being blocked. In Alaska, oil pipelines bringing raw petroleum down to the refineries from the Arctic oil fields had to be built with tunnels at key points to allow the migrating herds of deer to pass between their winter and summer ranges. In East Africa, the elephant population traditionally completed a great circular migration that took two years or more to complete. When large-scale farming enterprises developed, they cut across the migration route, preventing the elephant from following their centuries' old traditional migration. Instead, the elephants accumulated in the National Parks along the southern edge of the circuit, until their numbers were so great that they began to wreak havoc on the vegetation, destroying woodlands and changing the whole ecological balance in the area. Tragically, many animals died of starvation or had to be culled (➧ page 239).

▲ The landscape within the Arctic Circle provides few cues to guide the traveler. Lapp nomads rely on traditional routes in their annual spring migrations in search of pasture for their reindeer herds. Following the Chernobyl nuclear incident, the reindeers' main food, moss, has been poisoned, threatening the Lapps' existence.

▼ The millions of birds that pass through the Iberian and Italian peninsulas on their spring and autumn migrations have provided an easy source of food and objects of sport for the people that live there. Here a nightingale has been trapped in bird lime (a sticky exudate of the holly tree) that has been smeared onto a stick.

Interactions

8

The individual and the population...Predators, parasites, commensals...Energy flow and trophic levels...Food webs and chains...Nutrient cycles... Niches...Succession...PERSPECTIVE...The lynx and the snowshoe hare...Myxomatosis devastates rabbit population...Calcium cycling in a deciduous forest... Feeding adaptations of waders...Specialization and diversity...Colonists on a shingle beach

No creature that inhabits this planet does so in isolation from other individuals of the same and of different species. Every individual has an influence upon the others with which it comes into contact. In the case of individuals of the same species, such interactions are vital for the maintenance of social structure within a group and may be of considerable benefit to the species. The flocking of wading birds and their closely coordinated flight patterns may confuse predators, while the mutual preening between individuals in bat colonies helps to maintain sanitation and control parasites as well as strengthening social bonds.

A group of individuals of the same species is termed a population and for many animals, particularly social ones, the population is an important level of organization (◆ page 54). The size of a population in any locality may vary with the balance between births and deaths, immigration and emigration, which in turn may depend on conditions of climate and food supply. Breeding within the population provides a means of genetic mixing that is beneficial to the general vigor and adaptability of individuals. But if the population level falls too low such vigor may be lost and the resulting weakness can add to the problems of survival (◆ page 200).

▲ **Condors feed on carrion and they may have to travel long distances to locate a carcass. Hence the population density of condors is always low and the species endangered.**

▼ **Although elephants are slow-breeding animals, they have few enemies and their populations can become locally too large. The result is overgrazing and damage to vegetation.**

There are as many bacteria and yeasts living on a person's skin as there are people on Earth

When populations become too high for the resources of the environment to supply all their requirements, then safety valves come into operation. This may take the rather crude form of mortality, particularly among the very young, the old and the unfit (➧ page 229). Such a process is part of the operation of natural selection. Alternatively, there may be a wave of emigration. Central European birds such as nutcrackers, waxwings and lapwings irregularly irrupt westwards into the maritime regions of western Europe, especially when their food supplies are threatened by hard weather.

Among plant populations, such emigration is impossible and the results of high density are overcrowding, high levels of competition and consequent mortality. After a forest fire in temperate regions, high densities of birch seedlings may establish, but only a very few will survive to maturity and will add their seeds to the dormant assemblage within the soil, awaiting new opportunities.

The community

Populations do not exist in a biological vacuum, however. Each species population interacts with the various other species with which it comes into contact and this interaction can take a variety of forms. When an animal encounters another animal it may simply eat it. This is termed predation. Often the predator has preferred species among the selection of prey available (➧ page 91). So the oystercatcher preys

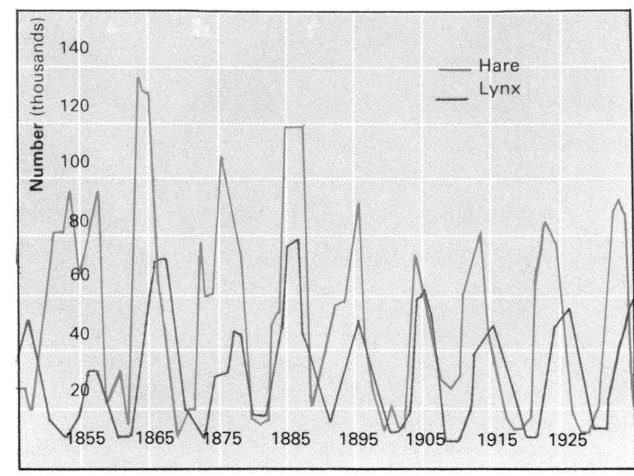

▲ Under conditions of low predation, when lynx populations are small, the snowshoe hare increases its numbers rapidly. But as this happens more food is available for the lynx, so its population begins to grow. Eventually it reduces the abundance of hares and then goes into decline itself. So the cycle continues.

▼ The snowshoe hare, a North American species, is here seen molting from its winter coat into the brown of summer. This coat change with season helps to protect it from the attentions of predators, but it remains a major item of diet for the lynx. In areas where snow is infrequent, some individuals remain gray-brown all year round.

Predator/prey interactions

When one animal feeds specifically on another, the two populations may be expected to behave in a coordinated way. As the prey species population increases, so will the predator's, for more food becomes available. But an increase in predator numbers may lead to a decrease in the density of prey, which in turn results in lower numbers of predators. In theory one might expect a repeating cycle of peaks and troughs in both populations, with the predator lagging slightly behind the prey.

The British ecologist Charles Elton (born 1900) found what he took to be an example of this when he examined the historical records of the Hudson Bay Company in Canada. The numbers of lynx caught over a 100-year period showed a distinct 10-year cycle. He interpreted this as a consequence of the oscillating relationship between the lynx and its main prey, the snowshoe hare.

But this explanation assumes that the lynx numbers are limited by hare populations and hare populations are in turn limited by predator density. In fact it is now known that the hare undergoes a 10-year cycle even in the absence of the lynx, perhaps in response to its own food supplies. So the model is only half true. The lynx does appear to decline in numbers when the snowshoe hare becomes scarcer. But even this limitation will only apply if the predator is strongly linked to one prey animal. On the island of Newfoundland, for example, when the hares become scarce lynx change their habits and move on to young moose. This prey switching ensures that the lynx population is not subject to erratic fluctuations in response to changes in the numbers of hares.

In the Arctic, a similar interaction exists between foxes, snowy owls and pomarine jaegers (predators) and various species of mice (prey).

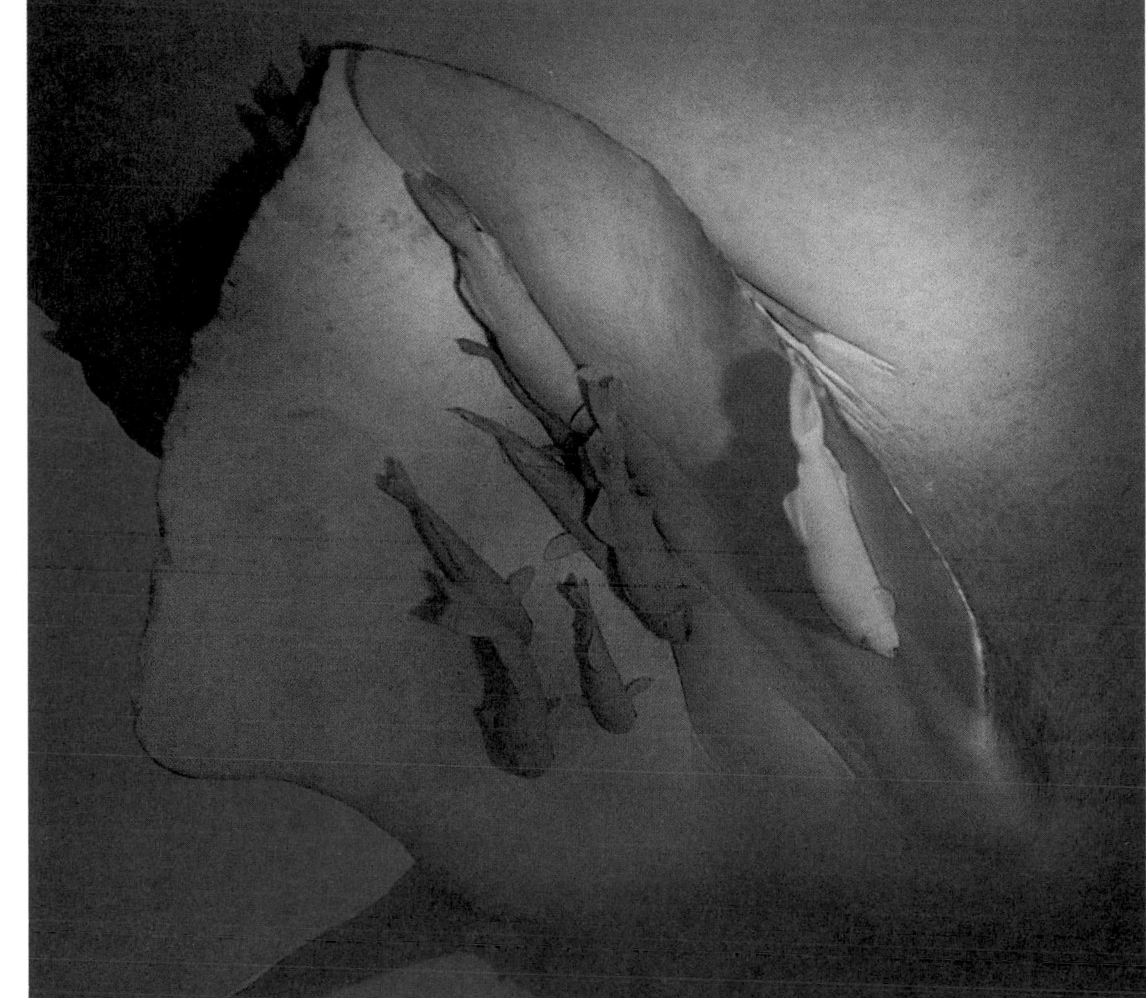

► *A remora, or shark sucker, is attached to the underside of a whale shark. The remora's dorsal fin is specialized to form a sucking disk by which it attaches itself. In this way it is carried around by the shark and probably benefits from the scraps of food left over after the shark feeds. The remora is a fish that is quite capable of swimming on its own, which it often does to prey on small fish. It chooses a wide range of hosts, including turtles, manta rays and even whales.*

▼ *Mature, healthy elm trees, like the one shown far below, have now become a rare sight in Britain and in North America because of the parasitic fungus, Dutch elm disease, which is carried from one tree to another by bark beetles. The fungus first causes leaves to wilt and then kills the tree, below, by destroying the veining.*

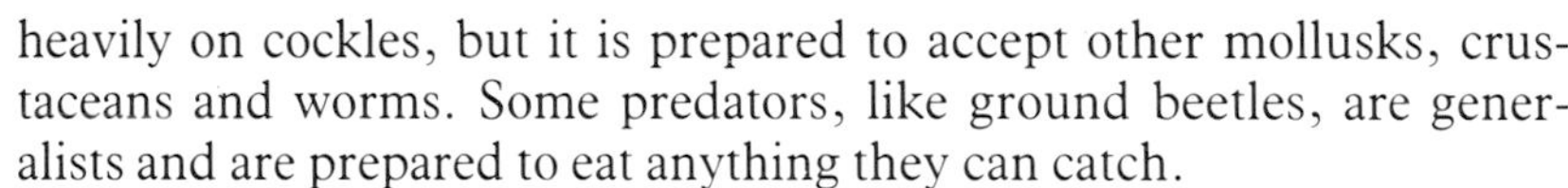

heavily on cockles, but it is prepared to accept other mollusks, crustaceans and worms. Some predators, like ground beetles, are generalists and are prepared to eat anything they can catch.

Some animals live off others without actually killing them. They are parasites. A few, such as fleas and ticks, spend their entire life outside their host (ectoparasites), while others, for example the tapeworm and various nematode worms, feed within (endoparasites). In a sense, animals that graze on plants are rather like parasites in that they take away some of the plant's tissue but do not usually kill their food source. Many microbes, such as bacteria, fungi and viruses, act in a parasitic way and infect both plants and animals. Those pathogenic microbes that kill their victims must be regarded as rather poorly adapted to such a way of life, but this is small consolation to the host. Often host and parasite have evolved together (coevolved) to form a balance between resistance on the part of the host and virulence on the part of the parasite.

Not all community interactions involve one species harming another. Some species benefit from the presence of others. The cattle egret, for example, often feeds in grassland where large herbivores such as cattle are grazing. The bulky grazers disturb insects, small reptiles and amphibians upon which the egrets feed. The bird may even hitch a ride upon the animal's back and in doing so may remove occasional ticks, so the benefits can be mutual.

The milkweed plant tolerates the depredations of monarch butterfly larvae, but not other animals, since the adult insects pollinate its flowers

Disease and populations

Most animal populations have evolved alongside microorganisms that are potentially dangerous and have developed a degree of resistance to their effects. During outbreaks of disease it is usually only the young, the old and the unfit that succumb. But when a new disease strain arises by genetic mutation or is brought into an area from outside, it can cause considerable mortality in a population. A salutary example is that of the European rabbit and myxomatosis. The disease caused by this virus was first observed in 1897 in Montevideo, South America, where it was responsible for the death of European rabbits kept for experiments in a hospital. It was found to be an endemic, but seldom fatal, disease of the native Brazilian rabbit and was clearly a virulent pathogen of European rabbits, which possessed no resistance.

Unsuccessful attempts were made to introduce the disease to an isolated rabbit population on a small island off the coast of Wales in 1936, with a view to testing its potential as a pest-control measure. In 1950, however, the disease was introduced into Australia and caused a 99.8% mortality among infected rabbits. Within ten years the effectiveness of the disease had dropped to about 90% as a result of the selective survival of more resistant rabbits, but it was still a very effective agent for population control.

In 1953 the disease reached Britain. Only at this stage was it demonstrated that the virus is conveyed by vectors such as fleas, and possibly also by mosquitoes. The lack of success in the 1936 experiments was apparently due to the absence of rabbit fleas on the Welsh island.

◄ The energy of sunlight is trapped by green leaves.

▼ Hippopotamuses have not always been welcome in African rivers, but it has been found that if they are removed the aquatic vegetation grows to such an extent that rivers become clogged up.

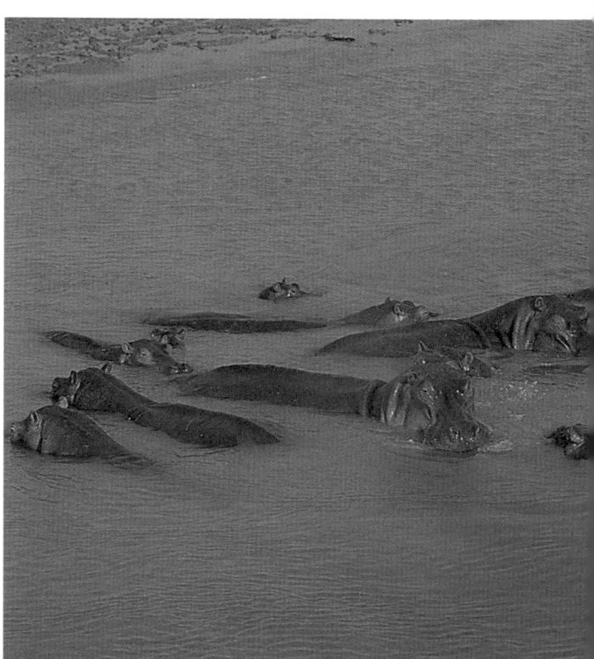

Even plants and animals have become coadapted, as can be seen in the great variety of pollination mechanisms in nature. The plant devotes a proportion of its hard-won energy to the supply of nectar or extra pollen for foraging insects, and advertises its generosity by showy petals and strategically placed markings to direct visitors (➧ page 206). Such energy investment is repayed by successful and often specific transport of pollen to other flowers of the same species. The animal's visitor, or vector, may be an insect, a bird or even a mammal, and it expends energy foraging for those flowers in which the reward warrants the time spent on the search.

So the community contains a whole range of interactions, from competition for limited resources and the resulting antagonism, to the cooperation of species in mutually beneficial relationships. The result is a complex system in which a change in any single component is likely to have repercussions through the whole organization.

The ecosystem

In 1935 the British ecologist Sir Arthur Tansley coined the term ecosystem to cover a concept of natural systems that goes somewhat beyond that of the community. The ecosystem comprises both the biological community and the nonliving physical and chemical

Energy flow in an ecosystem

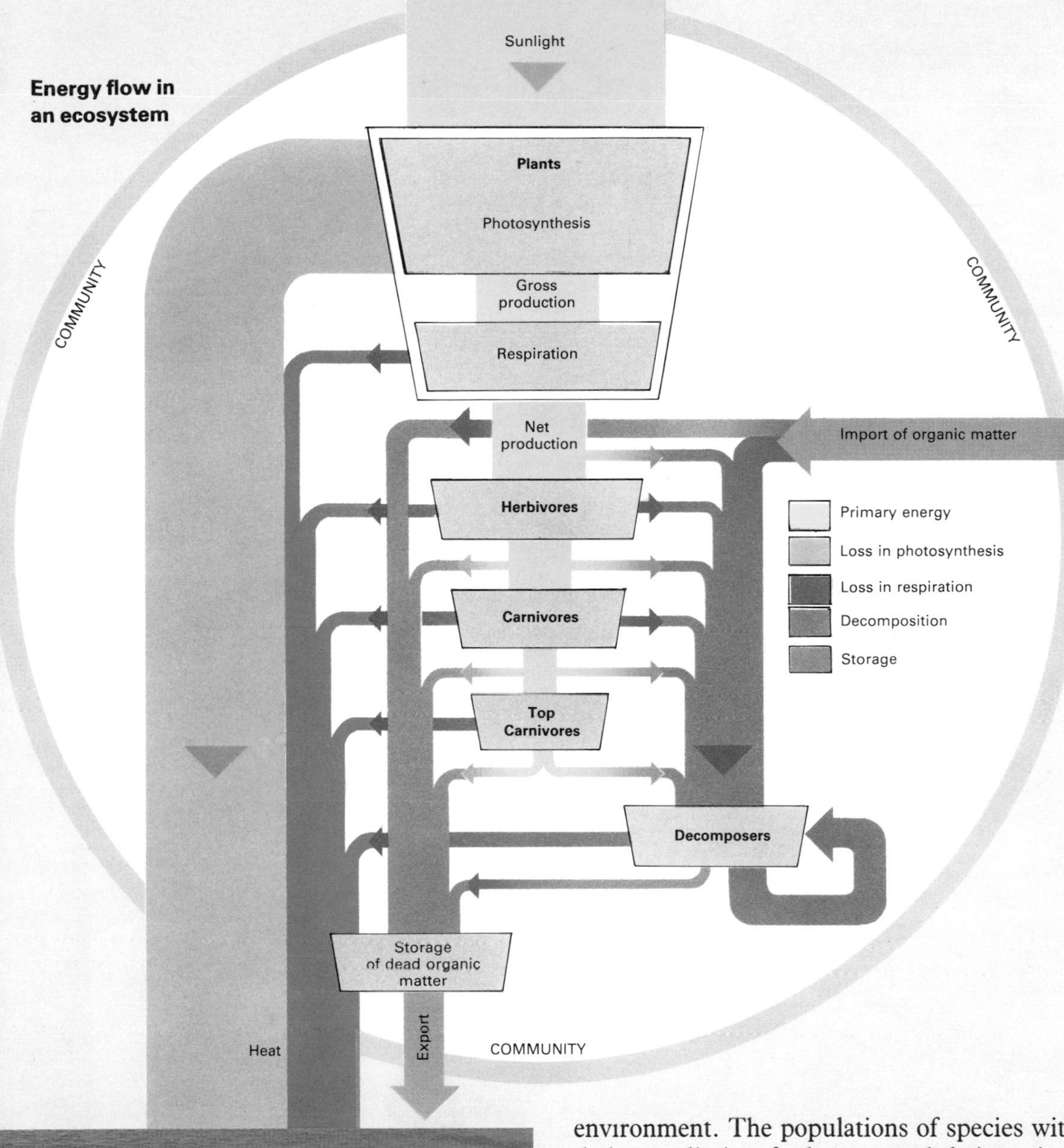

◀ All life is sustained by energy and almost all of the energy needed by living things is ultimately supplied by sunlight. Light energy is trapped by green plants and some of it is used in plant respiration. The remainder is then passed on to herbivores and carnivores; it is said to flow between these feeding, or trophic, levels. At each stage some energy is lost because no energy conversion is entirely efficient. Also each living organism uses up some energy in the work that it does, such as moving and seeking food, and this is lost as heat during respiration. Any spare energy is used up by the decomposers, the fungi and bacteria that utilize the energy of dead materials. Each group of organisms., such as the plant producers, herbivores, carnivores, and so on, can be thought of as temporary reservoirs of energy that higher trophic levels draw upon. It is possible, therefore, to construct an energy balance sheet for the ecosystem. In fact the picture can become more complicated as when some organisms feed at a number of different levels. For example, an omnivore may eat some vegetable matter, some fungi and a variety of meat, as do human beings.

environment. The populations of species within the community have their own limits of tolerance and their optimal requirements for various physical and chemical factors. Light levels, humidity, temperature and the availability of certain essential elements for healthy growth are all necessary for most living things, but each species differs in its precise needs.

The idea of the ecosystem can be applied at any level of scale. We can consider a pond as an ecosystem (➧ page 118), or a rock pool, a dead log, a forest, a clump of grass, or the entire globe. All can be studied from the ecosystem point of view.

Two of the most valuable ideas to emerge from the ecosystem concept are energy flow and nutrient cycling (➧ page 88). Each ecosystem has an external supply of energy, usually the Sun's energy, which is fixed by green plants (the primary producers), but some ecosystems import energy in the form of organic matter from other ecosystems. An estuarine mudflat, for example, has most of its energy brought in from outside by the flow of water. This energy is then available to the living organisms. Some is used up by the plants themselves in producing new tissues, leaves, wood, roots, flowers, nectar, pollen and seeds, and in respiration, which gives the plant the energy needed for the general running of its cell machinery (metabolism).

The individual is the first level of study in ecology and it is variation among individuals which is the raw material upon which natural selection and hence evolution works

Levels of study in ecology

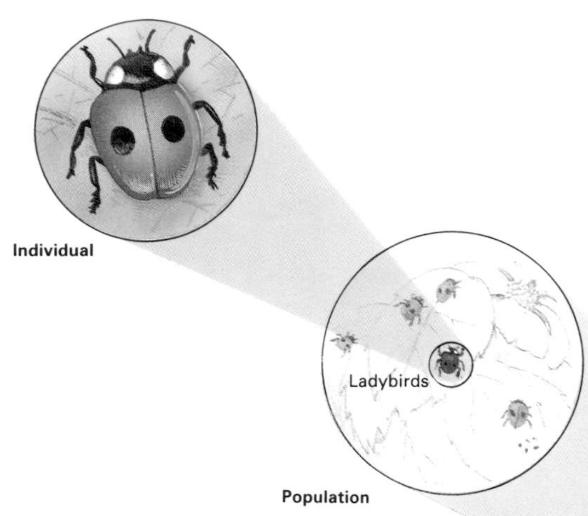

The flow of energy

Grazers obtain their energy from plants and build up energy-rich tissues which are available as a resource for exploitation by predators and parasites. So energy is passed through a complex web of feeders, with some being lost at each stage by the inefficiency of digestion and energy exchange and by being used up in respiration. Ecologists estimate that the average efficiency in the wild is about 10 percent at each step in the food chain. Those tissues which die and fall to the ground are consumed by detritivores and decomposers, which use up the residue of their energy content. In the end, all the incoming energy is dissipated or exported, unless some remains in the living tissues (biomass) as the ecosystem grows, or in the soil as humus or peat.

Thus energy flows in one direction through the system. It cannot be brought back to the beginning and reused. The situation with nutrients, or mineral elements, is rather different. Elements are needed to build up the living tissues of plants and animals and they are obtained

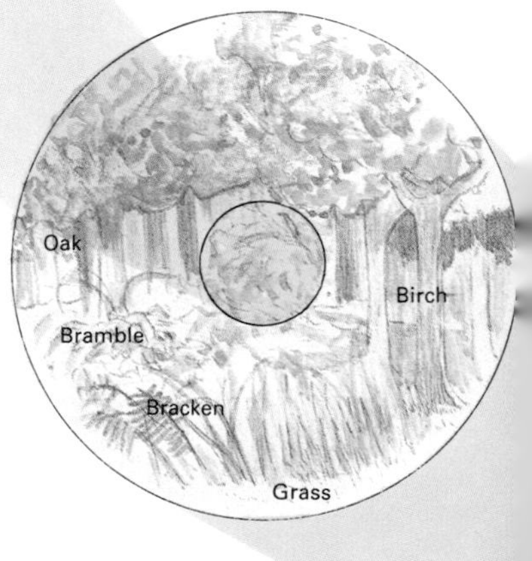

▼ ***Each plant may have several insects, birds and herbivores feeding upon it. These animals are eaten by a range of others, so a complex food web builds up. Here three plant species form the base of such a web and support four further feeding or trophic levels.***

The food pyramid

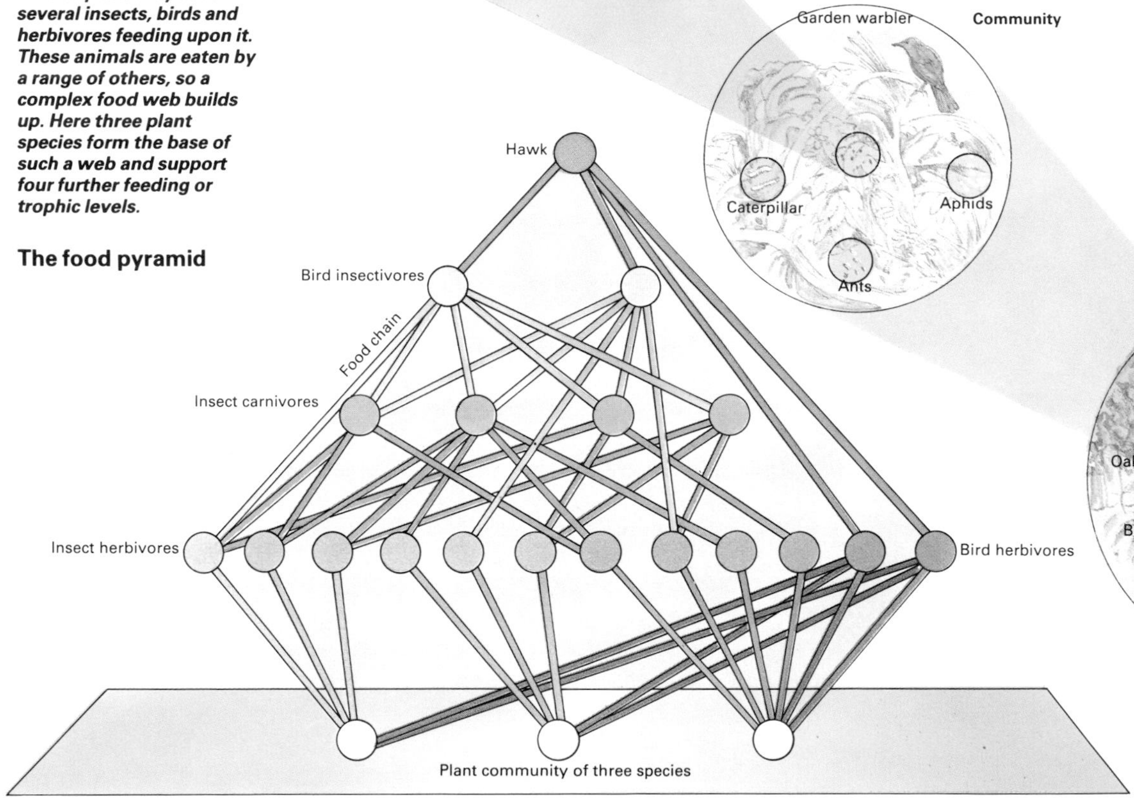

World primary production

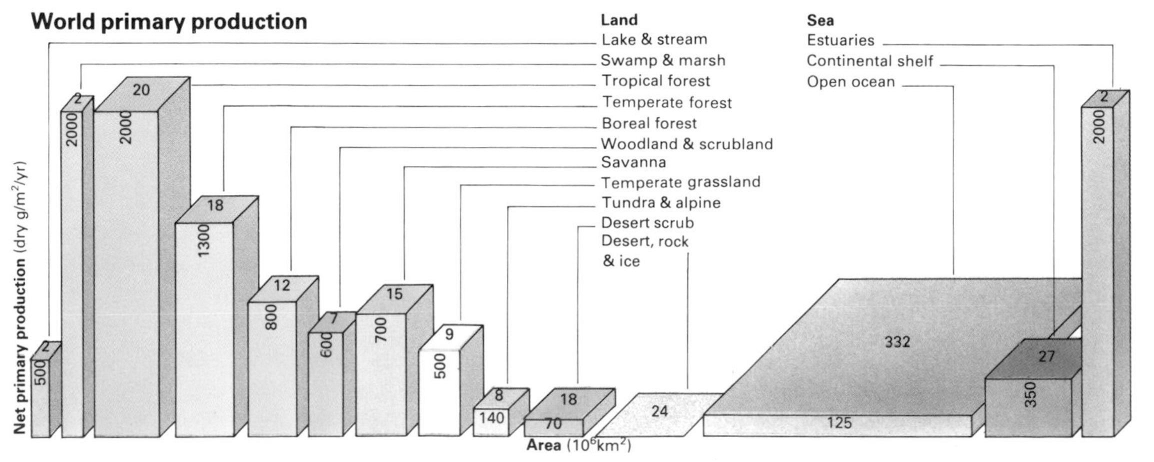

◀ ***The energy reaching the Earth from the Sun is fixed by green plants into usable food, but some types of vegetation are more efficient than others. The tropical forests are the most important energy-fixing systems, followed by the temperate and boreal forests and the tropical savannas. The oceans are not very efficient in terms of fixation per unit area – they resemble deserts – but their great area means that they nevertheless make an important overall contribution to global photosynthesis.***

ultimately from the nonliving world. Carbon, for example, is obtained by plants from the atmosphere as carbon dioxide and is built up into organic tissues with the aid of energy fixed from the sun. Nitrogen is taken up from the soil, as are phosphorus and calcium, and these too are combined into vital compounds; nitrogen, for instance, is needed in the construction of proteins, andd phosphorus is an essential element of both genetic material (DNA anf RNA) and an organism's energy-transfer chemicals (ATP and ADP).

When dead plant and animal material decays, the elements contained in them return to the non-living part of the ecosystem and from there can be reused by living organisms in a never-ending cycle (➧ page 88). Only deposition of an element in the deep ocean and its incorporation into rock can take it out of the cycle. But even this is not necessarily the end of the story, for over millions of years it may be elevated by crustal movements and be exposed to the process of weathering which will release it into the soil once again.

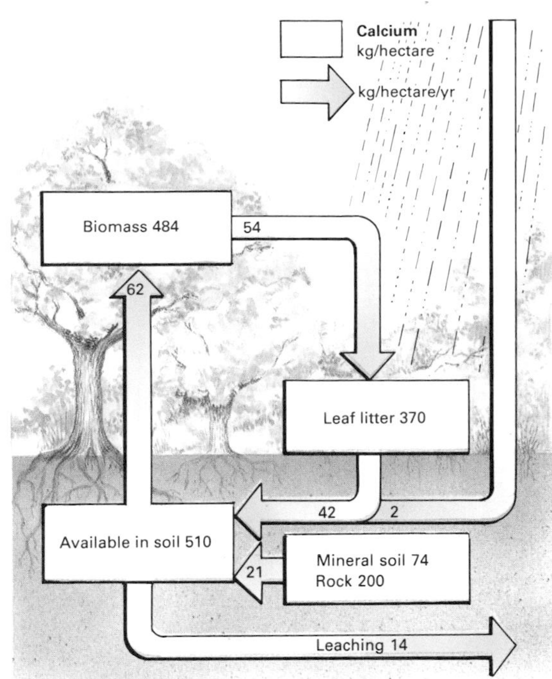

▲ *Inorganic elements, like calcium, cycle round the ecosystem. They enter as the rocks are weathered, or arrive as dust in rainfall. They are then taken up and stored in vegetation where they are available to animals. Leaf fall and decomposition completes the cycle.*

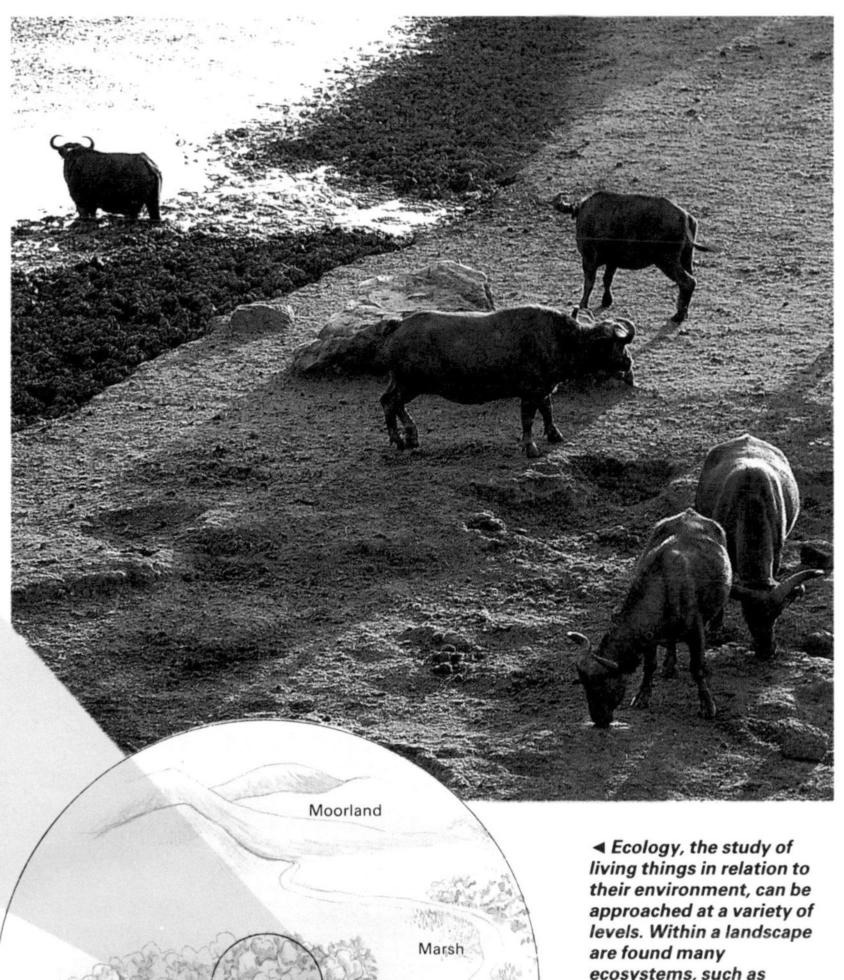

◀ ***Grazing animals depend largely on their plant food for the chemical elements they need for growth. But elements such as sodium may be in short supply in the vegetation because plants do not need them. Grazers, like these buffalo, make up for this lack by licking salt or eating soil.***

◀ ***Ecology, the study of living things in relation to their environment, can be approached at a variety of levels. Within a landscape are found many ecosystems, such as woodland. Within such ecosystems are communities of plants and animals. Each species has its own population contributing to the community. Populations consist of individuals of a single species. At each level of study different questions about the ways in which nature operates can be answered.***

Nutrient cycling in an ecosystem

The movement of an element round an ecosystem is well illustrated by a study of calcium in a deciduous forest in North America (see above). Calcium is necessary for all living things because it plays a part in regulating the activities of membranes in cells. In most vertebrates it is also needed for bone structure and in many gastropods for the formation of the shell.

There are two main ways in which calcium enters the cycle. First, from the weathering of rock, in which calcium-containing compounds are broken down and calcium released into the soil. Second from precipitation containing dissolved salts and dust from the atmosphere. The loss of calcium from the ecosystem takes place largely as a result of water draining through the soil and leaving via a stream. At this site it can be seen that more calcium is entering the system each year (23 kg/ha/yr) than is leaving it (14 kg/ha/yr). The remainder is accumulating within the soil or in the plant and animal biomass.

Within the ecosystem, about 99 percent of the calcium is in the soil and less than 0.5 percent in the living biomass. But of the calcium available for extraction from the soil, the vegetation manages to absorb about 12 percent, which shows how much calcium is in demand. Experiments in which forest has been cleared from watersheds have shown that the export of elements like calcium increase markedly. This is partly due to the release of elements from the biomass and also because the soil itself loses stability if the vegetation is killed or removed, and is eroded out of the ecosystem. The vegetation can therefore be regarded both as a reservoir of chemical elements and as a stabilizing influence upon the whole ecosystem.

Cycling of Chemical Resources

The Earth's chemical resources are in constant use and re-use within food chains and food webs (*page 86**). Elements such as carbon, nitrogen and oxygen are transformed from an inorganic to an organic state and back again in nutrient cycles.*

All life on Earth depends on oxygen. Animals, plants, fungi and bacteria use up oxygen from the air in respiration. Atmospheric oxygen is also used up when fossil fuels are burned. We depend on plant photosynthesis to replace this lost oxygen.

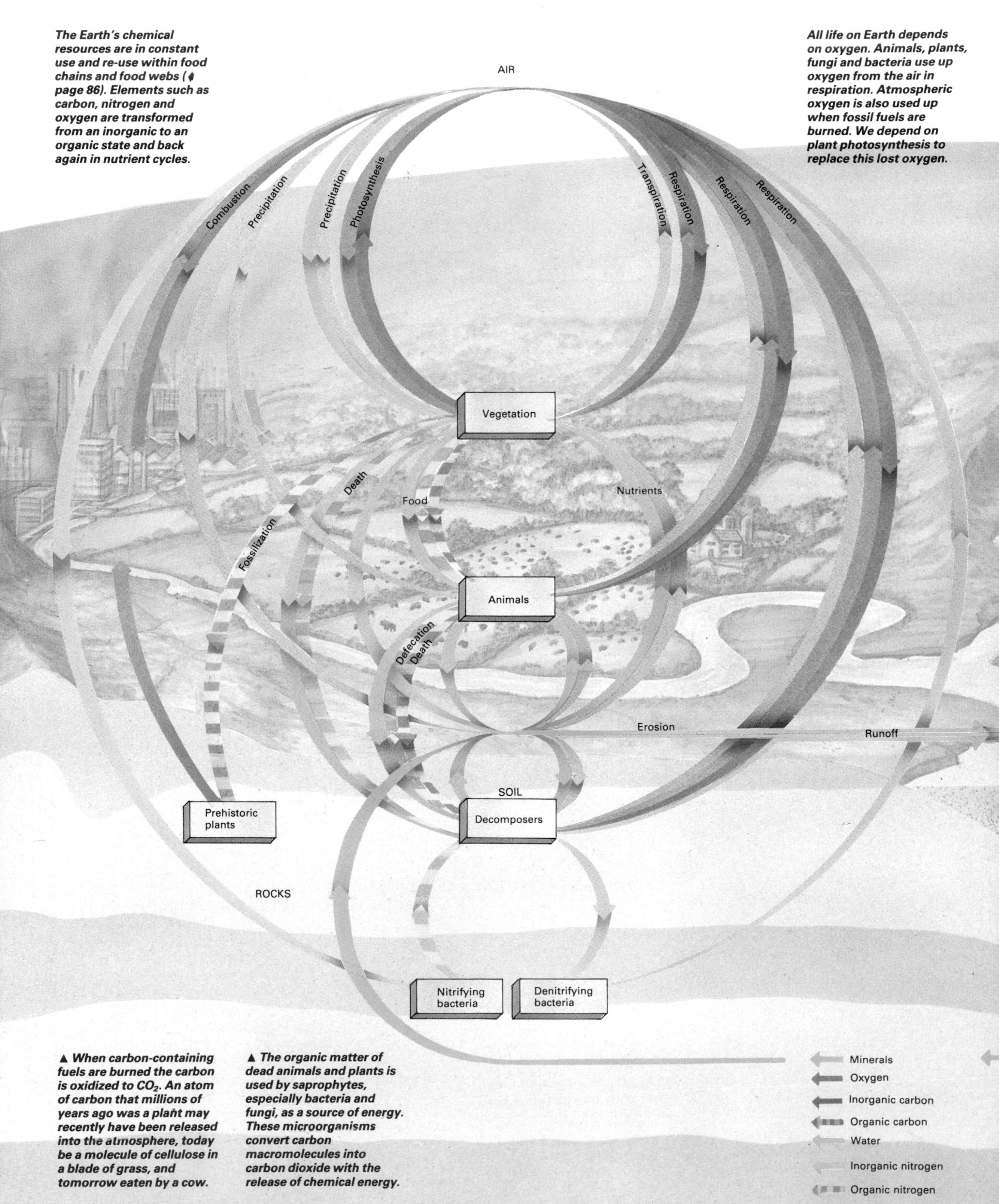

▲ *When carbon-containing fuels are burned the carbon is oxidized to CO_2. An atom of carbon that millions of years ago was a plant may recently have been released into the atmosphere, today be a molecule of cellulose in a blade of grass, and tomorrow eaten by a cow.*

▲ *The organic matter of dead animals and plants is used by saprophytes, especially bacteria and fungi, as a source of energy. These microorganisms convert carbon macromolecules into carbon dioxide with the release of chemical energy.*

Nutrient cycles exist on land and in water and the stages involved in both ecosystems are almost identical. Also, the two diverse ecosystems are closely linked. An upset of nutrient cycles in one produces a change in the other.

Limits to growth of aquatic organisms are dissolved oxygen and sunlight. In the atmosphere, oxygen is mixed with other gases and its concentration is quite constant. In aquatic ecosystems it is dissolved in water and concentrations vary considerably.

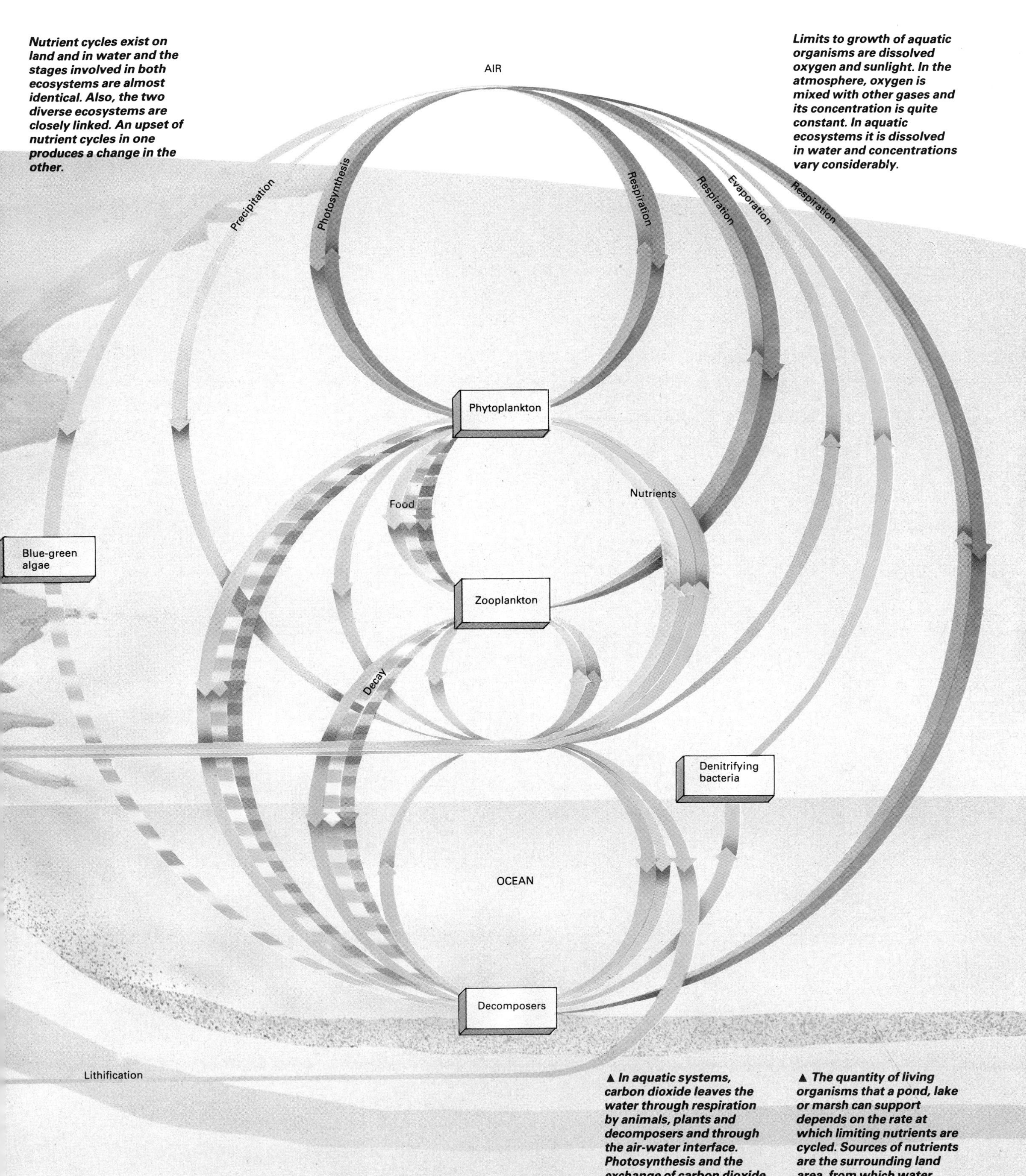

▲ In aquatic systems, carbon dioxide leaves the water through respiration by animals, plants and decomposers and through the air-water interface. Photosynthesis and the exchange of carbon dioxide and oxygen takes place primarily in surface waters.

▲ The quantity of living organisms that a pond, lake or marsh can support depends on the rate at which limiting nutrients are cycled. Sources of nutrients are the surrounding land area, from which water (run-off) drains into the body of water.

The Australian wombat and the American woodchuck inhabit the same niche and look similar but have evolved independently

The niche

Each species of plant and animal occupies a peculiar position in the community, both in a spatial sense (where it lives) and in a functional sense (what it does in the system to earn its living) (◀ page 86). This dual description of the role of the species was introduced by the British ecologist Charles Elton in 1927 and was referred to as the niche.

It is possible for the niches of two animals to overlap. For example, both shrews and moles will eat earthworms, but shrews feed mainly above ground whereas moles feed mainly underground, so the overlap is not a complete one. The barn owl and the kestrel both feed on mice, but the former hunts almost exclusively by night whereas the latter hunts by day. The more closely the niches of two species resemble one another, the more intense is the competition between them. Theoretically, it is impossible for two species to occupy the same niche in the same geographical area. The competition would be so strong that one or other species would be eliminated.

The competition resulting from niche overlap may result in a species being unable to attain its full potential. If you selectively remove one species of grass from a meadow, other species will soon move in to take advantage of the newly available space. Conversely, when the grey squirrel was introduced into Britain during the 1870s, the native red squirrel suffered a decline which has continued ever since. The North American species was more successful than the native species, particularly in deciduous woodland.

Such observations as these led the American ecologist Evelyn Hutchinson in 1967 to propose the distinction between a fundamental (i.e. ideally potential) and a realized (i.e. actually achieved) niche. The latter is smaller than the former and always lies within it.

It is possible for a natural resource to be incompletely exploited by the current inhabitants of an area, in which case one can propose that there is a vacant niche available for an appropriate animal.

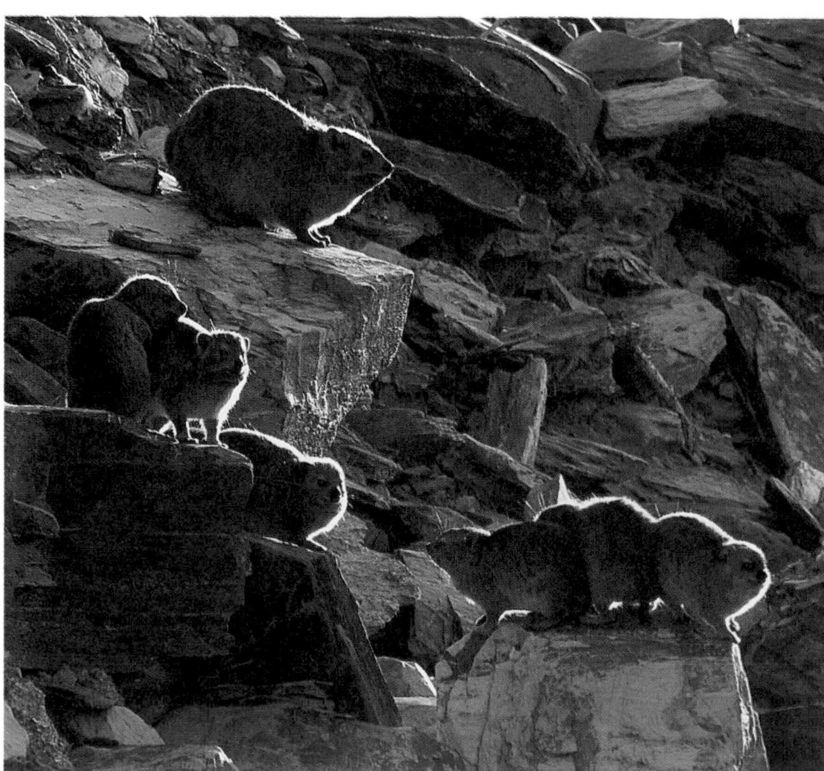

◀ The niche of the rock hyrax, seen here grouped together for protection, is that of a grazer within the unpromising habitat of rocky screes and boulders. It eats the meagre growth of plants and their seeds and may even resort to lichens. Rock crevices provide shelter for the colonies.

▲ Coral reefs provide opportunities for the development of a wide range of niches. The fish which shelter among corals find protection from predatory fish and may emerge at night to graze more widely. Their excreta fertilize the water and promote plankton growth on which the coral feeds.

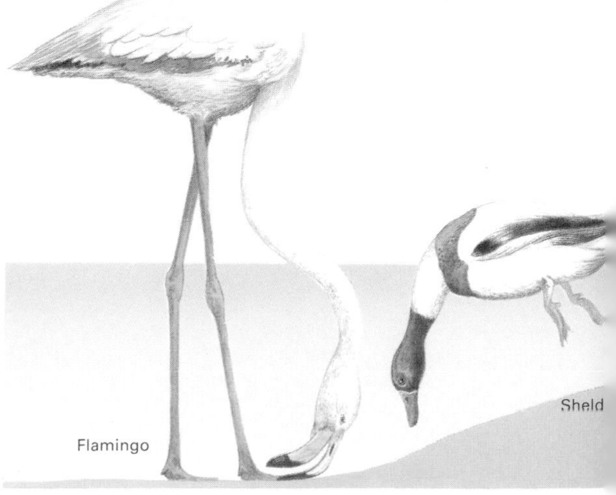

Niche specialization

In all habitats the resources available for exploitation are limited. A bird, for example, will seek adequate food as well as breeding space within a habitat, and any given individual is most likely to succeed if its requirements are slightly different from those of potential competitors. In other words, selective advantage is gained by those that find new resources to exploit or that can devise new ways of tapping those resources. Over the long term, therefore, evolution leads to the adoption of slightly different roles by the various organisms which go to make up an ecosystem.

An example of niche specialization is provided by the feeding adaptations of birds in shallow, saline lagoons, such as those of the Carmargue in southern France. The main food resource is the abundance of aquatic crustaceans living at the mud/water interface in the lagoons. But various techniques of hunting and adaptations of leg structure are necessary for the efficient exploitation of this resource.

Kentish plovers have fairly short legs (27-28mm tarsus) and feed mainly on the edge of lagoons, hardly entering the water. Oystercatchers (50-52mm) and redshanks (55-58mm) feed in the shallow water and may compete to some extent, but oystercatchers prefer small mollusks such as cockles, whereas redshanks stick largely to insects and crustaceans.

In the deeper water, birds with longer legs are better suited for bottom feeding, although redshanks can swim and often do so to extend their feeding range. Avocets (82-89mm) and black-winged stilts (125-127mm) can exploit the deeper water, wading out on their long legs and swimming when necessary. Shelducks can feed in yet deeper water, not because of their length of leg, but rather because of their ability to upend and use their long necks. Finally, flamingos, (tarsus length 259mm [female] to 323mm [male]), are able to exploit waters far too deep for the other wading species.

The outcome is that several species of bird are able to coexist on the same basic food resource in this habitat as a result of evolutionary specialization leading to a high efficiency of resource utilization within precisely determined limits. Specialization leads to more species within the community.

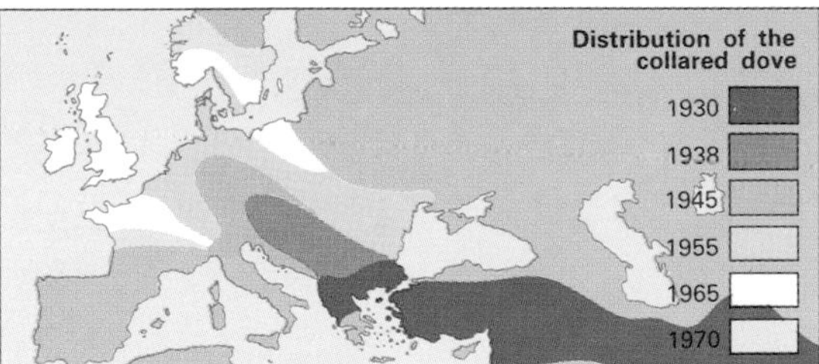

◄ The existence of a vacant niche can best be shown when an animal moves in to occupy it, as shown by the spread of the collared dove this century. Perhaps some genetic change fitted the species for a new role and allowed its expansion. (The spread of the collared dove has been arrested by the Atlantic.)

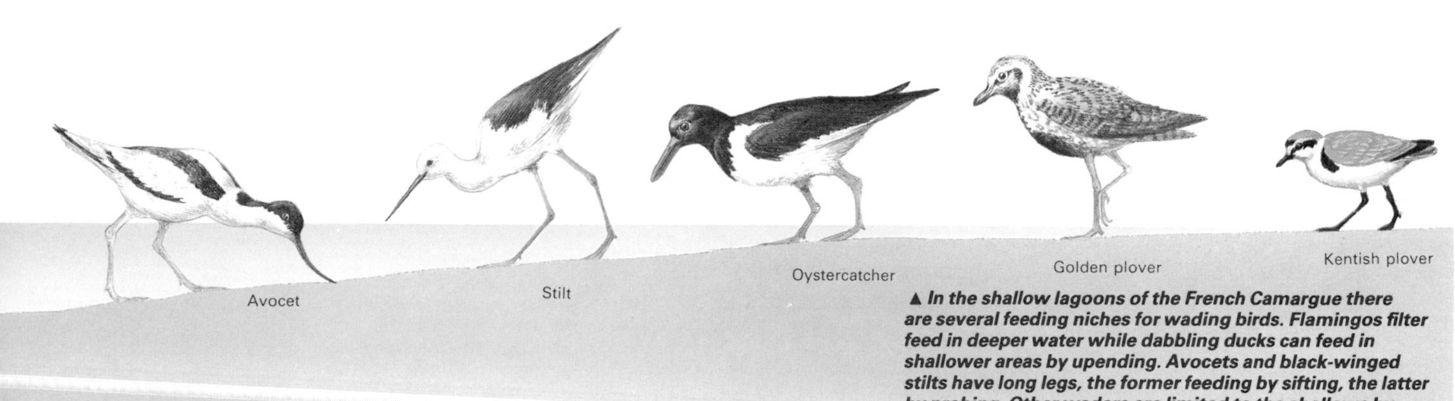

▲ In the shallow lagoons of the French Camargue there are several feeding niches for wading birds. Flamingos filter feed in deeper water while dabbling ducks can feed in shallower areas by upending. Avocets and black-winged stilts have long legs, the former feeding by sifting, the latter by probing. Other waders are limited to the shallows by their shorter legs.

It was the study of island life that led both Charles Darwin and Alfred Wallace to the theory of evolution by natural selection

◄ On the Galapagos Islands, off the coast of South America, isolation and a lack of predators has led to the evolution of some bizarre forms of life. One such is the giant tortoise much studied by Charles Darwin. Competition from introduced grazers, such as the goat could lead to its extinction (➧ page 197).

► Competition between animals for food resources means that any individual born with a new, advantageous feature may have a better chance of surviving, so that the new character is passed on to a greater number of offspring. This is how the giraffe developed its long neck and the ability to feed above other species.

Island biogeography

On islands, rather special conditions exist, since the arrival of new species depends upon such factors as how far they have to travel over the ocean and how large the island is. The chances of arriving at a small island are clearly less than those of encountering a large one. But the survival of populations on islands may also be in doubt. On a small island with small populations of species, extinction is far more common than on a large one. The overall diversity of species found on islands is thus determined by island size, distance from neighboring islands and, to some extent, by the diversity of habitat types found upon the island. These features of island biogeography have been studied in detail by the Americans Robert MacArthur and Edward Wilson, who have constructed mathematical formulae for rates of invasion and extinction on islands. Their theories have proved useful in the evaluation of nature reserves, for these also behave like islands in that they are fragmented habitats in the middle of an alien territory (urban or agricultural) (➧ page 196).

► Islands such as this, some 20km from Tahiti in the South Pacific, often have unique forms of life present on them, especially if they are very isolated. This is because the plants and animals which reach them by chance events are no longer able to interbreed with their original populations and they gradually become adapted to the peculiar conditions of island life in which they now find themselves. Land birds may no longer need to fly – indeed it could be a positive disadvantage, so the powers of flight are often lost. The takahe, for example, a rail unique to New Zealand, lacks predators and so has abandoned flight.

Diversity

Factors that lead to high diversity in an ecosystem include good growing conditions for plants, a variety of habitats and feeding opportunities for animals, barriers to widespread breeding, and general environmental stability. The more species that can be packed into a habitat as a result of specialization, then even more diverse the system will become. But when we consider plant and animal communities in conjunction, then certain problems arise. In a grassland, for example, a degree of grazing by sheep may serve to increase the diversity of the flora. This is a result of the sheep reducing the vigor of robust, fast-growing species which would otherwise dominate the community, and allowing a variety of smaller species to survive. The short turf sward produced, however, although rich in species, lacks complexity in its canopy architecture and so does not contain many microhabitats for invertebrate animals. These creatures reach their greatest diversity in tall grassland, even if it is poorer in plant species.

Birds, on the other hand, seem to prefer short turf as a feeding habitat. There are exceptions, of course, such as the corncrake, which prefers tall grass meadows. But more species are generally prepared to feed in short grass rather than tall. This may be because their prey is easier to catch in such a situation, or it may reflect the ability of the feeding birds to keep an eye open for approaching predators such as hawks and cats.

In the Coto Doñana National Park in Spain, a cyclic succession of dune ridges, damp sand, grass and sedge, invading umbrella pine and mature pine forest, has been going on for at least 13,000 years.

Succession on the shingle beach

Shingle beaches form as a result of the prevailing drift of coastal currents depositing lengthy strips of pebbles and other large particles, particularly during storms. They are inhospitable habitats for plants and animals since water drains freely through them as a result of their large particles. The soil usually contains little available organic or mineral material.

Plant colonists, such as the sea purslane and the yellow horned poppy have a marked impact on the general conditions of the habitat in that they bind the mobile pebbles together with their deep, penetrating roots. Their shoots and leaves create a complex microclimate, shielding the beach surface from some of the intensity of radiation, temperature variation and desiccation. The dead organic matter produced in litter provides an energy resource for microbes and a true soil begins to develop. Lichens grow and, with some of the microbes, help increase the nutrient resources of the habitat.

But stability and improving conditions may lead to invasion by more robust species of plant, such as hawthorn and gorse. These shrubs soon out-compete the early colonists which succumb in the newly shaded environment. The colonists, by their very presence, have modified their surroundings so that their continued survival becomes less likely.

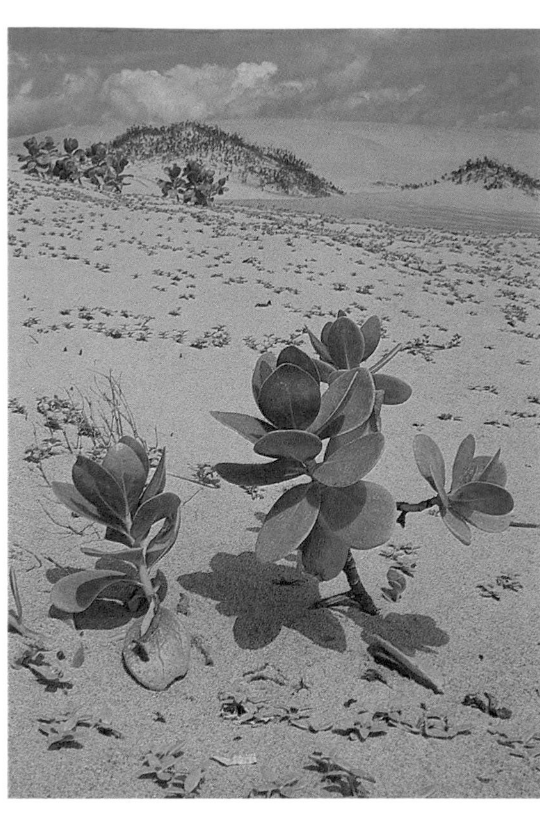

◄ Maritime dunes, like these along the east coast of South Africa, are dry, unstable habitats which are colonized by sea-distributed drought-tolerant plants and creepers that are able to stabilize the soils. In the middle-distance more stabilized dune hummocks can be seen.

► Volcanic eruptions produce soils which are rich in many nutrient elements, such as phosphorus, sulfur, sodium and potassium, but are low in organic matter and are free-draining and dry. Colonization by plants adds organic litter and improves their water-holding capacity.

▼ Retreating glaciers leave behind some types of habitat which are particularly unpromising, like these bare rock faces. Lichens grow on the rock and their dead matter accumulates in crevices forming a soil that can be colonized by mosses. Succession has begun.

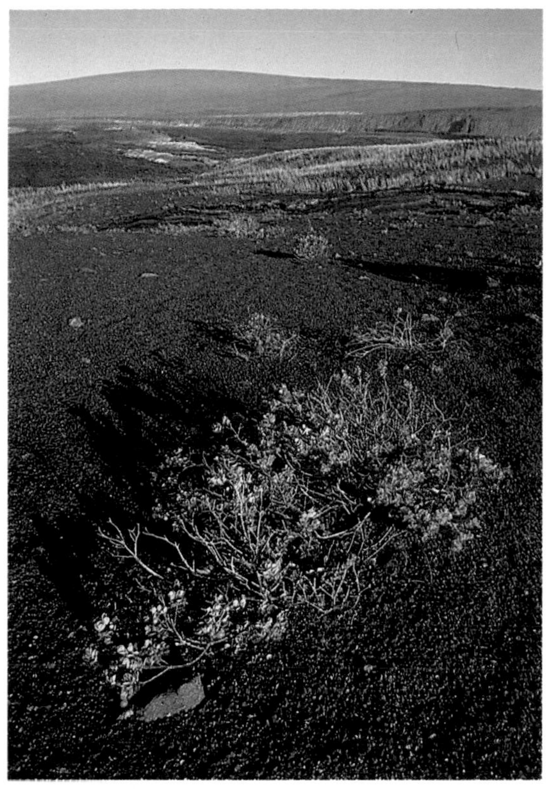

The growth of the ecosystem

F. E. Clements, the North American botanist, in 1916 put forward the formal idea that vegetation undergoes a series of predictable changes in the course of its development, leading ultimately to a stable equilibrium state, or climax. He called this developmental process succession, and it was his view that within any given climactic situation all such developments would proceed towards the same end-point. The idea of convergence was based on the premise that the climate ultimately determines the nature of vegetation, so no community will achieve stability until it has arrived at the most efficient vegetation type for that climatic condition (▶ page 174).

Clements allowed the possibility, however, that certain factors could stop the process before it reached its natural conclusion. Shallow, dry soils could result in scrub or grassland developing within a forest zone (subclimax). On the other hand, unusually favorable conditions, such as a river valley within dry prairies, could result in the development of woodland (postclimax).

One of the main ideas underlying succession theory is that of facilitation. This entails the physical alteration of the environment by the presence of one species which permits the invasion of other species. The presence of marram grass in a sand dune, for example, slows the surface wind, allowing sand grains to be deposited and so encourages the growth of dunes. The consequent alteration of microclimate (more moisture, smaller range of temperature, less desiccating wind) and soil conditions (less salinity, more organic matter, better water retention), offer opportunities for certain other plants to establish themselves which could not have survived on the open dune surface. Eventually forest may develop.

In recent years some aspects of this theory have been challenged. Some ecologists feel that other factors, such as the sheer chance arrival of seeds of a plant, are of greater importance than facilitation. What we regard as successional stages may simply reflect the length of time it takes certain species to mature. This would account for the herb-scrub-forest development one often finds in the succession following the abandonment of old fields. Yet the biomass of the ecosystem takes several hundred years to reach its maximum development and in doing so it passes through recognizable communities of plants and animals.

▼ Open bodies of water provide another habitat in which succession takes place. This lake in the Upper Amazon of Peru is formed from an old river meander, or ox-bow, and is being invaded by floating and emergent aquatic vegetation. It will gradually fill up as organic litter accumulates in the form of mud. Eventually, as the water becomes more and more shallow, flood-tolerant trees will invade and the forest canopy will close over the former pool.

See also
Communities of the Land 97-116
Communities of the Water 117-128
The Last Great Hunt 181-190
The Shadow of Extinction 195-208
Threats to the Environment 209-224

The question of stability

The final, or climax, stage of succession is said to be stable in the sense that no further change in the overall community of organisms is detectable within the human lifetime. Over geological time, as climate itself changes, the general vegetation is also modified. Thus in the temperate zone over the past 11,000 years, the vegetation has developed from open tundra, through birch/pine forest, to mixed deciduous woodland (➧ page 102). This may seem a slow process, but in terms of the generation times of trees, it is quite rapid.

Stability is therefore a relative term. What concerns ecologists rather more than long-term changes in vegetation with climate is the ability of communities to resist disturbance by human beings. As human populations expand and the demand for living space and resources increases, we place greater demands upon natural habitats. How well are they able to withstand the onslaught?

At one time it was believed that the complex structure and diverse interrelationships of communities such as tropical rain forest would provide them with an inbuilt mechanism for resisting disturbance. If one species suffered, there were always plenty more to take its place. But this has proved to be entirely wrong. The more complex the relationships in a community, the more fragile and sensitive that system becomes. Disturbance of one link in a complicated net of interactions may result in the whole structure and composition of the community being disrupted (➧ page 212).

As we consider the range of communities present over the Earth, their delicate balance must be borne in mind. Some habitats have taken thousands of years to develop. Once destroyed they can be regarded as lost forever.

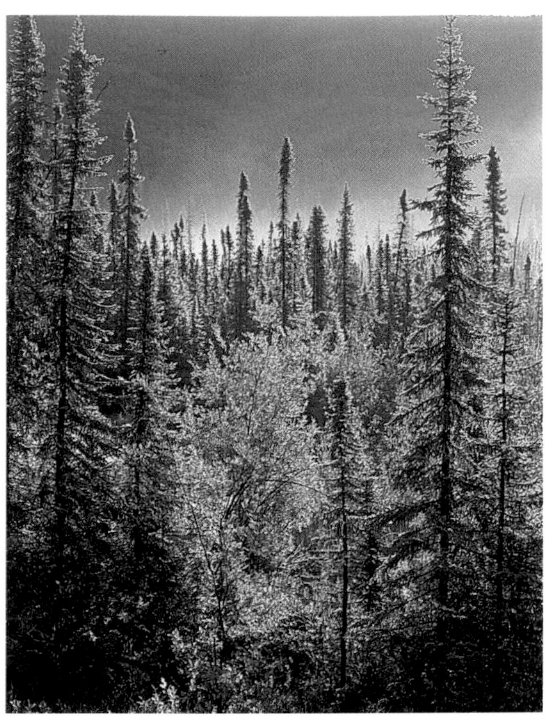

▲ Spruce forest may appear stable, but it is subject to regular wind-blow and fire. Natural fires in North America occur about every seventy years.

▼ Fire is much more serious in tropical forests such as here in Brazil. When burned by people it may take centuries to recover, and it may not do so at all.

Communities of the Land

9

The effect of latitude on climate...Similar climates produce similar habitats...Ecology of different types of forest, grassland and desert...PERSPECTIVE...The influence of circulating air masses...The floral mosaic ...Life at different levels...Adaptations to extremes of climate...Variations on photosynthesis...Mountains are like islands

The Earth's surface is constantly being bombarded with energy in the form of light and heat radiating across space from the Sun. Not only is this the source of almost all the energy used by living things on the planet, it is also the driving force behind climate. And it is climate, in turn, which ultimately determines what types of communities can develop in different parts of the globe.

Two main components of climate directly affect the performance of living things: warmth and water. All the biochemical processes going on in the cells of plants and animals are governed by special protein catalysts, the enzymes, and these molecules will only operate within a fixed range of temperature. Generally this lies between about 15° and 30°C. Warm-blooded animals, of course, can avoid some of these environmental limitations by keeping up their body temperature. But they can only achieve this if their energy (food) supply is adequate to invest in this way.

Many plant and animal tissues consist of over 90% water, so a regular supply of this commodity is vital for life. Some animals control their body temperature by means of water loss, and most plants suffer considerable water losses from their leaves as they open the pores (stomata) in order to take in carbon dioxide from the atmosphere. The amount involved varies between living organisms; those adapted to drier habitats often have mechanisms which avoid undue loss (➧ page 107), such as plants that only open their stomata at night, and reptiles which excrete crystalline uric acid to save water.

The communities of the Earth are thus governed by the availability of adequate temperature and water supplies. But these commodities are not evenly distributed over the surface of the globe.

Wind circulation

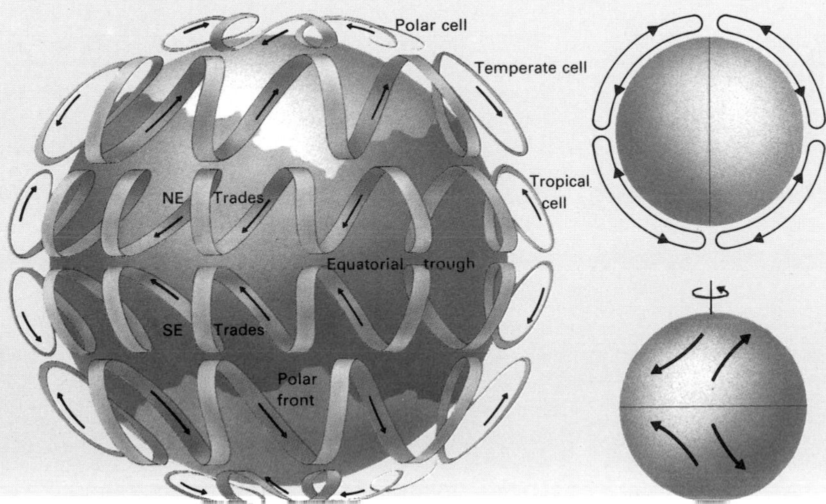

Patterns of climate

The equatorial regions receive the most intense supply of energy. The energy heats the ground surface and gives rise to convection currents in the atmosphere. Hot, moist air currents are pushed upwards and, as they rise, they cool and their water load condenses and falls as rain. So the equatorial regions are characterized by year-round high temperatures and heavy rainfall.

As the air masses rising over the equator are pushed polewards, they cool and descend. Their descent is most concentrated over latitudes 20-30° north and south of the equator, and these regions experience the high pressures created by the falling masses of air now dried as a result of the water shedding over the equatorial zone. These dry regions harbor the world's deserts. Between them and the equator is a broad transition along a gradient of increasing wetness.

Some of the falling air is deflected polewards, where it meets the cold, descending polar air on its way towards the equator. When the two airflow patterns meet in the temperate zone, a mild but wet climate is created. Once again, the region between the hot arid zone and the mild temperate zone is intermediate. It is characterized by hot, dry summers and mild, moist winters.

The polar regions receive the least incoming radiation. The air descending over the frozen poles is both cold and dry. This means that the Arctic and Antarctic areas receive very little precipitation – they are cold deserts.

Between the poles and the temperate zone lies the boreal region (so-called because of the Greek word for the north wind, "boreas"). This is also a cold zone in winter, but has warmer summers and generally more precipitation, especially as snow.

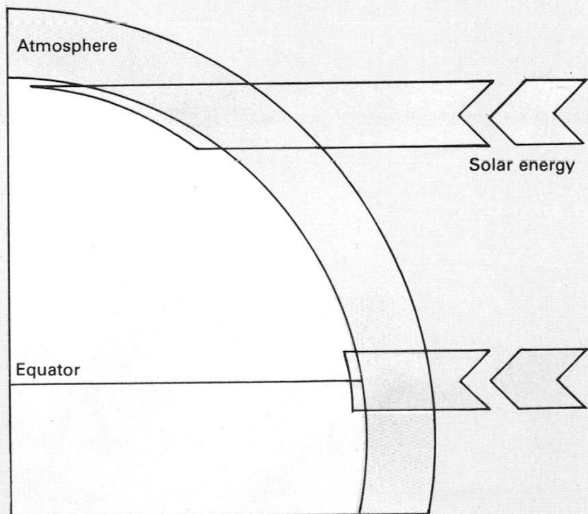

▲ Energy arriving at the Earth from the Sun is unevenly distributed over the surface of the globe. At the equator it passes through little atmosphere and is spread over a small area while the reverse is true over the poles.

◀ The general circulation pattern of the atmosphere is complicated by the rotation of the Earth. Rising equatorial air masses are deflected westwards and produce a spiral motion as shown here. The meeting of temperate and polar air masses produces a zone of turbulence and instability in the atmosphere, resulting in an unstable climate in mid latitudes.

Tropical rain forests cover less than 10% of the land surface but account for some 30% of the Earth's biomass

The world's biomes

Much of the Earth's land surface is covered by a blanket of vegetation and this greatly alters the amount of energy actually reaching the ground. This means that the soil surface may experience quite a different type of climate in terms of temperature and humidity from the one expected in such a geographical setting. The vegetation creates its own microclimate.

The nature of the living community depends very largely on the type of architecture created by the vegetation, but this is itself subject to the general, or macro-, climate of the area. In the equatorial zone, for example (▶ page 100), plants experience neither drought or frost. Losses of water from the canopy are of no consequence since water is in plentiful supply. No delicate buds that are held aloft are in danger of being chilled, so tall growth is unimpeded. The scale and complexity of the botanical architecture thus created provides a vast assortment of microsites and opportunities for a wealth of animal and microbial dependants.

Seasonal climates bring their problems. When one time of year is unfavorable for growth, either because of drought or cold, only hardy vegetation can survive. The diversity of plant life is lower and the canopy construction is consequently simpler. In extreme situations, such as a very limited wet season, only scrub vegetation or grasses may persist, resulting in a very different landscape with a much simpler structure and pattern of microclimates.

The major types of vegetation produced under different climatic conditions and the assortment of animals that they support are termed biomes. They form convenient units within which to study the pattern of living communities over the land surface of the Earth, and range from tundra to tropical rain forest.

Major world biomes

◀ *Frost is a serious problem for plants because the formation of ice crystals inside cells can cause the rupture of its delicate membranes. The danger is greatest when the plant tissues are thawing; a quick thaw causes the most damage and can kill the plant.*

The architecture of vegetation

The climate of an area places certain demands upon the plants and animals that are able to survive there. These various stresses may be extremes of heat or cold, of drought or wetness. They may occur only at certain times of the year and form an "unfavorable season", like the cold winter of temperate North America or the dry winter of northern India. It is often these extremes which determine just what plants and animals can, or rather cannot, live in a region.

Vegetation prevents a proportion of the Sun's energy reaching the soil surface. A layer of scrub in the chapparal, for example (▶ page 110), may experience dry, scorching conditions on its surface, yet underneath in the shade temperatures are much lower and the humidity considerably higher. For this reason many animals, especially small ones such as invertebrates, are very dependent upon the architecture of the vegetation at a site.

The vegetation, however, may be exposed to the

▲ **Vegetation is broadly controlled by climate, especially by temperature and precipitation. But the pattern of climate over the world is complex because of the effects of ocean currents, continental land masses, atmospheric circulation patterns and belts of mountains. So the final pattern of vegetation is not a simple zonation with latitude, but a complicated mosaic. This has led to an equally complex development of vegetation types, or formations, and these in turn support different kinds of animals. The major global units of flora and fauna are termed biomes.**

full force of the climate, with the exception of those plants that hide in the shade of others. The most vulnerable part of a perennial plant is usually the bud, as this contains the new potential growth. If it is damaged, growth and development of the whole plant is set back. The bud is usually enveloped in protective scales, but these may prove inadequate against the blown ice crystals of the arctic or the intense heat of the desert. Under these conditions the buds need extra protection.

In the moist tropics there are no such climatic rigors to be coped with and buds can be vaunted with impunity. The seasonal drought experienced in the tropics as one moves away from the equator, however, imposes restrictions on growth. Leaves may fall in the dry season and buds need protection from drying. The height and structural complexity of the canopy is thus diminished.

In the arid zone this effect of drought becomes more pronounced and prolonged with the result that the tree form of growth is not generally successful (▸ page 106). Low growth with reflective surfaces is more appropriate, or the capacity to survive longer periods of drought as a seed. Bulb and corm plants are also well suited because their buds are buried underground. Vegetation is often sparse and the protection afforded to animals limited.

In the moister temperate zone the tree comes into its own once again, but the buds need protection against the cold of winter. Herbaceous plants in this region often keep their buds at ground level where they escape much of the chilling wind. Finally, in the arctic and antarctic latitudes, the winter coldness of the air above the ground and the permanently frozen layers below the ground exert a combined stress on vegetation. The most efficient answer is the dwarf evergreen shrub or the cushion plant in which the buds are held in tightly packed masses just above the soil surface. These form the physical setting in which the animal component of the community can develop (▸ page 112).

Tropical rain forest

The equatorial region, with its consistently high temperatures (in the lowlands often between 20 and 28°C), its very considerable rainfall (between 2.5 and 4m, but sometimes reaching 10m a year), and the lack of seasonal variation, combine to make this biome the richest of the Earth's resources, both in the number of species it harbors and in the sheer mass of living material it supports. On average, each square meter of ground may have as much as 80kg of living plant and animal material above it (◀ page 86).

The rain forest has a complex structure with many layers of canopy. Often one can discern a series of towering emergent trees with a height of about 40m, a second and continuous canopy at about 20m and a third layer at about 15m. This structured world of branches and leaves is so productive and varied that many of its animal inhabitants never need to descend to the forest floor throughout their lives. The diversity of the trees themselves means that some species will be flowering and fruiting throughout the year, so there is never a shortage of forage for the herbivores.

Plant-feeders are therefore abundant in the canopy and include insects, mammals and birds. The insect life is so varied that the majority of species have yet to be described. In a recent study of the insect life on a single tree species over 600 new species of beetle were discovered. The butterflies are almost equally represented, their larvae consuming the endless supply of new herbage being produced and the adults adding a wealth of color to the canopy fauna. These adults can often be seen along the banks of rivers where they collect mainly to take up inorganic salts from the muds. Sodium in particular may be in short supply up in the canopy and is needed especially by the males for the production of their sperm packages.

The fruits of the forest canopy provide an energy resource for many animals. Toucans and hornbills, for example, are fruit feeders especially in the emergent and high canopy layers. The characteristic bill of the toucan seems excessive as a fruit gathering structure and its bright colors suggest that it serves a sexual function in courtship. Bats are another group of animals which are very typical of the forest canopy. In the forests of Panama bats represent almost half of the mammal species present. Most feed on fruits, as do many of the monkeys of the canopy, like the colobus monkey of Africa.

Some bats, and birds such as hummingbirds, rely on the nectar from flowers for their energy. They expend much energy as they hover to obtain this, but reap the reward of an energy-rich food.

In an environment where plant-eating animals are so numerous, it would be surprising if there were no predators to avail themselves of this food supply. Insectivores abound, especially among the birds, where species such as the night-hunting potoo and the day-active motmots exploit the insect resource. Some birds seek bigger prey, such as the crowned eagle of Africa, which feeds mainly upon monkeys.

Not all of the plants in the canopy have their roots in the forest floor. Many, such as epiphytic orchids and bromeliads, spend their lives rooted in pockets of humus that accumulate in the branch crevices of the large trees. Some trees even begin life this way as it saves them from the problems of establishing themselves in the dark, dank underworld beneath the canopy. Many figs use this strategy, starting life as an epiphyte and later sending down roots to the forest floor, often strangling their host tree in the process.

The layers of the tropical forest

▼ The boa constrictor inhabits the floor and the lower part of the canopy, feeding on small mammals.

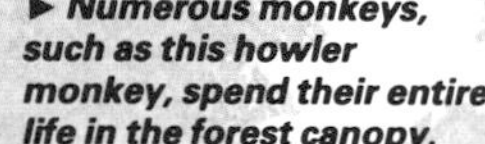

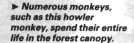

► *Numerous monkeys, such as this howler monkey, spend their entire life in the forest canopy.*

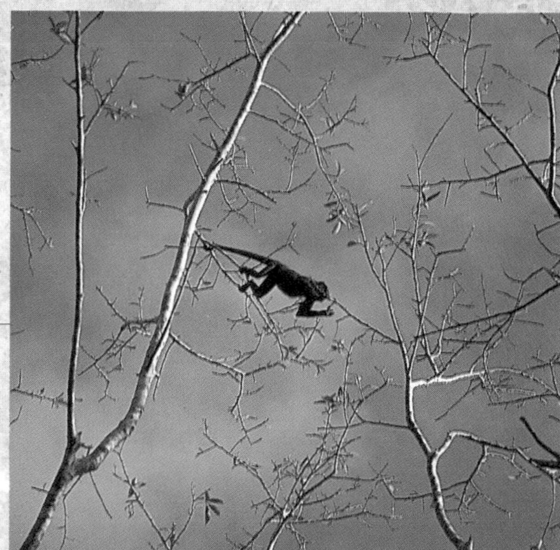

Upper storey

Middle storey

◄ *Many bats feed upon fruit, like this one, seen carrying a fig and assisting seed dispersal.*

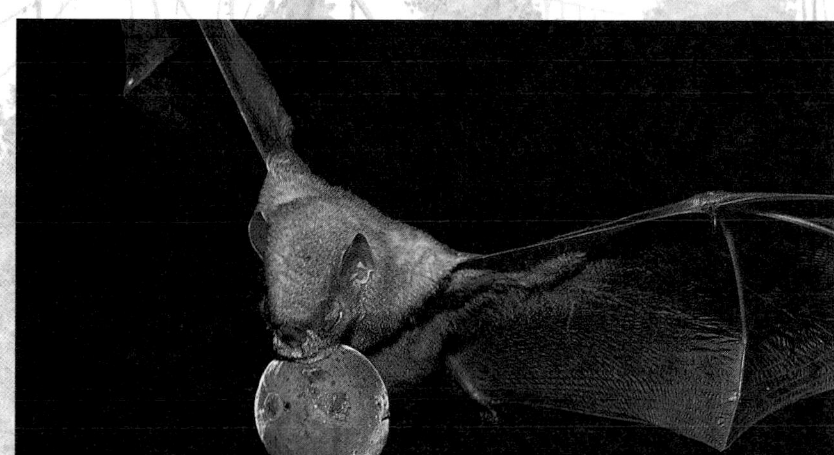

► *The coati is a social mammal that hunts for insects and lizards on the forest floor.*

Understorey

◄ *Insects, like these red stainer bugs, feed upon fruits that have fallen to the forest floor.*

Life on the forest floor

Here light levels are low, except for the rare fleck of sunlight that succeeds in penetrating the leaf cover above. The soil and the air are permanently moist and the temperature stable and warm.

Few green plants are able to grow on the floor because of the lack of light, unless the canopy has been disturbed by felling or by the death of a tree, when a surge of young growth makes the vegetation temporarily impenetrable. Some orchids have adopted a saprophytic way of life, feeding upon the decaying plant matter that falls like unceasing rain from the canopy.

But most of this organic rain is consumed by the hordes of animal detritivores and the armies of fungal and bacterial decomposers that occupy the forest floor. Little in the way of litter actually accumulates on the ground despite its abundant supply from above, for the rate of decomposition in the moist warm conditions is very rapid. In this way all the mineral nutrients demanded and used by the luxuriant vegetation, such as calcium and nitrogen, are released back into the soil from which they are soon reabsorbed by the plants to be cycled through the canopy once again (◄ page 88).

The coast redwood of Californian temperate forests can grow more than 100m tall and some specimens are more than 2,000 years old

Temperate forests

Just outside the tropics, mainly on the eastern seaboards of continents, are the temperate rain forests. They occur only where conditions are warm, oceanic and wet, and here the broadleaved evergreen type of tree thrives since there is no period of drought or frost to interrupt its growth.

In all of these forests, the cool, moist floor becomes covered by a thick layer of undecayed humus. The lower temperatures of these temperate regions means that the invertebrate animals and the microbial decomposers cannot perform as efficiently, so some of the organic produce of the trees remains to build up on the floor (◀ page 85). But the small animals inhabiting the litter and slowly degrading it form a food resource for insectivores that haunt the forest floor. Among these is the New Zealand kiwi, which has lost the ability to fly during the course of evolution. It has nostrils in the tip of its long, curved bill and seeks out its food by smell in the forest detritus. In the canopy above, birds such as the tui and the bellbird feed upon insects, fruit and nectar. They are particularly fond of the honeydew produced by mealybugs on the leaves, and the secretions of sugar also provide a food source for fungi, especially sooty molds, growing on the leaves.

In the cooler temperate areas, the winter may be sufficiently cold to restrict the growth of plants and this leads to a season of dormancy for many creatures. Under such conditions, there may be advantages in the possession of deciduous leaves, since the presence of leaves in winter is of no energetic value. Hence the development of the temperate deciduous forests of eastern North America, eastern Asia and much of Europe. The structure of these woodlands is simple; there is a tree canopy and a lower layer of shrubs. Beneath this is a fairly rich layer of herbaceous plants and often a ground cover of mosses beneath, for the light penetration, especially in spring and autumn, is adequate for the growth of shade-adapted plants.

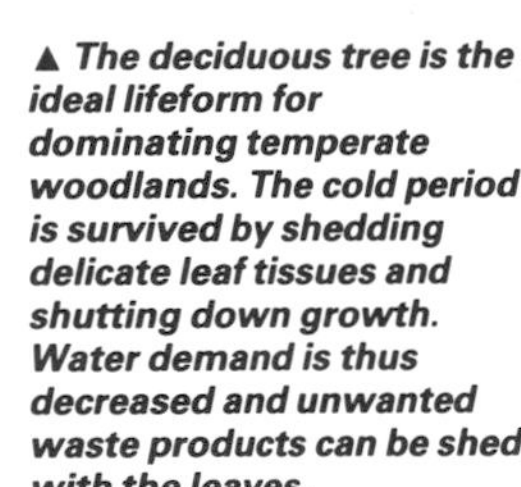

▲ The deciduous tree is the ideal lifeform for dominating temperate woodlands. The cold period is survived by shedding delicate leaf tissues and shutting down growth. Water demand is thus decreased and unwanted waste products can be shed with the leaves.

◄ Complex food webs develop beneath the forest canopy. Here an omnivorous brown rat is being ingested by a snake, in this case a carpet python. Up to five or six trophic levels can be recognized in the energy flow patterns of forest communities.

► Large herbivores are not as numerous in forests as in grassland. Red deer are most successful in forest glades rather than under a heavy mature canopy. Stone-age people opened up the forest to encourage these game animals even before agriculture had been developed.

Coping with winter conditions

Winter is cold enough to present problems to many of the resident animals. Food supplies become scarce and some, like the hedgehogs, meet this problem by entering a sleeplike torpor in which their cell biochemical activity drops to extraordinarily low levels. They simply tick over in this state of hibernation until spring brings a new supply of food.

Some herbivorous birds and mammals, like jays and squirrels, do not hibernate but lay up stores of food for themselves which they hunt out and consume during the winter. These activities assist in the dispersal of many plant species, particularly trees.

Detritivore animals and microbes in the soil receive a large input of new food in the autumn. Beneath this thick layer of litter the temperature rarely falls below freezing, so they are able to maintain some activity throughout the winter. But their populations increase greatly as the soil warms up in the spring and their burst of activity releases a flush of inorganic nutrients into the soil where they are urgently needed by the roots of the plants as they begin their spring growth (♦ page 87).

A variety of evergreens and conifers

In the Northern Hemisphere, the temperate rain forests of Japan have evergreen oaks and beeches, while in the Southern Hemisphere, the forests of New Zealand can boast of some of the largest of all trees, the coniferous kauri pines, which were used by the Maori peoples for the construction of huge war canoes. With the kauris are found the evergreen southern beech trees that cover much of New Zealand.

In South America it is the monkey puzzle which is the dominant tree, while on the eastern side of North America the swamp forests of Florida are of a temperate rain forest type and these are dominated by the swamp cypress. Although conifers, these trees do lose their leaves during the winter. On the west coast are the other giants among trees, the redwoods, which occupy a humid coastal strip on the seaward side of the Rocky Mountains. The most massive tree in the world, the giant sequoia, forms natural forests in the Sierra Nevada.

Forest food chain

Each spring, a deciduous tree puts out new leaf tissue and in this early stage of leaf expansion the leaves are most vulnerable to insect grazers. As the season proceeds there is a buildup of chemicals, such as tannins in the leaves, which deter insects because of their unpalatability. Insects therefore abound in early summer and form a food resource not only for resident insectivores but also for large numbers of migrant birds. These take advantage of the summer food supply and long days to breed in the temperate regions (♦ page 76).

But the grazing intensity in temperate deciduous forest is really very light. Only about 1-4% of the productivity is consumed by grazers. Most of the energy supply goes straight to the decomposers. Among the larger grazers, deer, squirrels and wild pigs are the most important. Predators include wolves, foxes, badgers and owls (♦ page 82).

In continental parts of the boreal biome winter temperatures can fall to −60°C

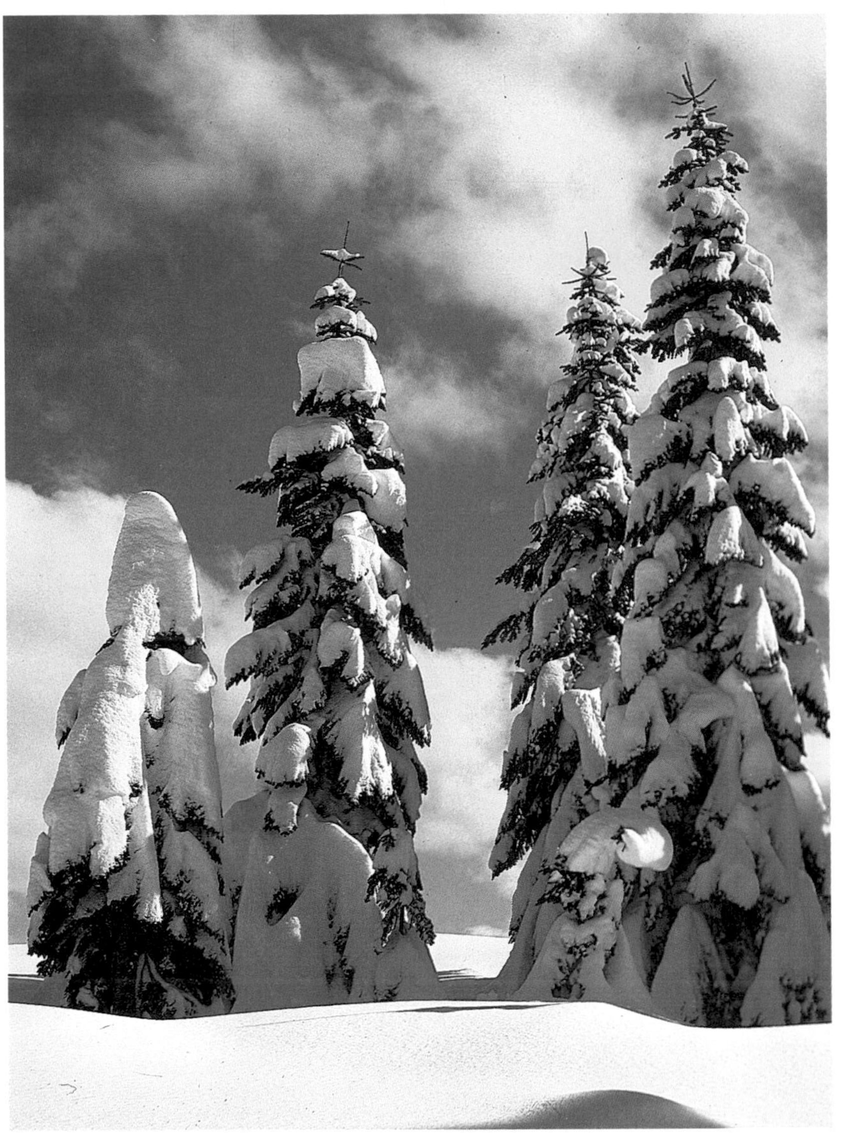

◄ The steeply conical shape and the narrow needle-leaves of spruce trees allow them to shed the snow that would otherwise weigh down and eventually break their branches.

Boreal forest

Forming a belt around the Northern Hemisphere continents to the north of the temperate woodlands and grasslands are the boreal forests. Their absence from the Southern Hemisphere is simply the result of a lack of any land in the appropriate latitude. The dominant life form in the vegetation is the evergreen coniferous tree, and the structure of the forest is even simpler than that of the deciduous forest, for there is no shrub layer present. Beneath the dense tree canopy, light penetration is often poor, but a layer of dwarf shrubs and a ground cover of mosses and lichens is usually found.

Here, the winter is more severe than further south and snowfall is heavy. The restricted summer growing season is a time of feverish activity for animals as well as plants. Population expansion in the invertebrate grazers may lead to the complete defoliation of areas of woodland. Trees such as spruce and birch are prone to population explosions among certain moth caterpillars. Insectivorous migrant birds, such as warblers and flycatchers, exploit these resources for the brief breeding season before returning south for the winter (◆ page 76).

One curious feature of the reproduction of coniferous trees is the fact that the cone takes several years to mature and during this time

The economics of leaf production

In the boreal forest the evergreen habit dominates as it does in the equatorial regions. But here its success depends largely on the long life of a leaf. The building of a leaf demands a certain investment in energy and nutrients, and if it can be made to last for two or three years then it proves more economical, especially when soils may be poor in the required nutrients. In these northern regions with short growing seasons, however, the main advantage of the evergreen leaf is that it can quickly begin to photosynthesize as the temperature rises in the spring and the days increase in length. It takes deciduous trees much longer to expand a new leaf canopy and get going in spring. But despite this, some deciduous trees, especially the birches and, in Siberia, larch, do survive well in boreal regions and usually extend further into the Arctic than their evergreen counterparts.

► A top carnivore of the boreal regions is the peregrine falcon. It hunts in open areas and feeds mainly on birds like the golden plover that it has killed here.

▲ The larvae of various species of beetle attack the bark of trees in the forest.

► In podsol water leaches nutrients, minerals and organic matter and deposits them lower down.

The formation of podsol

As conifer needles senesce, die and fall to the floor, they build up in a thick mass of undecomposed humus on the surface of the soil, which may render the vegetation subject to periodic fire. This organic build up is quite unlike the temperate and tropical forests, where the organic detritus is rapidly incorporated into the soil and broken down (◆ page 101). The main reason for this is that the chemistry of the dead leaves, particularly the polyphenolic compounds, make them distasteful to many soil invertebrates, such as earthworms. Add to this the generally low temperatures on the dark forest floor, the wetness of the habitat and hence the poor oxygen penetration into soils, and clearly conditions are not ideal for the decomposer. Within the soil itself, the seepage of these chemicals down the profile leads to the development of a distinct layering, called podsolization.

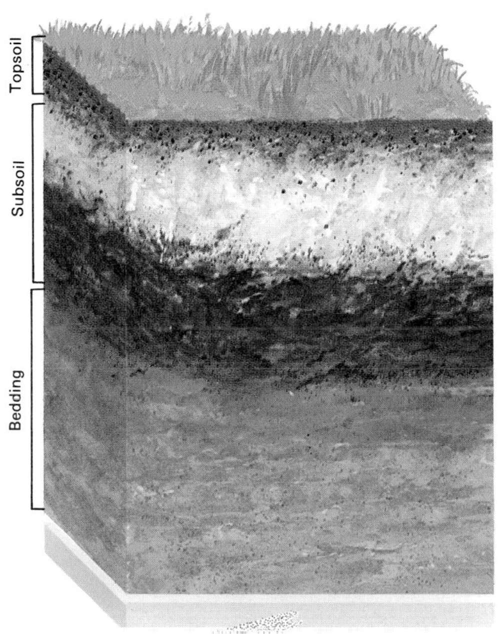

the seeds are developing within its protective scales. These cones therefore form one of the most reliable sources of food in summer and winter, and both mammals, like the squirrels, and birds, such as crossbills, take full advantage of this fact. The crossbill has evolved into a distinctly specialist feeder, its curved, overlapping mandibles being ideally suited to prising open pine cones. If the cone stocks run out in the winter, this bird has no alternative but to erupt in large flocks from its homelands and seek more southerly pines. Squirrels are less specialized feeders and many species are quite omnivorous. They may act as predators and eat the eggs and young of birds. But in their turn they are preyed upon by pine martens and goshawks.

On the forest floor, hares and deer are important grazers and these are preyed upon by wolves, lynx and brown or grizzly bears. In the wetter regions, beavers make a considerable mark upon the forest landscape in the damming of rivers and felling of trees. The boreal regions are rich in wetlands and the summer forest is alive with blood-sucking insects, especially mosquitoes. These feed upon the blood of mammals, both grazers and predators, and may then be themselves consumed by insectivorous birds. Thus it can be seen that here feeding relationships form complex, interlinking webs (▲ page 86).

Since many desert plants have extensive root systems, the below-ground biomass may be greater

Desert and desert scrub

Some of the most stressful conditions of the Earth's surface are experienced in the high pressure tropical belt where descending air masses bring no water to hot, parched lands (◀ page 97). These are the desert latitudes, found both north and south of the equator and reaching their most extreme form where air currents move predominantly offshore, as in the Namib of Africa and in the western part of Chile, or where high mountain ranges protect the desert from oceanic winds, as in the Gobi of central Asia.

Summer temperatures under cloudless tropical skies can climb as high as 50°C and the surface sand can almost reach the boiling point of water. The problems which beset the residents in such sites are twofold, therefore; how to keep temperatures low enough for living tissues to survive and function, but how to achieve this without excessive loss of scarce water.

Some of the most important and abundant of desert grazers are the small mammals – gerbils, jirds, ground squirrels and jerboas. These live in underground, labyrinthine tunnels, particularly beneath desert shrubs, and here they shelter from the heat of the Sun and emerge in the evening and at night. Many feed upon fallen seeds, which are relatively abundant since many desert plants rely on a good seed production as an insurance against the recurrent catastrophes of life in mobile sand, and where next year's rain may never come.

Larger grazers include the oryx and many species of gazelle, together with such animals as the wild sheep and wild ass of the Middle East and Asian deserts. The birds of such extreme habitats may spend much, or even all of their time on the ground. They include coursers, bustards and ostriches. Or they may be dependent on flight to take them to sites of water availability, as in the case of the sand grouse.

Invertebrate detritus feeders, like millipedes and woodlice, feed upon fallen herbage and are preyed upon by insectivorous birds, beetles and lizards. These in turn are eaten by larger predators, such as foxes, coyotes, and wolves. Among the top carnivores of the desert regions are mountain lion, caracal, cheetah, and various other members of the cat family (▶ page 247).

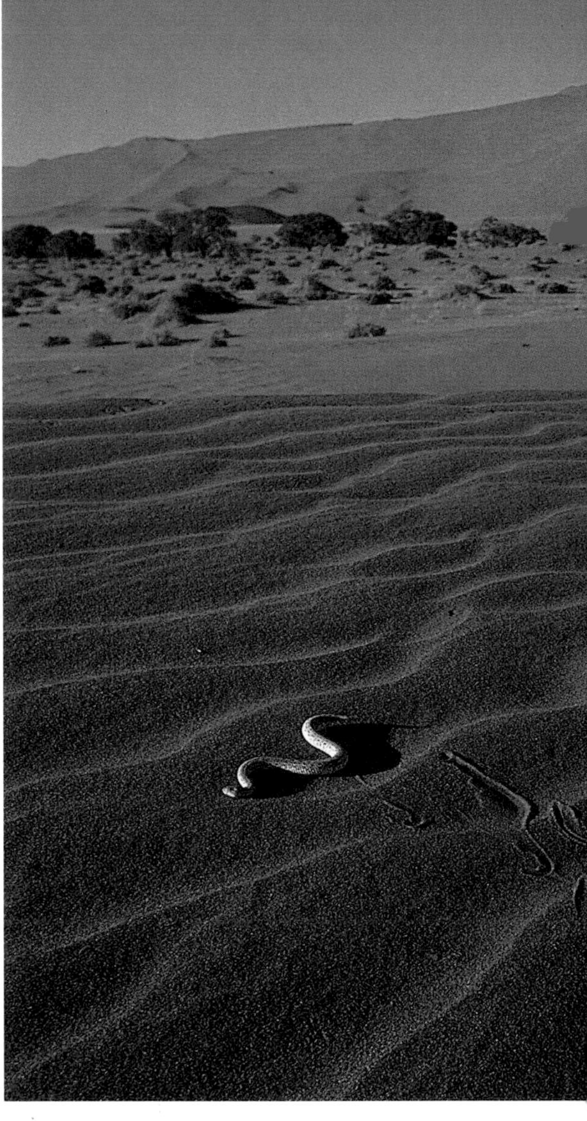

▲ The mobile sands of the Namib Desert are blown into ripples and ridges. Locomotion for animals can be difficult in the sand and the sidewinder snake has developed a distinctive sideways motion to overcome this problem. Sidewinding allows the animal to keep much of its body off the dangerously hot sand.

◀ A characteristic invertebrate carnivore of the desert is the scorpion, which immobilizes its prey by means of the stinging tail. Scorpions shelter beneath stones during the day and emerge to hunt in the dusk. Worldwide in deserts there are more than 2,000 species of scorpion. Some can reach 16cm in length.

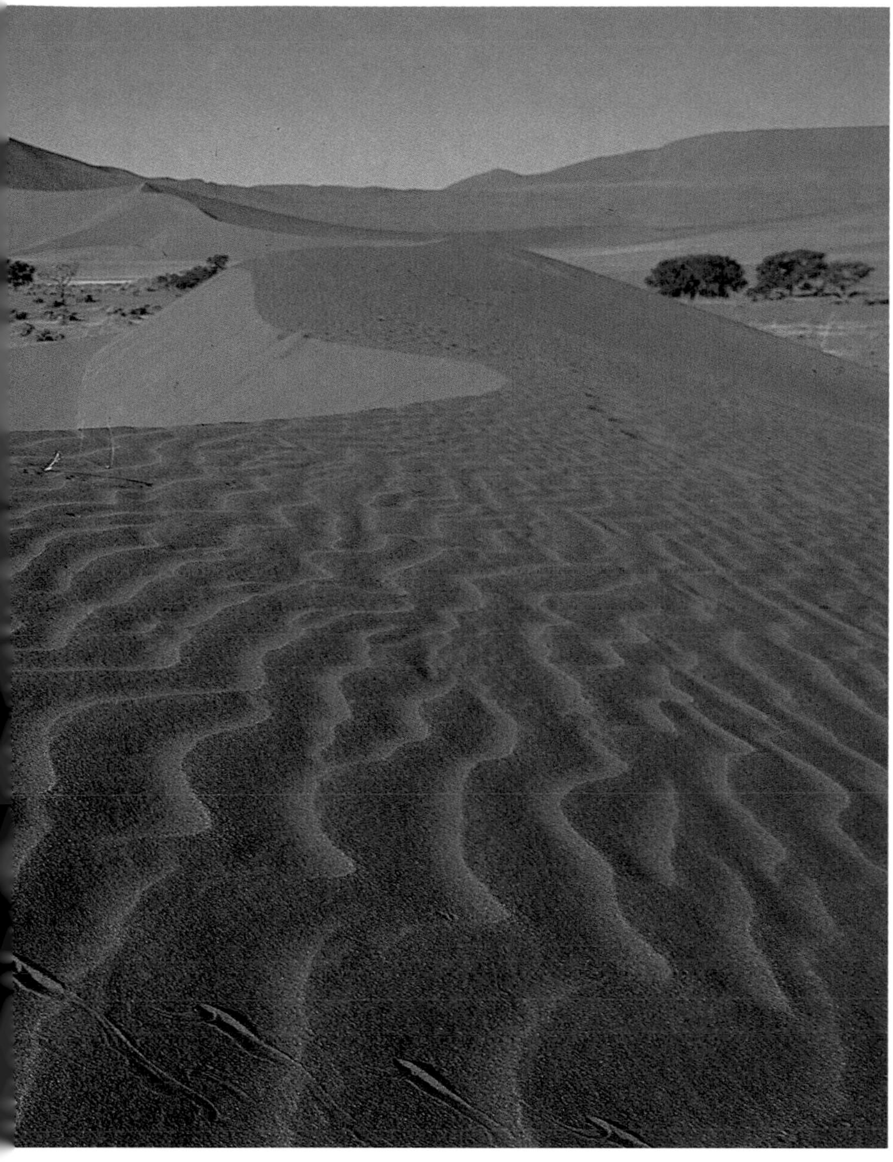

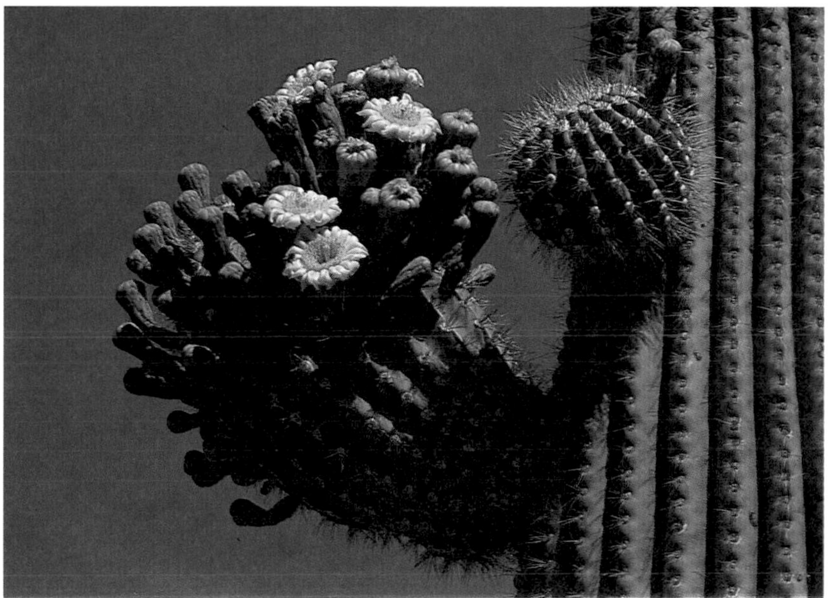

Desertification

*In many respects the desert ecosystem is a relatively simple one in that there is a low density of living things. But it is also finely balanced and very sensitive to disturbance. The imposition of an additional grazing stress, such as occurs when nomadic pastoralists spend a long time with their flocks in a single area (**▶ page 160**), may result in the total destruction of the sparse vegetation cover and the loss of the meagre resource base upon which the desert ecosystem depends. It is this type of overgrazing that leads to the spread of low-productivity desert, so-called desertification (**▶ page 220**), commonplace in parts of North Africa. This destruction of habitat often results from political and sociological pressures which have encouraged the settlement of nomadic peoples.*

Plant strategies for survival

Since plant life is the primary source of energy in any ecosystem, the existence of life in deserts is determined by whether photosynthetic plants can overcome these problems. Even in the hottest place on Earth, like Death Valley, California, some plant life does survive. Biochemical studies have shown that these plants have enzymes in their cells which can still operate at very high temperatures. They often also have shiny, reflective surfaces which absorb only a small proportion of the incident light. Their leaves are often reduced in area, or may be totally absent, the swollen green stems taking on the function of photosynthesis.

Some photosynthetic plants spend almost all their life beneath the sand, only sending fruits above its surface to aid dispersal. Others, particularly tiny lichens and algae, live within the crystalline structure of the rock where they avoid the worst effects of a desiccated environment.

The greatest problem facing a plant in the desert is how to photosynthesize, which involves the absorption of carbon dioxide from the atmosphere through pores (stomata) in its surface, without losing too much water to the atmosphere through those selfsame pores. Some of the most successful plants, like the cacti of North America, have adopted a modified form of photosynthesis in which they only open their stomata at night, so taking in carbon dioxide at a time when the temperature is lower and hence water loss less problematical. The carbon is stored temporarily in the form of organic acids in the cell sap and then is passed through the normal photosynthetic process during the day.

But the growth of such plants is slow and the animal life which can be supported by low productivity is very limited. Only in the dry stream beds (wadis) which periodically fill during freak storms is the density of vegetation improved, and here a low scrub may develop, as it also does in the somewhat damper areas of the desert margins.

◀ Cacti are highly adapted desert plants in which photosynthesis takes place mainly in stems rather than leaves. This reduces water demand, as does their ability to open their pores for the intake of carbon dioxide only at night. Their stems are often ribbed, enabling them to expand and contract with water intake or loss. Flowers arise from specialized pits on the stems known as areoles

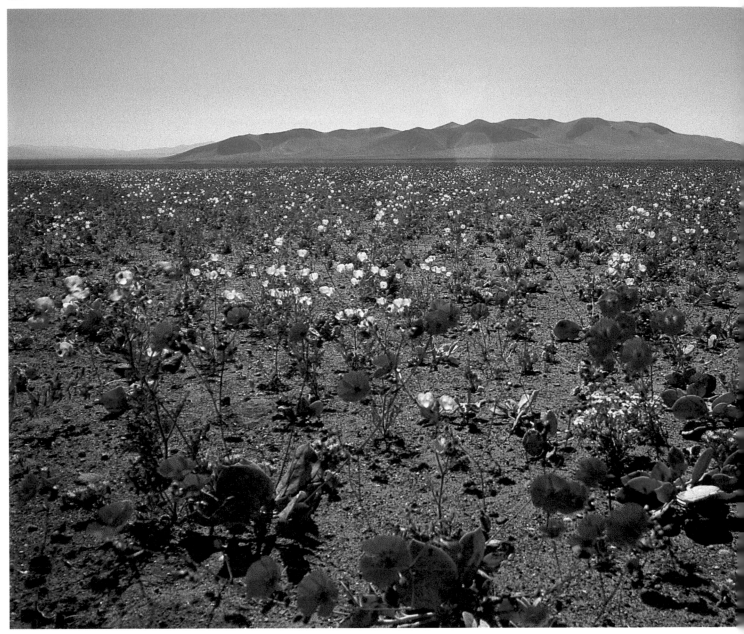

Specialized grasses

In the savanna, conditions are not as dry as in deserts but many of the resident grasses have developed a system of photosynthesis resembling that of the cacti and succulents (▶ page 107). They can accumulate carbon dioxide from the atmosphere by an efficient, though temporary, fixation process that is so rapid it limits water loss through their stomata to the minimum. They then transport the organic acids produced to specialized cells around the veins where it is refixed in a more permanent form. These grasses are some of the most productive plants known in the world and among them are some that we have brought into domestication, such as maize and sugarcane (▶ page 153).

Exploiting the resources

Giraffes feed mainly by browsing on the low branches of trees, so they do not compete with animals that feed at ground level. The eland will also browse, but cannot reach as high as the giraffe. As the dry season progresses and nutritive food becomes scarcer on the drier uplands, many grazers move down onto lower pastures. The robust buffalo are often the first to invade the damper valleys, where they can cope with the really tall grasses. Then comes a succession of herbivores that feed on the poorer and shorter grasses. By the time these have moved through an area, the shorter growing broadleaved plants are exposed, and these form the main food resource of smaller grazers such as Thomson's gazelle.

Vultures

One consequence of the high density of large herbivores and their predators is the availability of discarded carcasses following a kill. Here the scavengers take over, most conspicuous among which are the vultures. Vultures fall into three distinct ecological groups: the large gregarious vultures, like the griffon in Africa and the white-backed in India, the large solitary species, such as the black vulture, and the smaller species with less robust bills, like the Egyptian and the hooded vultures. These types attack a carcass in sequence, with the aggressive black vulture taking precedence at a kill, the large gregarious types coming next and the small species mopping up the remains.

◀ *The supply of water limits plant growth for much of the year in these dry areas, and the coming of rain releases dormant bulbs and seeds, causing the dry land to blossom and take on a meadow-like appearance. Mechanisms to survive prolonged drought are a vital requirement for plants in these regions. They must also complete their lifecycle and set seed quickly before the coming of the dry season and the onset of renewed drought.*

Savanna

In the climatic zone between the seasonal tropical forests and the desert scrub is found an extensive area of grasslands with scattered trees. These are the savanna grasslands. Much of Africa is occupied by this biome, as is a large area of India, though the intense impact of human agriculture has modified much of the natural vegetation (➧ page 220). There is some controversy concerning the degree to which human management, mainly in the form of burning, has modified and perhaps even created this vegetation type. The long drought period, usually in winter, is broken by a summer wet season in which new life comes to the savannas. It is in the dry season, however, that these areas are most subject to burning and this factor may play a large part in suppressing the invasion of woodland.

The grasses that dominate the savannas are massive, often over 2m tall. Among them are patches of scrub and trees, particularly acacias and the remarkable baobab, with its swollen trunk rich in stored water; it is often raided in times of drought by such animals as elephants.

With this great resource of productive herbage, the tropical grasslands have been habitats in which many species of large, herbivorous animals have evolved. In the East African savannas alone, there are about 40 species of antelope apart from other large grazers such as buffalo and zebra. The reason why so many species can be supported at this trophic level is partly the high productivity, but also the degree of specialization in feeding habits found among them. This specialization takes two forms; either the grazers have very specific tastes and limit their grazing to a few species of plant, or they feed in a particular area at only certain times of the year and hence reduce the overal grazing load.

Any plant litter that evades consumption by all these grazers falls to the ground where it is soon attacked by the detritivores, the most conspicuous of which are the termites. These build towering nests in the savanna in which they cultivate gardens of fungi (➧ page 54). In this way they can gain access to those parts of the plant tissue that they cannot digest on their own.

The great range of herbivores offers opportunity for a similar variety of specialization among the carnivores. Lions hunt in packs and have limited speed and stamina, so need to approach their prey closely and rely on panic to catch them. Leopards hunt singly at night, while spotted hyaena are also nocturnal but hunt in packs. A range of smaller carnivores, such as genets, civets and mongooses, feed on smaller prey animals (➧ page 56).

▼ *Grevy's zebra are capable of grazing on poor grass forage. Their trampling opens up the savanna and allows other grazers to take advantage of the lower growing herbage. The topi, for example, moves in to graze upon short mat-forming grasses, while the wildebeest tackles upright stems. All aspects of the ecosystem's productivity are partitioned and exploited by the army of grazers.*

◀ *Open areas of water within the savanna region are often rich in wading birds. The East African lakes are famous for their flamingoes, that aggregate in their thousands to feed on the algae-rich waters. Here are sandhill cranes from North America, which are most typically associated with the tundra but which can also be found in wet areas of the dry grassland region. The lakes form a staging point along one of the world's busiest migration routes – southern Africa to northern Europe*

In a concerted effort to save the wisent of East European grasslands, in the 1920s the six remaining individuals were brought together and bred successfully such that today there are about 2,000

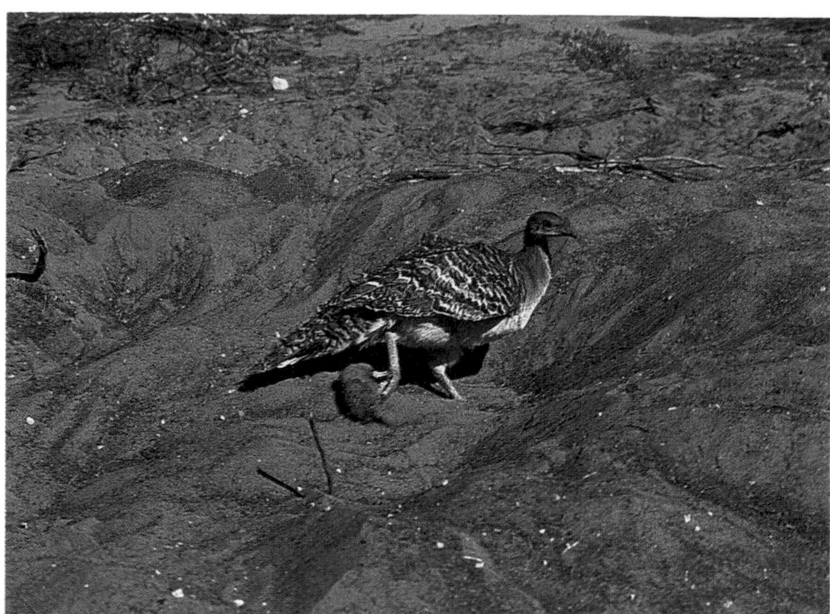

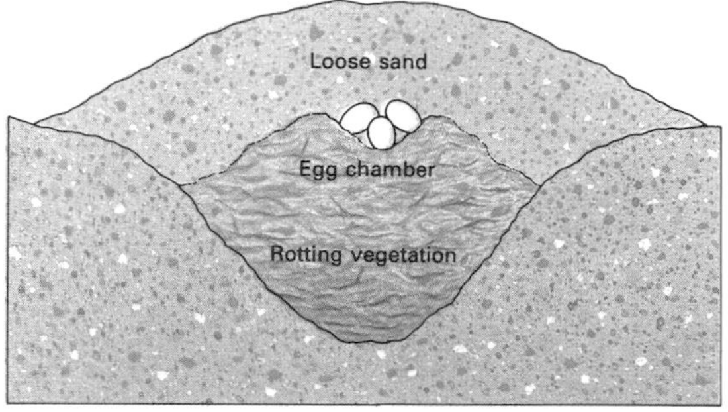

► The prairie grasslands of North America have been very badly damaged by humans. Intensive agriculture and overgrazing have had widespread damaging effects, but some areas are conserved in a fairly natural state and still support considerable numbers of some of the native grazers such as these pronghorn antelope.

◄▼ The Australian mallee fowl builds its nest in open scrub or grass and lays the eggs on a large pile of rotting vegetable matter that generates heat as it decomposes. The egg chamber is covered with sand so that the adult bird does not need to concern itself with the incubation of the eggs.

Temperate grassland

In the continental interiors of the temperate zone, summer temperatures rise to high levels and rainfall is lower than in the more oceanic areas. The continental influence also means that the winters become very cold, especially in some of the Asiatic regions where monthly minima can reach as low as −15°C. As a consequence of the summer drought, the growth of trees is restricted to the wetter regions around rivers and most of the landscape is dominated by grassland. In North America these grasslands are called prairies, in Asia steppes and in South America pampas.

Although grasses form the dominant plants, there are many plant species that grow among them, especially on undisturbed, virgin prairie. Unfortunately, the temperate grasslands have proved so attractive to agriculturalists that little undisturbed habitat now remains (♦ page 238), and many of the cultivated regions have been mismanaged and have suffered severe soil erosion.

The abundant supply of palatable plants makes the temperate grassland an ideal habitat for herbivorous animals. Most conspicuous in the days before human hunting reached excessive levels was the bison of the North American prairies and its equivalent, the wisent, of the Eurasian steppes. The large size of these animals is of great advantage since it permits the development of a capacious gut in which the digestion and absorption of the herbage can take place. Various antelopes also exploit the temperate grassland, such as the pronghorn in America and the saiga in Asia. The South American pampas have their own grazers, the pampas deer and the guanaco. It is in South America also that large, flightless birds occupy the grazing trophic level, namely rheas. These seem to prefer broadleaved herbs to grasses, but also eat insects. Horses are another typical group of grassland animals, but the native American species became extinct and the horse was then reintroduced by the Spaniards.

In the soil, the activity of invertebrates is considerable during the summer and there is a rapid incorporation of organic detritus. This is why the prairie soils are so attractive to agriculturalists. However, the evaporative movement of water towards the soil surface in summer can lead to salt accumulation (♦ page 213).

Drought – and overgrazing – specialists

Among the dominating grasses, the specialized photosynthetic system found in the tropical savannas (♦ page 108) is less common in the temperate zone. Those grasses that operate in this way gain their greatest advantage in the hot, dry conditions which prevail in full summer. So it is after the normal grasses have completed their full growth that the drought specialists come into their own. They grow most efficiently in August and September. Those species that conduct normal photosynthesis grow most successfully in the relatively cool, moist conditions of early summer.

On their southern edges, these Northern Hemisphere grasslands become increasingly dry in summer, and in parts of the Middle East, such as western Iran, the steppes merge into the desert scrub of the semiarid zone (♦ page 106).

An additional pressure upon the vegetation is the intensive grazing, and many of the resident plant species are well adapted to withstanding this stress. Two main strategies have evolved: either the plant may be palatable, in which case it needs to be able to recover from grazing damage, or it can become unpalatable as a result of building up distasteful or poisonous chemicals in its tissues. Grasses have adopted the former strategy. They have their growing points at the base of their shoots, so removal of the top part by herbivores does not result in irreparable damage.

*The alternative strategy has been adopted by plants such as the joint pines (*Ephedra *species). Their toxicity is such that they can be a serious problem for pastoralists in these regions. Toxins in such plants may be produced as organic compounds by the plant itself, or may be selectively absorbed from the surroundings. This is the source of toxicity in the locoweed* Astragalus *of the American prairies, which accumulates selenium into its tissues out of the soil. The concentrations built up are sufficient to affect the nervous system of any grazing animal, causing apparent madness – hence its common name.*

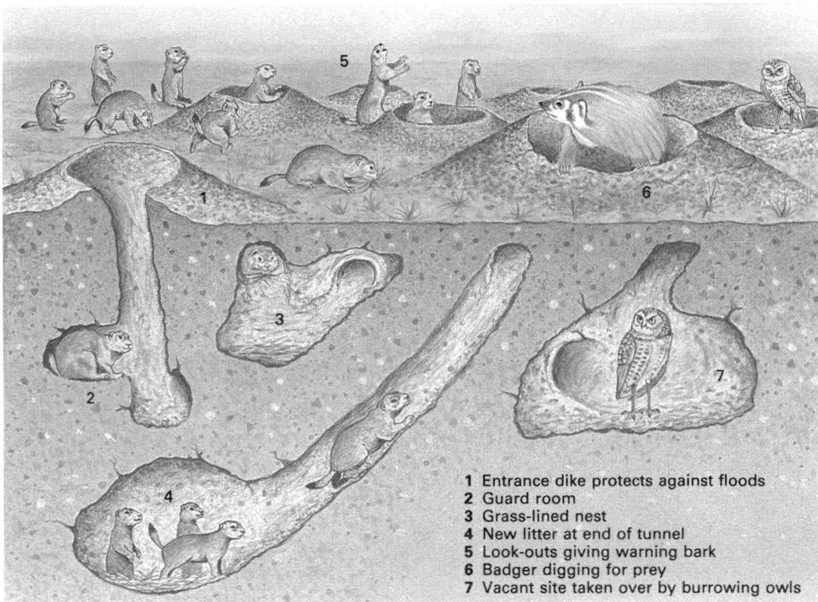

Burrowing to safety

Escape from predators on the open grasslands can present a problem to small mammals, and many of them, like the prairie dog and its Asian equivalent, the souslik, use burrows to avoid the buzzards, eagles, coyotes and foxes that prey upon them. Even some birds, such as the burrowing owl, nest underground because of the absence of available above ground sites.

Burrowing itself creates new habitats in the prairie. When the American badger digs up virgin prairie for example, it opens up the soil for many weed species that may come to dominate the area. If, however, it disturbs manmade prairie, dominated by bluegrass, it may serve to disrupt this domination and allow other species to enter the community, thus increasing the plant diversity.

◄ Prairie dogs are really vegetarian ground squirrels that inhabit the North American prairies. They are social rodents that build extensive tunnels in which they avoid their predators. They are easy to observe because they are diurnal and can often be seen sitting upright and alert for buzzards by their burrows.

In summer, blood-sucking flies make the boggy areas of the tundra uninhabitable to warm-blooded animals, including humans

Tundra

The far north of the two major continental blocks in the Northern Hemisphere (Asia and North America) is clothed with tundra. This biome bears a vegetation type that is as highly adapted to stress as that of the desert, though stress of a rather different kind. In northern Europe there is only a narrow band of tundra, for the tropical water currents of the Gulf Stream Drift keep these high latitudes warmer than might otherwise be expected.

The word "tundra" is one of the few Finnish words to find its way into the English language. The Finnish "tunturri" refers to the bleak, treeless fells of Lapland that lie just to the south of the true Arctic tundra zone. As in the case of the boreal forest, there is very little land in the Southern Hemisphere that lies in the appropriate latitudes for the development of tundra, but some of the oceanic islands, such as South Georgia and the Falklands, bear a tundra vegetation.

The climate in the tundra is cold for much of the year. In Alaska the mean monthly temperature in winter is often below −30°C and in Siberia below −50°C. The fact that the sun does not rise in some areas during midwinter adds gloom to the freezing conditions. But in summer, the very long days, coupled with much warmer temperatures, lead to a high degree of biological activity during the short growing season. Precipitation is low at all times of year in these high pressure, polar zones; figures of 150-300mm are common, and these compare closely to those of desert regions. But the generally low temperatures mean that evaporation is minimal, so water is rarely scarce, except on rocky and sandy ridges, where plants such as the opposite-leaved saxifrage often have to cope with extreme drought in summer.

The main features of the tundra vegetation are its low stature and the predominance of perennial plants. The only "tree" species that survive here are the dwarf birch and dwarf willows, which seldom exceed a meter in height.

Invertebrate animals abound during the short summer season, but spend much of the year in a dormant condition as eggs. Their emergence in summer, together with the fresh growth of the plants, brings many visitors to the tundra from the boreal zone (◀ page 104). Migratory reindeer (caribou) join the resident musk oxen to graze the new shoots. Their breeding season is timed to coincide with this abundance of food. Such timing is achieved by their perception of increasing daylength which affects the development and function of their reproductive systems. With these grazers come their predators – wolves, wolverine and brown bears – which ensure that only the very fittest of the young calves survive.

Some of the resident mammals, like lemming and hares, also reproduce rapidly in favorable conditions. Sometimes, in the case of lemmings, this leads to the famed migrations of large populations seeking new grazing as resources become consumed. Predators of the small mammals, for example the snowy owl, Arctic fox and rough-legged buzzard, derive great benefit from these times of plenty.

There is a further resource, however, from which the tundra benefits, namely the sea. Carnivores such as the resident polar bear or the migrant Arctic tern (which travels all the way from the Antarctic for its summer breeding in the arctic tundra) depend upon the sea as their prime energy resource. The bear feeds mainly on seals and the tern on small fish, both of which are part of the marine food webs. It is the plankton of the oceans (▶ page 122) and not the vegetation of the tundra that supports these members of the Arctic fauna.

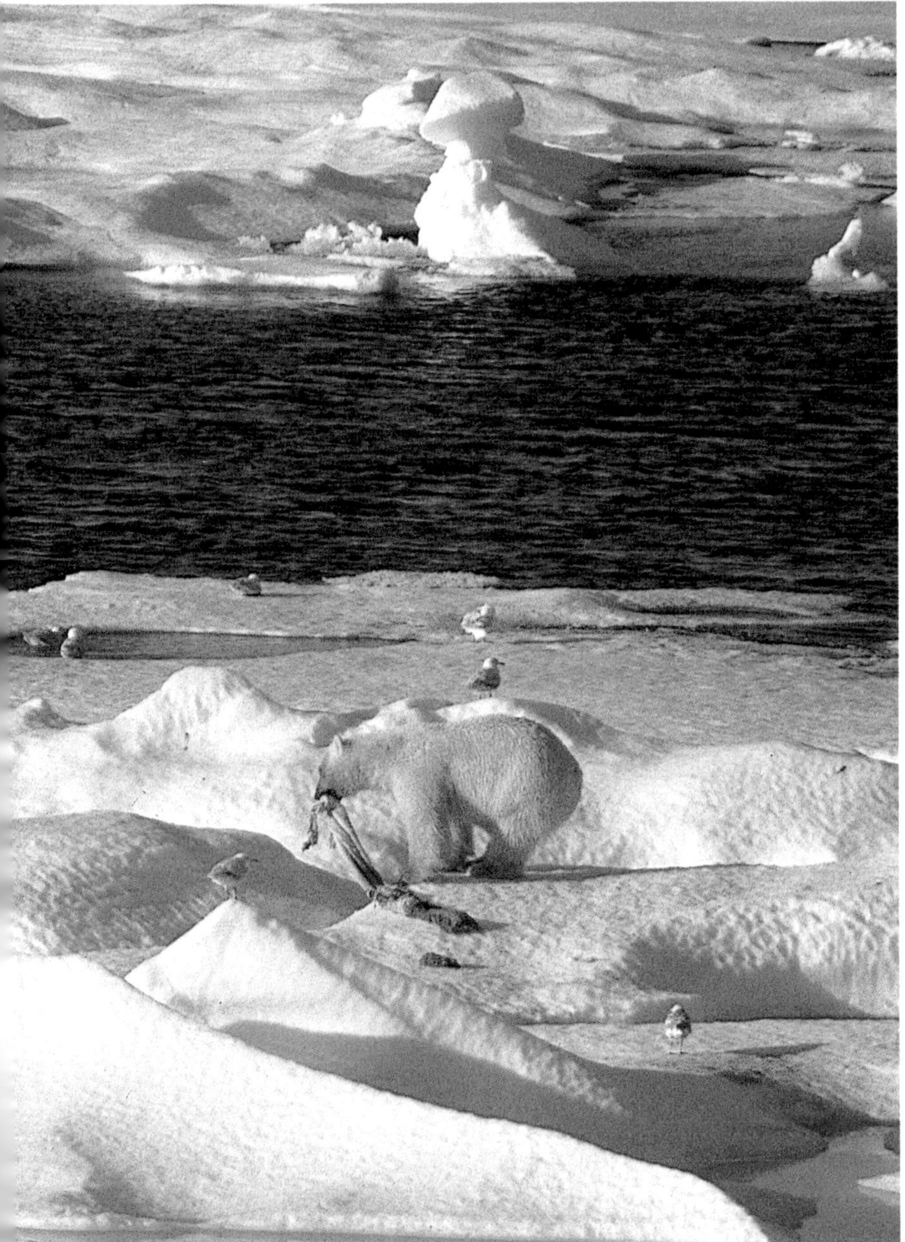

Adapted to the harsh elements

The main problem plants have to face is the blast of the high winds in winter, driving crystals of ice across the landscape and scouring all in its path. The lack of snow means that there is little protection for the vegetation. Added to this is the problem of a permanently frozen soil. Only the upper layers thaw out in summer. Below this is the permafrost. Under these conditions tall plants and bulb plants, as well as animals, suffer severely and are therefore very scarce. Perennials that have a dwarf, cushion form and fairly shallow, fibrous or spreading roots (chamaephytes) are the most successful. Among these are various species of the heath and crowberry families.

◄ Polar bears are the top carnivores of the northern Arctic tundra. They feed upon seals and walrus and build up layers of fat that carry them through a long winter of hibernation beneath the snow. The female bear even gives birth to its cubs beneath the snow cover in winter.

▲ One characteristic of tundra plants is their low, cushion-forming stature. This moss campion plant has bright flowers to attract the pollinating insects in spring. It may suffer drought in the long, rainfree days of summer, and ice-blasting during the winter.

◄ The tundra is a productive biome in summer and many animals migrate from further south to take advantage of the available grazing. Among these is the caribou, or reindeer. Vast herds of these animals migrate to the tundra to breed and are preyed upon by wolves.

► Many birds also migrate into the tundra for the productive summer. These white-fronted geese, which breed in the Arctic, are an example. These are herbivores, but many other birds come to feed on the abundant insect life of the tundra in summer.

Temporary residents

The abundance of plant and insect life proves attractive to large numbers of migrant birds, some of which, like the greylag, white-fronted, snow and Canada geese (♦ page 76), may have spent the winter far to the south in the warm temperate regions. The Arctic warbler, an insectivore that breeds in scrub-tundra from Siberia to northern Norway, winters in the forests of southeast Asia. Thus many of the animals that exploit the resources of the tundra are mere visitors. Apart from the food it supplies, the tundra also provides the long days which are so valuable, particularly to insect-eaters, that need all the time available to feed and rear their young.

Most mountain butterflies are darker than related lowland species, which allows them to absorb the maximum of heat from the Sun's ultraviolet rays

Mountains

High mountains create conditions that have much in common wherever they occur. Elevated land creates a climate of its own. Not only does temperature fall with altitude (about 0.5°C fall with every 100m rise in altitude), but warm, moist air currents shed their water as they cool, thus increasing precipitation at high elevation. Because of the cooling effect, the ascent of a mountain in the tropics is not unlike a journey towards the poles. One passes through communities of plants and animals resembling those of the biomes of cooler climates.

The Himalayas, for example, lie only just outside the Tropic of Cancer, and their foothills, the Siwaliks of northern India, are clothed with seasonal monsoon forest. Above 1,000m a heavily disturbed scrub zone leads into temperate deciduous forest of oak and rhododendron, superficially resembling the woodlands of eastern North America or western Europe. Only the presence of exotic fan palms breaks the image. At higher altitudes, around 2,000m, the presence of coniferous trees becomes more noticeable. The deodar cedar gradually assumes dominance, together with pines and firs. Such a community is reminiscent of the boreal zone, but the presence of Hanuman langur monkeys reminds one of the hotter lowlands.

Above about 3,400m the coniferous forests give way to an alpine scrub of birch, juniper and more species of rhododendron, ultimately leading to the permanent snowline above 4,500m. These high altitude scrub and grassland communities are the equivalent of the Arctic tundra and the cushion growth form of the vegetation confirms the similarity of the climatic conditions. But there is one major climatic difference, which results from the nearness to the Equator. During the day, especially in summer, the Sun is almost overhead (◀ page 97), which results in high daytime temperatures that fall to low levels at night. The diurnal fluctuation in temperature is thus much greater than in the high latitudes.

High day temperatures lead to a rich insect fauna, but these animals risk being swept away by high winds at such altitudes. This has led to an abundance of flightless species. However, the same winds bring food into the alpine tundras, for other insects are blown up from the warmer lowlands and are chilled, killed and deposited on the snowbeds to be consumed by the expectant residents.

▲ Melting snow and ice on high mountains provide the water for rivers, which flow fast in these regions and denude the landscape by eroding soils. The braided patterns of streams and sand banks of these alpine regions provide an important habitat for plants and animals. Many wading birds breed in such country. In the summer months, insects abound, many of them flightless; they travel with the winds.

▼ Ascending a tropical mountain (right of diagram) involves passing through climatic zones that are similar to those found in higher latitudes (below, left), and the vegetation is also zoned in an equivalent fashion. On Mount Kilimanjaro, for example, the tropical forest gives way to montane forest with a distinct bamboo zone, then to heath, alpine, and finally glacial ice. However, here daytime temperatures on the peak far exceed those in the Arctic.

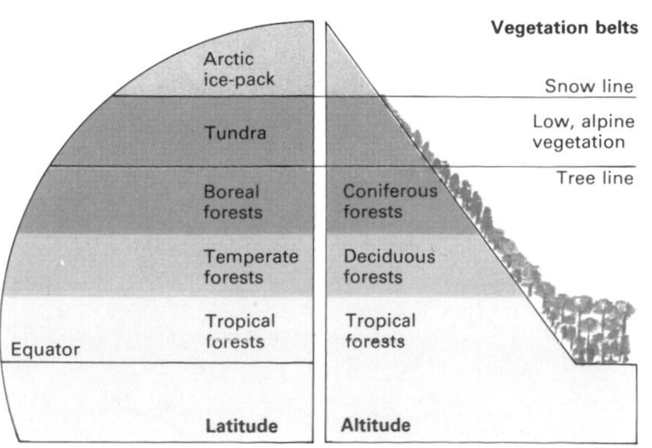

▲ Isolated on the subalpine region of Mount Elgon in Kenya, familiar plants, such as the groundsels (Senecio species) and lobelias, have evolved stumpy tree-like forms.

◄ Animals at high altitude, like these ibex, face many problems. Low temperatures require thick, insulating coats, and the low atmospheric oxygen results in difficulties in absorbing sufficient oxygen into the bloodstream.

Endemics and relicts

Tropical mountains are islands of coldness surrounded by an inhospitable sea of heat, and this has led to a distinct isolation of their plants and animals. Apart from strong-flying birds such as the condors and lammergeier, there is often no opportunity for animals to migrate from one mountain top to another. Nor is it likely that plant seeds will be carried over the distances often involved. The result is that mountain life has evolved in isolation, with little of the genetic mixing by interbreeding that keeps a species uniform. As a result, some strange organisms have developed, such as the giant senecios of the East African mountains around 4,000m. These relatives of the temperate ragwort have stout trunks topped by tufts of leaves. Many such mountain plants are endemics, being restricted to only the single site where they have evolved.

Other organisms, though, have had a wider distribution in the past and now are confined by habitat or climatic change. The dung beetle Aphodius holdereri *is now restricted to a few peaks in the high Tibetan plateau of the Himalayas but has been found as a fossil during the last Ice Age as far west as the British Isles. Such species are termed relicts and they, like endemics, are frequently encountered on mountains.*

See also
Animal Migration 71-80
Interactions 81-96
The Resources of Trees 173-180
Threats to the Environment 209-224
Saving Habitats 225-234

◀ ***The cold of a high-altitude winter induces hibernation in insects. Here is a communal winter roost of ladybirds. In low temperature they lack the energy for movement.***

▲ ***The inaccessibility of mountains provides a refuge for animals whose range is becoming restricted by human persecution, such as the cougar of the Rockies.***

A number of mammal species are found only in mountain areas and some, like the European Alpine marmot, spend more than half of their life asleep as they await the summer season of growth and reproduction. The conservation of energy during hibernation is achieved by dropping body temperature to about 3°C, and the heart may beat only two or three times a minute. Not all mammals hibernate, however. The mountain sheep, such as Dall's sheep of Alaska and the snow sheep of Siberia, spend the whole year active. Energy conservation is achieved by means of a thick coat of hair, which is developed to perfection in the Angora goat and the chinchilla, a rodent from the high Andes. The large mammals of mountain areas, such as the sheep, are preyed upon by wolf and leopard, both of which live at or above the treeline.

Communities of the Water

10

The ecology of lakes and ponds...Bogs, marshes and mires...Temperate swamps and saltmarshes... Mangroves...The marine habitat...Estuaries...Rocky, sandy and muddy shores...PERSPECTIVE...Trophic levels in a lake...Feeding strategies in birds...The most diverse fauna in the world...Energy supplies to inhabitants of sea and shore....Coping with tidal changes

Freshwater bodies

Less than 3% of the world's water is fresh, and of this, the bulk is frozen in the ice-caps and glaciers. Some of the fresh water is circulating in the atmosphere as cloud and ice crystals. The remainder lies upon the surface of the earth as lakes and ponds, is draining away to the sea in the form of streams and rivers, or is soaking through soils and rocks, forming subterranean aquifers. The surface water-bodies support much life, but what types of community develop depend upon such variables as the chemistry of the water and the speed at which it moves.

The energy supply for animals living in fresh water may come from an outside source (▸ page 85), such as the vegetation overhanging a pond. Dead leaves and detritus may lead to the formation of muddy sediments upon which animals and microbes may feed. But in most water bodies there is also an input of energy from the resident plants in the system. These may be free-floating plants of microscopic size – the phytoplankton – or they may be larger plants (macrophytes), either floating freely or rooted in the bottom sediments.

Clearly the existence of suspended populations of tiny plants can only be maintained if the motion of the water body is slow, as in lakes. In streams, the fast-moving water soon carries any microscopic life away unless it is attached or sheltered by pebbles or rooted plants. In the lakes, however, there is a constant process of photosynthesis and replication taking place among the unicellular algae, especially the desmids, diatoms and chlamydomonads in the surface layers of water where light levels are highest. This productivity reaches its height in the spring, when the lakes "bloom". The growth of algae may become apparent to the human eye, especially if filamentous green algae like *Cladophora* are involved.

These populations of microscopic producers are preyed upon by minute grazers, the zooplankton, and it is the activity of these little harvesters that prevents the phytoplankton becoming very dense. Their intensity of grazing is so great that in some lakes it has been calculated that the waters are filtered through the guts of zooplankton about four times each day. Yet still the productivity continues. Many of these planktonic grazers are small crustaceans, like *Cyclops* and *Daphnia*, the water flea, which also breed at prodigious rates when conditions are suitable.

Two possible fates may befall the plankton communities. Either they may be consumed by higher predators, such as fish, or they may die and settle to the bottom of the lake where they join the accumulating sediments. The fish that feed upon the crustaceans often maintain a larger biomass than their prey, in sharp contrast to terrestrial communities, where the food materials are always present in greater bulk than the feeders (▸ page 86).

Productive lakes, unproductive rivers

If a lake is sufficiently shallow (less than about 3m), aquatic macrophytes may float freely or may be rooted in the muds. The presence of these more bulky plants, such as water lilies and pondweeds (Potomageton species), slows the flow of water through that part of the lake. This, coupled with their high organic productivity, may increase the rate of sedimentation in these regions. The slower flow of water also favors invertebrate animals, such as pondsnails, which feed not only upon the macrophytes themselves, but also upon the coating of sedentary algae and filter-feeding microscopic animals that cover their leafstalks and undersides of leaves.

*In contrast to lakes (and ponds), the fast-flowing rivers and streams are something of a desert. Vegetable detritus becomes available to animals only when it enters the relative shelter of crevices between pebbles. Here it forms a food resource for crustaceans such as the water fleas (*Gammarus*) that need the high oxygen content of fast-flowing water to survive. Some fish, such as trout and lamprey, also prefer this more active, oxygen-rich environment.*

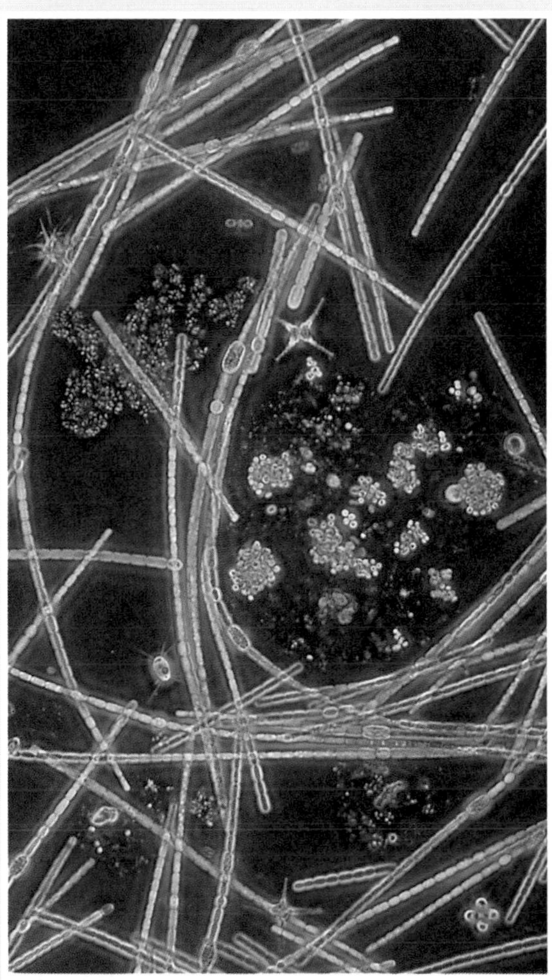

▲ Microscopic algae, some single-celled and others colonial, are the main primary producers of the open waters of lakes and the sea, forming the basis of many aquatic food webs

The dam-building activities of beavers have been instrumental in creating many of the tiny lakes of Canada, Alaska and Scandinavia

Feeding interactions

It is the very rapid population turnover in the plankton that allows some middle and top carnivores to outweigh species lower down in the trophic levels and ponds. Plankton-feeding fish are in turn consumed by predatory fish or by birds, such as herons, kingfishers or saw-billed ducks like the goosander. When such ducks are eaten by terrestrial predators, for example the fox or the peregrine falcon, then the energy they contain may have passed through as many as six levels, which is an especially long food chain (alga-crustacean-plankton-feeding fish-predatory fish-duck-raptor).

Those planktonic individuals, that evade predation long enough to die naturally, add their residual energy to the muds of the bottom, where it is efficiently consumed by detritivores such as mollusks and worms, and by the fungal and bacterial decomposers. The inorganic nutrients thus released become available to the phytoplankton once more. But this recycling is of less importance in this community than in terrestrial ones (◀ page 88), for there is a constant replenishment of nutrients in the flowing water, bringing the leached materials from catchment soils. Some dead material survives the attentions of the decomposers, such as the silica shells of diatoms, and these build up on the floor of the lake, which thus becomes shallower with time.

Food web in a temperate lake

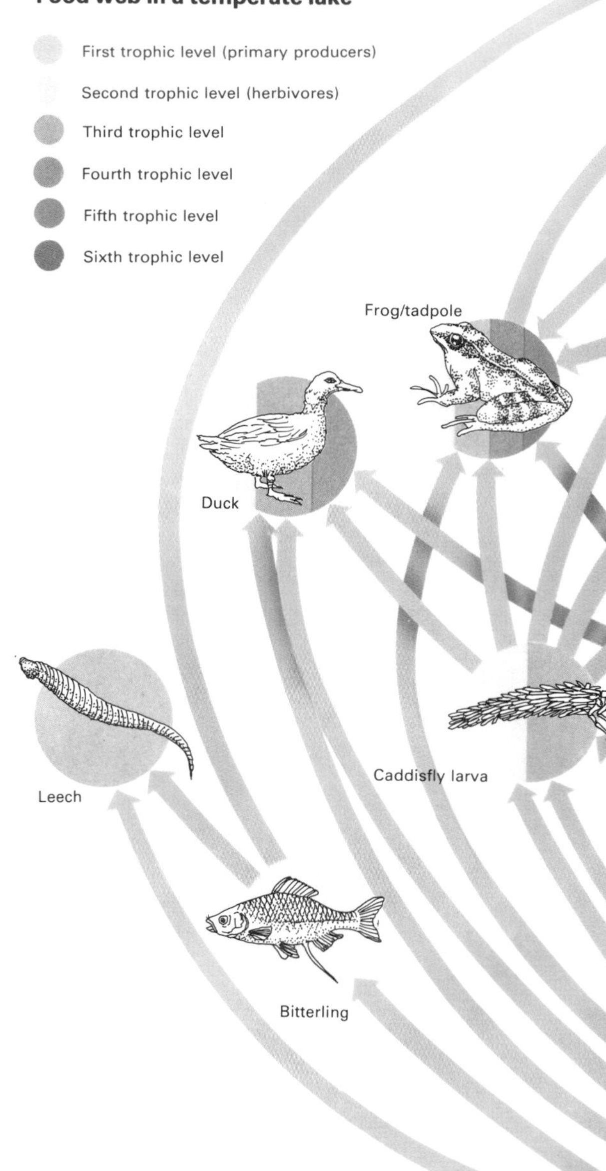

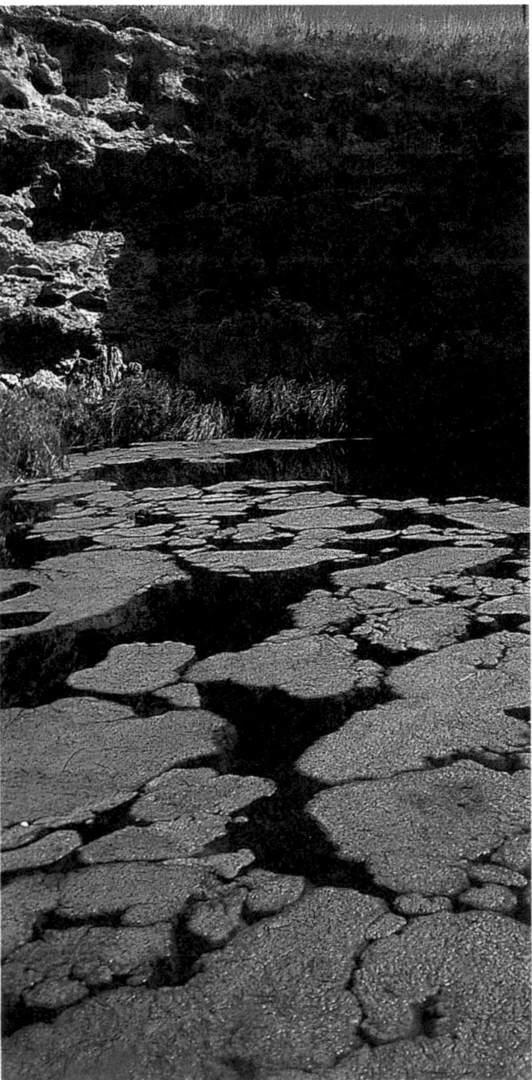

◀ Some algae, particularly the filamentous ones, may form dense masses, or platforms that float to the top of water because of the oxygen they produce during photosynthesis. Rapid growth of algal masses can be harmful because as they die they decompose and use up the oxygen in the lower layers of water that have been cut off from the atmosphere by the platform.

▼ Excessive growth of water plants is usually controlled by the grazing animals that feed upon them. The great pond snail, seen here with an egg rope attached to its shell, is an efficient consumer of the primary production of freshwater plants.

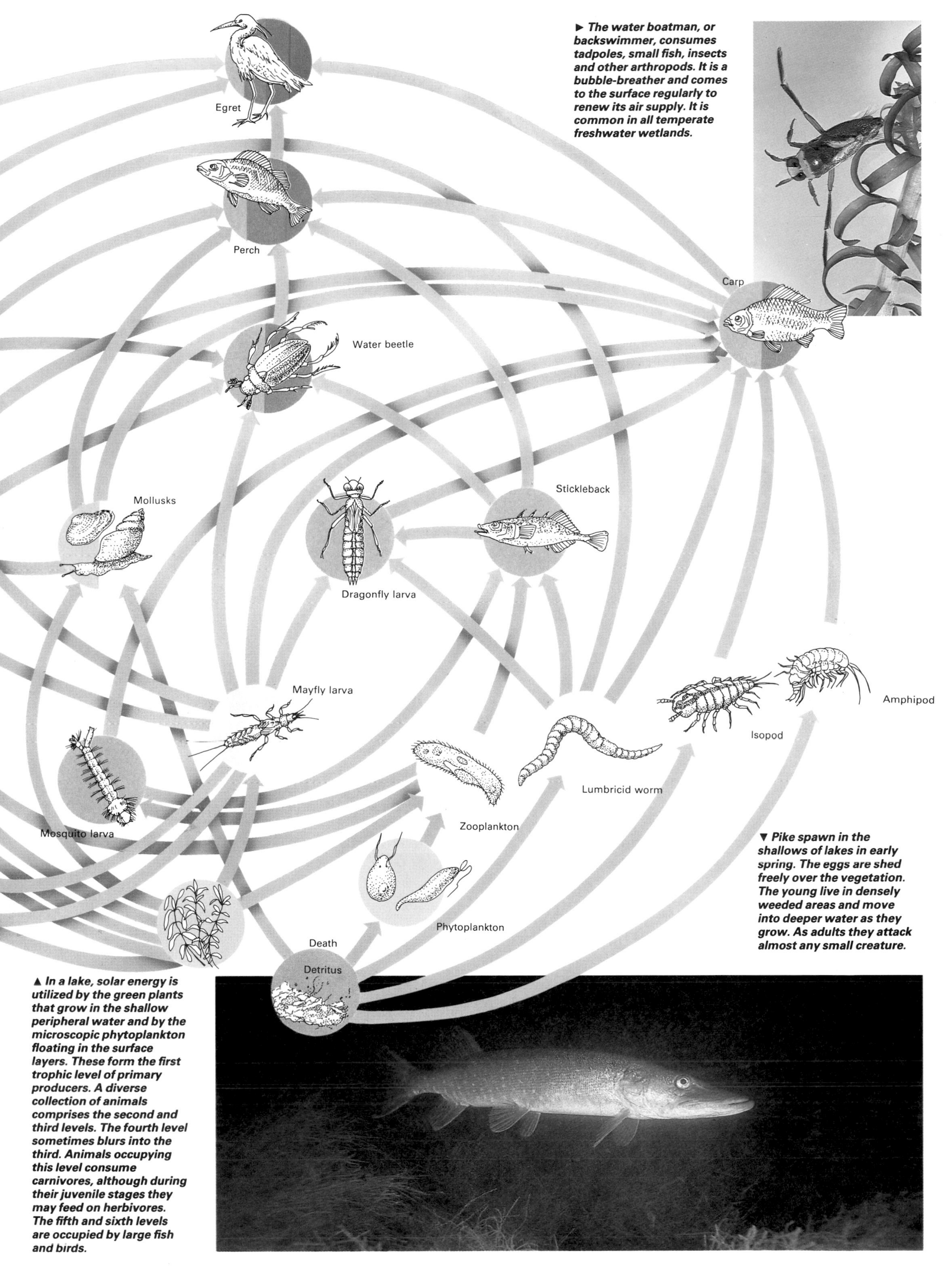

► *The water boatman, or backswimmer, consumes tadpoles, small fish, insects and other arthropods. It is a bubble-breather and comes to the surface regularly to renew its air supply. It is common in all temperate freshwater wetlands.*

▼ *Pike spawn in the shallows of lakes in early spring. The eggs are shed freely over the vegetation. The young live in densely weeded areas and move into deeper water as they grow. As adults they attack almost any small creature.*

▲ *In a lake, solar energy is utilized by the green plants that grow in the shallow peripheral water and by the microscopic phytoplankton floating in the surface layers. These form the first trophic level of primary producers. A diverse collection of animals comprises the second and third levels. The fourth level sometimes blurs into the third. Animals occupying this level consume carnivores, although during their juvenile stages they may feed on herbivores. The fifth and sixth levels are occupied by large fish and birds.*

Swamp vegetation in the Amazon basin has to cope with a fall in water level of 10m or more between June and December

Freshwater wetlands

Around the edges of lakes and ponds and wherever water accumulates on the surface of the soil, peat-forming habitats may form. Peat consists of the undecomposed remains of living organisms, chiefly plants, which collect in sites where conditions are too wet to permit full decomposition (◀ page 85). Where water stands, the microbial populations are deprived of oxygen and many of them are unable to thrive. Some anaerobic types manage to operate effectively, but the overall rate at which detritus is broken down becomes slow and some of the organic remnants, especially the less palatable bits such as wood and fibre, survive as peat. In some sites, like valley mires, the accumulation of peat further restricts drainage and so encourages more surface wetness and more peat development.

Along the fringes of lakes, tall, emergent aquatics form swamps in which the surface of the water is always above the sediment surface. The presence of plants such as reeds and cattails slows the flow of water and leads to increased sedimentation, so swamp vegetation leads towards fen in which the summer water table is below the soil surface. Swamp vegetation provides a habitat in which animals can easily emerge from the water by climbing stems of plants, and this attracts the attention of insects like dragonflies in which the adult is terrestrial but the larva is aquatic. The female lays her eggs under water while perched on an emergent shoot, and the larva can climb up such shoots on its completion of the aquatic phase of the animal's life history. Many other invertebrates, such as grazing gastropod mollusks, also make use of this easy means of access between air and water.

Swamps are among the world's most productive habitats and they support a very large invertebrate biomass. This is exploited by numerous predators, including many birds and amphibians, especially frogs. Among the top carnivores are alligators, crocodiles, snakes, large raptors and mammals such as the American mink.

Sphagnum bogs

As the soil surface rises with the accumulation of peat, less flood-tolerant plants can invade, and trees such as willow and alder are often among these. The course of succession (◀ page 94) can thus lead into wet woodland or carr. There is one type of moss, however, which can alter the whole direction of the succession. This is the bog moss *Sphagnum*. Although of diminutive stature, this moss can absorb mineral elements out of its wet environment and leave the water acid and poor. By doing this it gains a great competitive advantage over other plants, which cannot tolerate conditions of such nutrient poverty. Even the trees, willow and alder, are not immune to the subtle strategy of this moss, and the study of peat stratigraphy, which records the past history of peatland habitats, shows that wet carr woodland has often been converted into sphagnum bog by the invasion of this moss.

Sphagnum bogs are found in both the Northern and Southern Hemispheres, but are most abundant in the boreal regions of Canada and the Soviet Union. They are rich in certain insects, particularly mosquitoes, but their low levels of calcium mean that they cannot support snails. Frogs are still frequent, however, and these are fed upon by cranes, which are the major avian predators of the northern bogs. In Eurasia, the strange cries of the common crane are a distinctive feature of the boreal peatlands, whereas in Canada it is the sandhill crane which occupies this niche.

◄ *Sphagnum bogs are remarkable ecosystems in which a rapid buildup of peat can elevate the surface layers above the influence of ground water. The vegetation then becomes fully dependent on the rainfall and receives little in the way of nutrient input from this quarter. Elements like nitrogen and phosphorus become especially scarce and some plants overcome this problem by catching and digesting insects that are relatively rich in such elements.*

► *Amphibians like toads are dependent upon wetland habitats for breeding. Their eggs are without shells and are sensitive to drought.*

▼ *The curlew both feeds and breeds in wetland habitats.*

Feeding and nesting strategies

In freshwater habitats birds abound. Some, such as sedge warblers, are insectivores. The snail kite of the Florida Everglades has a long, hooked upper mandible with which it can extract gastropods from their shells. Egrets, storks, bitterns and herons are all very fond of frogs, as well as fish and other birds. Large raptors include the marsh harrier in Europe and the northern harrier in North America.

All of these birds must solve the problem of nesting sites in such wet habitats. Reed and sedge warblers weave their cup-shaped nest into the swaying reed stems. Many herons and night herons abandon the swamp and nest in neighboring trees. Bitterns make low platforms of dead reeds which float on the water, and the harriers build large mounds of reed litter deep in the shelter of the most extensive beds.

▼ *Wetland birds display a variety of feeding methods. Flying insects are best caught on the wing, while wading birds exploit aquatic food in either shallow or deep water, depending on the length of legs and bill. In open water, swimming wildfowl may dabble or dive, a strategy also used by kingfishers. Predatory birds hunt mainly on the wing.*

1 Yellow wagtail catches insects on the wing
2 Snipe probes in wet mud for worms
3 Heron catches fish and frogs in the shallows
4 Mallard often upends for vegetable and animal matter
5 Shoveller sieves plants and invertebrates from the water surface
6 Tufted duck submerges for animal and vegetable matter
7 Kingfisher hovers then dives into water for small fish
8 Marsh harrier catches small birds on the wing

The Oceans

▶ ***Microscopic plants, phytoplankton, are the main producers of the open ocean and are preyed upon by microscopic animals, the zooplankton.***

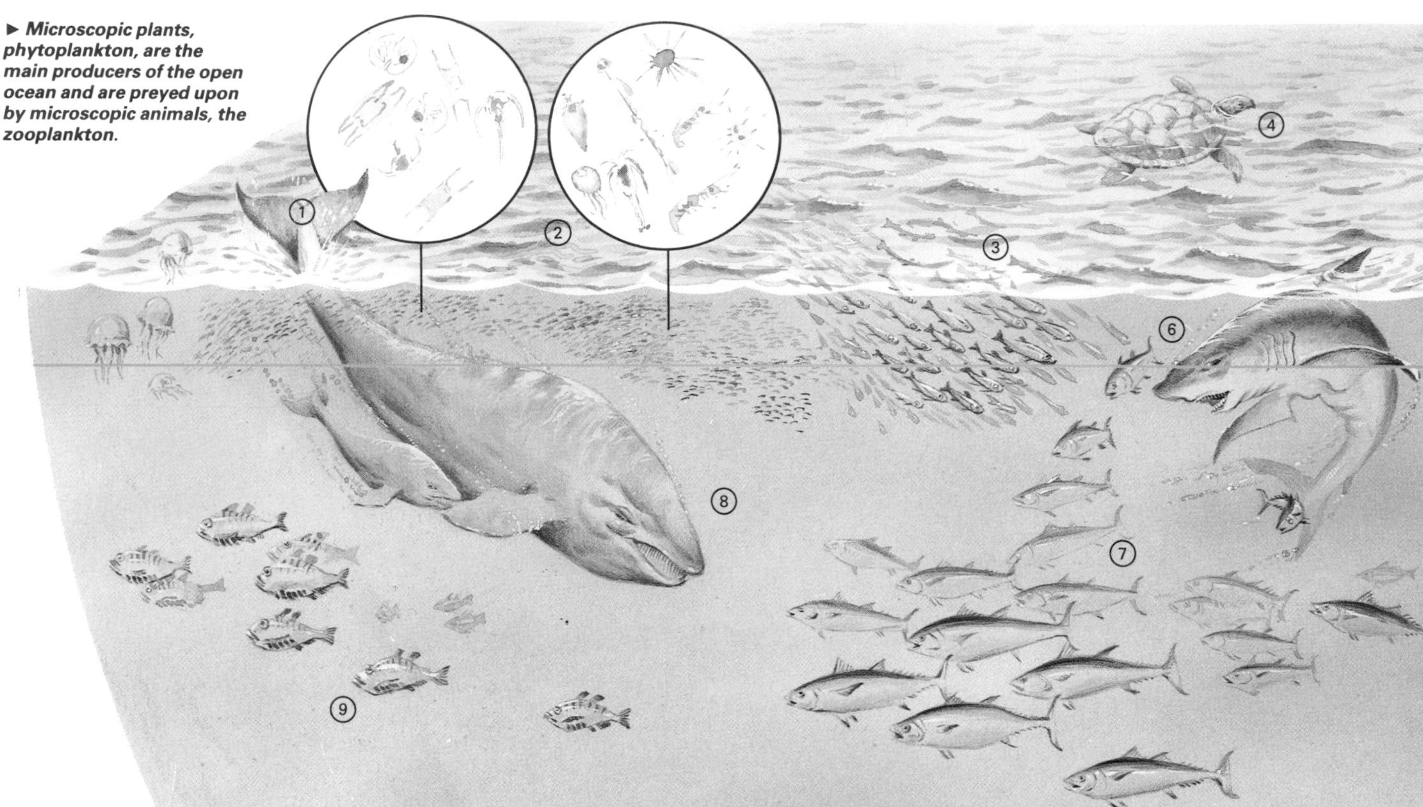

The pelagic community

Plants and animals which occupy the free ocean waters are said to be pelagic. They drift with the ocean currents, though some, like fish and cetaceans (whales and dolphins), are able to swim and therefore determine their movements more precisely. These are the nekton, as distinct from the drifting plankton. They may even migrate seasonally to ensure continued food supplies. Marine turtles belong to the pelagic community, but move onto land to lay eggs. They may migrate thousands of kilometers to achieve this, like the green turtle which travels from its feeding grounds off the coast of Brazil to Ascension Island in the mid-Atlantic to breed (♦ page 72). Perhaps its ancestors have been doing this for millions of years, since that time when the continents of Africa and South America first began to pull apart and the distances involved were quite small.

Within the pelagic community the plankton supports a range of feeding levels. Some of the animal life, the nekton, is able to swim in search of food. Many fish, like herring, mackerel and pilchard feed directly on the plankton, as do many of the great whales (♦ page 189). Others, like damselfish and sharks feed on larger prey. A large variety of seabirds adopt a pelagic habit and seek out land only for breeding. Among these are various species of albatross that are unable to breed until about 15 years old. Some seabirds, like the Manx shearwater are known to cover immense ranges in their feeding activities. Such birds may also be responsible for a very considerable transfer of nutrient elements to the shore around their breeding and roosting grounds. The guano deposited by the seabirds of Chile have proved important sources of phosphorus for fertilizer.

1 Phytoplankton
2 Zooplankton
3 Anchovy
4 Green turtle
5 Dolphin
6 Shark
7 Bluefin tuna
8 Grey whale
9 Hatchet fish
10 Squid
11 Lantern fish
12 Oarfish
13 Giant squid
14 Deep-sea jellyfish
15 Skate
16 Brittle star
17 Deep-sea shrimp
18 Angler fish
19 Tripod fish
20 Sea cucumber

▲ ***The deep sea angler fish dangles a luminous lure to attract its prey – other small fish. This carries with it the risk of attracting enemies as well.***

► *The coral reef is among the most productive of all marine habitats. But all animals here are ultimately dependent on the microscopic algal plankton.*

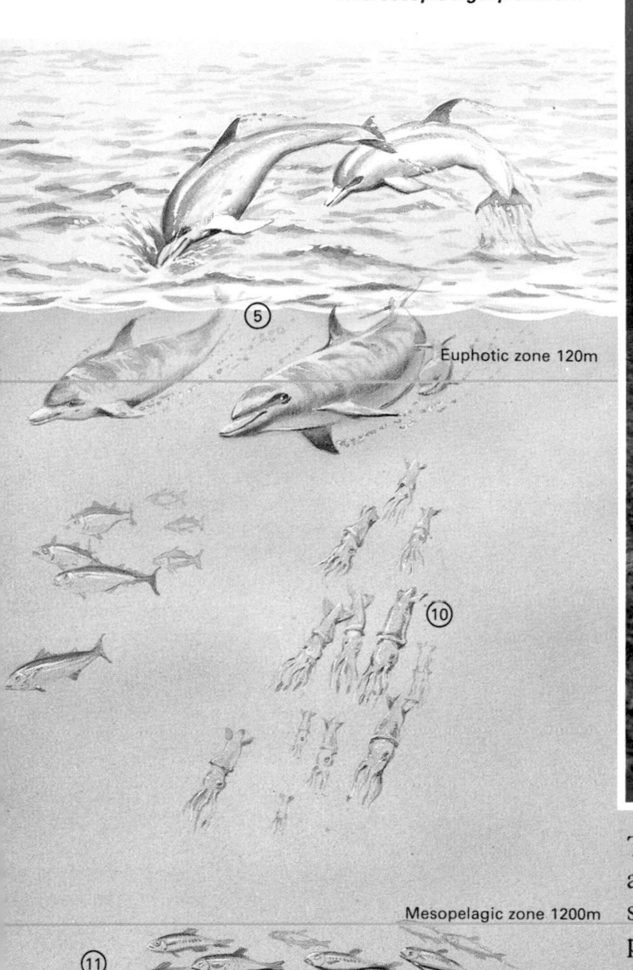

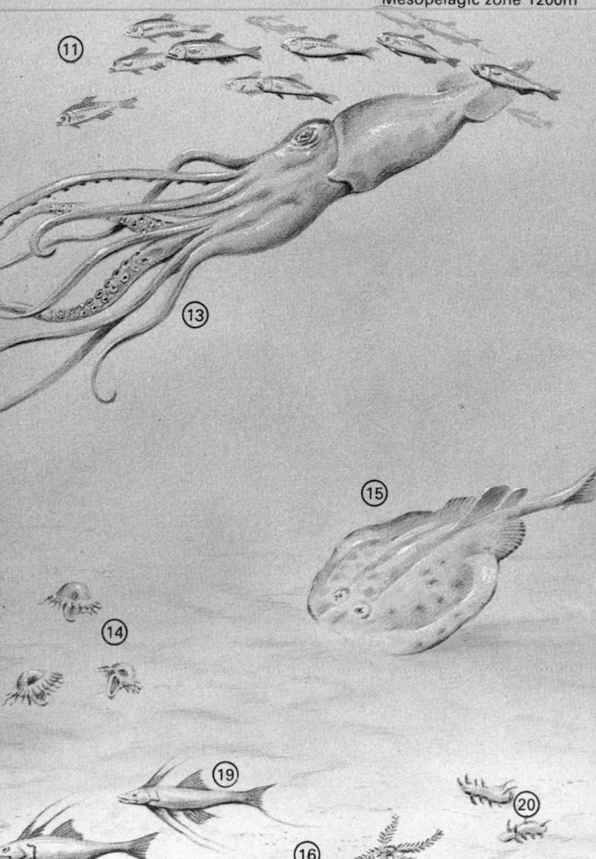

The bulk (70%) of the Earth's surface is covered by the oceans and about 97% of the world's water is salt and lies within them. In the surface layers – the upper 100m where light penetrates – the phytoplankton photosynthesizes and supplies the energy resource for the ocean community. However, the total amount of production per unit area is really rather low (on average less than 200g carbon/m^2/yr). The oceans are thus effectively wet deserts, but they are so large that this productivity still makes a substantial contribution to the Earth's total production (◀ page 86). It is this which makes it possible for people to harvest over 60 million tons of marine life annually (▶ page 181).

One of the main factors limiting productivity is the availability of the required inorganic nutrients, particularly phosphorus. In northern waters, as the light levels increase in spring, there is a burst of phytoplanktonic activity, but soon the levels of nutrients in the waters become limiting and by summer the population growth rate has declined markedly. There is often a second peak in production in the autumn, when winds cause turbulence in the water, mixing up the deeper layers with their nutrients. The polar waters are kept productive throughout the summer by the upwelling of deep water currents.

In the deeper waters, where light does not penetrate, the sole energy source is the rain of dead plankton from above, together with those fish which feed upon this detritus.

On the ocean floor, the organic rain forms a soft ooze. This is rich in decomposer bacteria, as well as detritivores, such as mollusks and worms. There are also large populations of monstrous worms up to 3m in length, together with large bivalves living around the vents of submerged volcanoes. The plankton rain is unlikely to be adequate to support such big animals in such high numbers. There are, however, species of bacteria which can gain energy from oxidizing the sulfur given out by the vents, and it is this productivity which is believed to be the basis of the food web in these remarkable abyssal communities.

Pollinators

Once plants with insect-pollinating systems have become established in a saltmarsh, then in summer very large numbers of pollinating vectors, particularly bees, become attracted to the habitat from the land. Grazers, such as rabbits, are also tempted onto the marsh, but these terrestrial species need to be able to escape at high tide. Invertebrate grazers like grasshoppers are mainly confined to the upper levels of saltmarshes where flooding is infrequent and does not normally cover the entire shoot systems of plants. Here too are the breeding grounds of many species of bird. In the muddy, saline soils and in the permanent pools of the marsh it is an essentially marine fauna which survives, from microscopic crustaceans to scavenging shore crabs.

Grazing such saline vegetation must impose considerable stress upon the kidneys of mammals, but on the other hand the salty habitat is relatively poor in parasites. It is because of the absence of liver fluke that the saltmarshes of England became valued as sheep grazing areas in former times.

◄ Where a river enters the sea it brings both inorganic silts and organic material from the land. The organic matter provides an important source of energy for the animals living there.

▼ Estuary muds are rich in small crustaceans which are eaten by wading birds. Black-winged stilts probe for their food, but the avocet swings its bill to catch prey in the water.

▲ **Flowering plants, like sea lavender, grow along the edge of the sea. For migrant insects, such as the painted lady butterfly, they are a welcome source of food.**

▼ **When constantly flooded, the roots of plants suffer from lack of oxygen. Mangrove trees differ in their tolerance to flooding depending on their roots.**

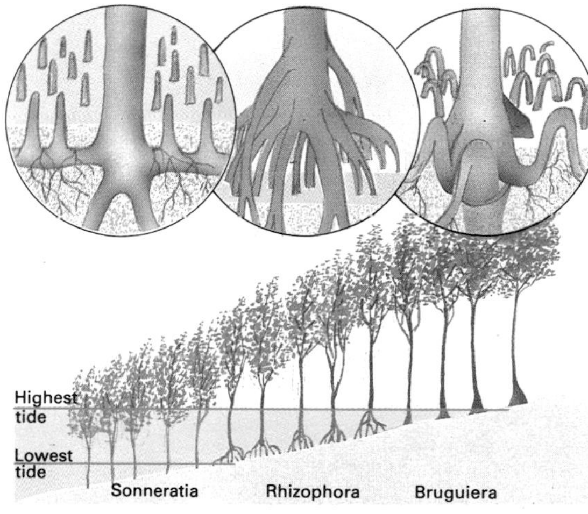

Saline wetlands

Whereas the communities of rocky shores (➧ page 126) are largely comprised of marine organisms, those of muddy shores usually have a distinctly terrestrial element within them. This is particularly true of the plant life, for as the mud levels begin to rise they are often colonized by higher plants rather than marine algae. In the temperate areas the vegetation develops into salt marsh and in tropical regions into mangrove swamp.

The detritivore communities living in the imported muds of estuaries have a rather unstable existence, for the water currents are constantly shifting and altering the topography of mud banks. They attain greater stability when colonized by plants such as eel grass. This fine-leaved flowering plant can grow on muds that are left uncovered only for short periods of time, close to the low tide mark. They are an important source of food for brant geese in winter and European brant goose populations were badly affected when eel grass declined following an epidemic disease earlier this century.

Mud stabilization is even more effectively achieved by the invasion of more robust species of flowering plant, usually the glasswort (*Salicornia*) and the chord grasses (*Spartina* spp.). Most of the glasswort species are annuals, so the degree of stability which they can impose is limited, but the chord grasses have extensive root systems and they are perennial, so their influence is more lasting. Their shoots filter the mud and silt from moving waters and add to the rate of sedimentation, so raising the soil surface and reducing the amount of time spent submerged in salt water. This facilitates the invasion of other, less tolerant species of plant, such as the thrifts and sea lavenders. All of these plants have to face the problem of regulating their salt contents and they overcome this by having salt excreting glands which actively pump the excess salts out of their tissues.

The most characteristic plant of tropical saline wetlands is the mangrove. There are many different species with varying capacities to tolerate flooding, but all are robust shrubs or trees that become rooted in anaerobic salty muds. One of the main problems for such trees is obtaining enough oxygen, and many species have specially adapted roots that grow upwards out of the mud and exchange gasses in the air above. Another difficulty in the shifting muds is the establishment of seedlings. Some species have distinctive, arrow-shaped fruits that fall from the tree when ripe and stick vertically into the mud. In this way they avoid being swept away by the tide before they can germinate and establish an anchoring root system.

Coping with a scarcity of oxygen

The mudskipper lives in the mangrove swamps of the Indo-Pacific region. It is a fish rather like a goby but which is able to emerge from the water on the surface of the wet muds at low tide. Its eyes are placed on top of its bulbous head and it is constantly on the alert for predators, which it avoids by a sudden straightening of its bent tail, enabling it to leap over half a meter. Its most remarkable feature, however, is its ability to breathe not only through its gills, but also through its wide mouth and throat, absorbing oxygen from the air in a most un-fishlike manner. Perhaps it is the scarcity of oxygen in this type of habitat which has led to this piece of adaptive evolution.

◀ **Mudskippers among mangrove roots**

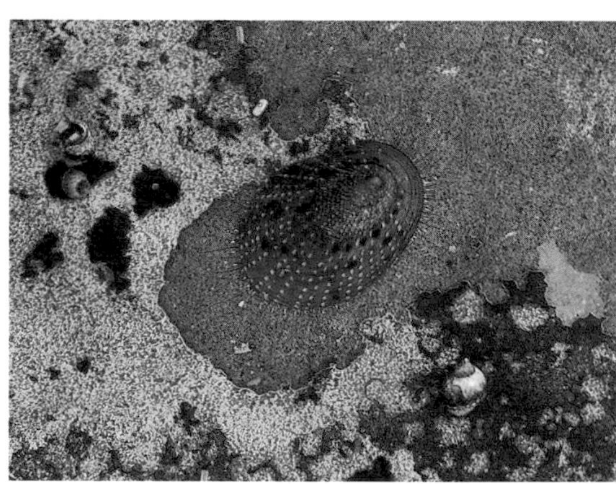

▲ Feeding in the open ocean, these albatross need dry land to nest on. Cliffs are most convenient and may become crowded with breeding birds.

► In rock pools the algae which coat the surfaces of rock are grazed by limpets which wander around rasping with their rough tongues.

◄ Another source of food in the rock pool is provided by small swimming creatures such as fish, and these are caught by the stinging tentacles of sea anemones.

Life in the eulittoral

On many rocky shores the eulittoral, like the sublittoral, is dominated by brown algae, but here the wracks, or fucoids, do best. These are generally smaller than the kelps, but more desiccation-tolerant. Some red and green algae also grow among them, sometimes taking advantage of the greater bulk of the browns to gain protection from the drying winds at low tide. Many animals adopt the same strategy; herbivorous periwinkles and predatory dogwhelks creep beneath the mounds of gelatinous weed to survive their periods of emersion.

Other animals are sedentary, like the barnacles that coat the bare surfaces of rock. These do not graze upon the large algae, but act as filter feeders, extracting plankton from the waters which immerse them during high tide. At low tide they close the hard valves of their shell and protect themselves against desiccation. Limpets adopt a similar way of life, clamping firmly onto the rock during low tide, but they graze the surfaces of the rock, scraping off colonies of minute algae and settling sporelings when the tide is in.

► The rocky shore usually shows a distinct zonation pattern of plants and animals. Each zone contains a different community of organisms which are best fitted to cope with the degree of drought and competition experienced there. Sublittoral zone species are least adapted to drought-tolerant since they are emersed for only short periods of time at low tide.

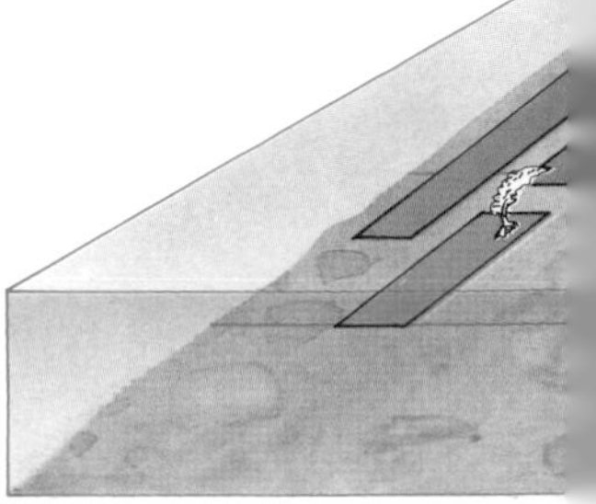

The shores

The zone where the land meets the sea is one in which many difficulties are experienced by living organisms. Some of the time the zone is immersed in sea water and becomes part of the oceanic system. Then the sea withdraws and it undergoes the rigors of a terrestrial existence. Most of the inhabitants of the intertidal region have their origin in the sea and it is the emersion, or exposure to air, which is most stressful. But some, like the birds that feed at low tide and the flies that lay eggs among the seaweed, are terrestrial organisms that now exploit an opportunity in an inherently difficult habitat.

The action of waves on the land results in erosion, and where the sea is bounded by hard rocks this may result in a scouring to produce smooth rock faces and boulders, the rocky shore. The material eroded from such shores comes to rest in other areas, sometimes along shores where the wave action is not so severe, or the geology is softer. Here the inorganic particles are deposited to form a sandy shore. There are also areas where rivers from the land surface debouch into the sea, carrying not only the inorganic products of erosion but also organic matter from the primary producers of the land. These materials usually accumulate in the mouth of estuaries, particularly where they are protected by sand or shingle bars. They form mudflats (◀ page 124).

Zonation on the beach

There are three main zones that can be discerned on most rocky shores: the sublittoral zone, which lies below the normal low water mark, the eulittoral zone, which occupies most of the intertidal strip up to the upper limit of barnacles, and the littoral fringe, which extends approximately to the limit of the highest tides. It is not possible to generalize completely about the distances over which these zones spread as they extend higher up those shores that are exposed to severe wave action. In very exposed shores the littoral fringe and even the eulittoral may extend well above the strict limit of high water. The splashing effect of waves keeps these upper levels unusually moist.

The sublittoral zone is the least stressful and is occupied by an essentially marine community of organisms. It contains the highest biomass of benthic algae, often in the form of large members of the Phaeophyta, the browns. These are some of the world's biggest seaweeds, the kelps, which can attain many meters in length. Unlike land plants, they lack supportive woody trunks and stems since they are kept upright by their buoyancy in the water. Some have air bladders to aid flotation.

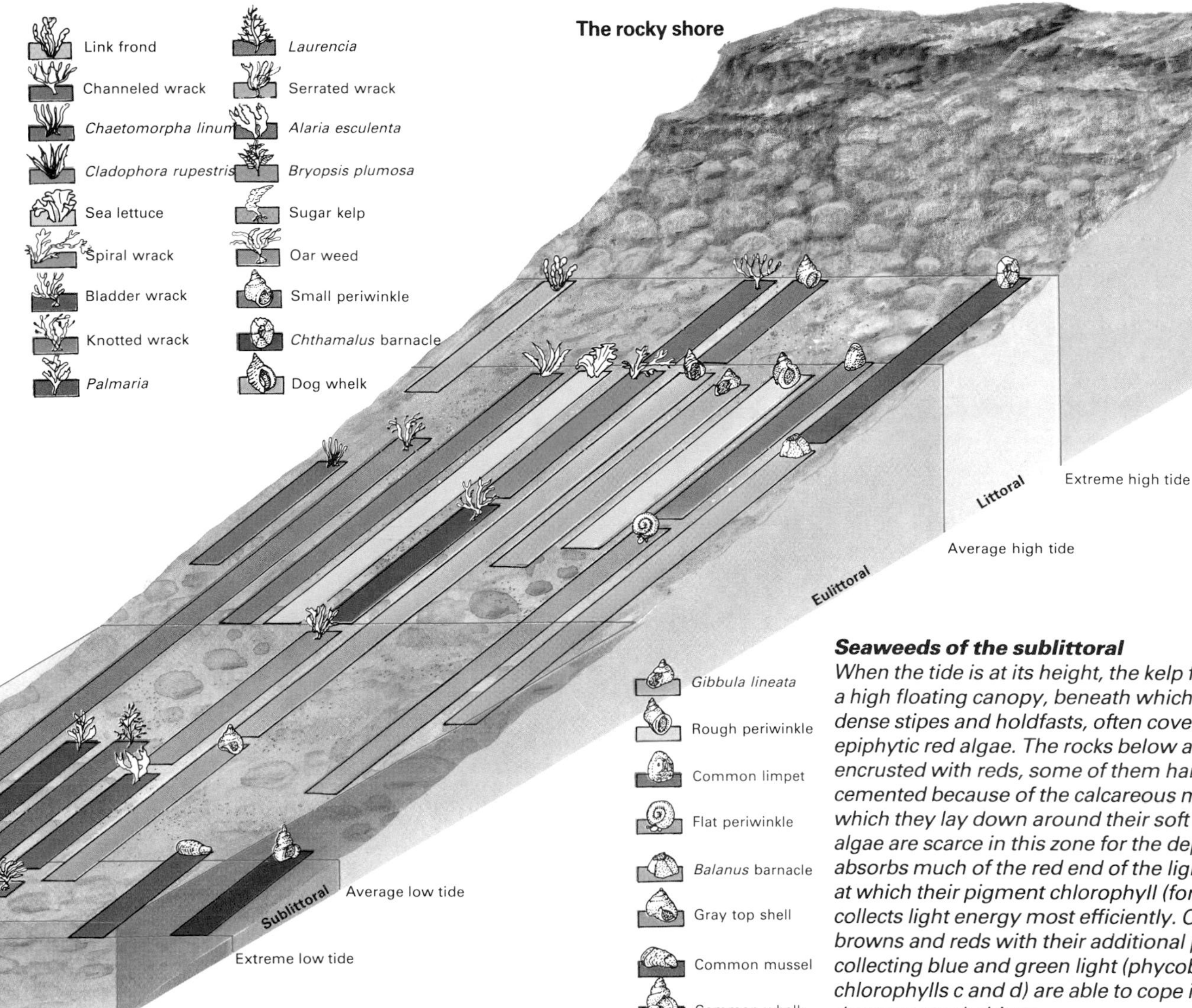

Seaweeds of the sublittoral

When the tide is at its height, the kelp forests have a high floating canopy, beneath which are the dense stipes and holdfasts, often covered with epiphytic red algae. The rocks below are also encrusted with reds, some of them hard and cemented because of the calcareous materials which they lay down around their soft cells. Green algae are scarce in this zone for the depth of water absorbs much of the red end of the light spectrum at which their pigment chlorophyll (forms a and b) collects light energy most efficiently. Only the browns and reds with their additional pigments for collecting blue and green light (phycobilins and chlorophylls c and d) are able to cope in these deeper water habitats.

See also
Interactions 81-96
Communities of the Water 117-128
The Shadow of Extinction 195-208
Threats to the Environment 209-224
Saving Habitats 225-234

Rocky and sandy shores rely upon two main sources of energy for their maintenance. One is the production of benthic, attached algae, and the other is the pelagic phytoplankton that each tide brings within reach. In the case of the mudflat, there is an additional source of energy in the organic detritus brought down from the land.

In all three shores there is some differentiation of the community type which occupies the different levels above the low water mark. This is most marked in the rocky shore, which is also generally the most diverse in the life it supports (◆ page 126).

In sandy and muddy shores, the instability of the substrate means that there is little opportunity for the larger seaweeds to obtain a footing. Some smaller photosynthetic organisms do survive, however, such as diatoms, which often form an orange-colored coating on the mud surface. But the bulk of the energy in these situations is brought into the ecosystem by the sea water, either as plankton or as detritus.

The main difference between muddy shores and sandy ones is the degree to which sea water drains out of the substrate and air enters into it during low tide. Sand is free-draining, so there is adequate oxygen movement into the pores as it drains. In mud, however, the pores are much smaller and water is held in them by capillarity. The waterlogged interstices become deprived of oxygen and only those few animals which are tolerant of anoxia can survive here. One such is the amphipod crustacean *Corophium*, a detritivore which often attains great population densities in muddy sediments.

Like the rocky shores, sandy and muddy shores also prove attractive to birds which feed upon the invertebrate animals. Many of them, like the whimbrel and curlew, are equipped with long bills with which they probe the soft sediments in search of worms and crustaceans. Others, like the terns, hunt small fish and sand eels in the pools and along the shallow sublittoral of these gently sloping shores.

Filter feeders and sand eaters

Many animals that live on sandy shores take advantage of the plankton and detritus input by filtering the waters and extracting energy-rich particles. Bivalves, such as clams and cockles, operate by drawing waters through their body and extracting all potential food material. These are the filter feeders. Other animals ingest the sand together with its content of organic detritus. The lugworm, for example, makes a U-shaped burrow and constantly consumes the sand that pours into one entrance and defecates it out of the other to produce the characteristic worm casts.

▲ Crabs are generalist feeders, mainly relying upon detritus and carrion. They inhabit all kinds of shore, though this spider crab is most typical of deeper water.

▼ Open ocean organisms such as jellyfish may find themselves in an alien environment when they become stranded on sandy shores. They can die in an hour from desiccation.

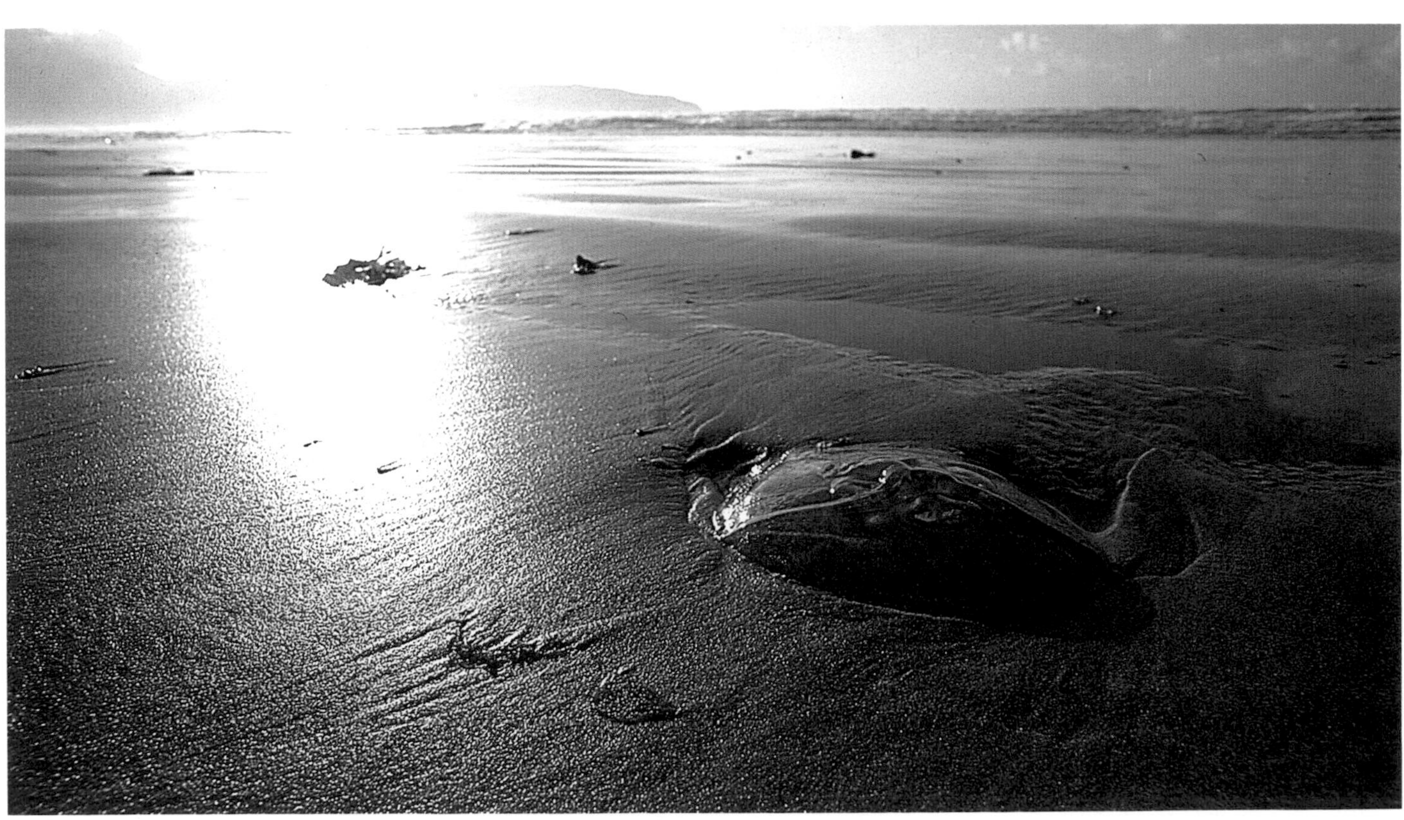

An Ape Called Man

11

From* Australopithecus *to* Homo*...How we differ from all other animals...Controlling our environment... Human achievements...Living in harmony with our fellow creatures...*PERSPECTIVE*...Man the farmer... Population explosion...The evolution of language... How we use our evolutionary advantage

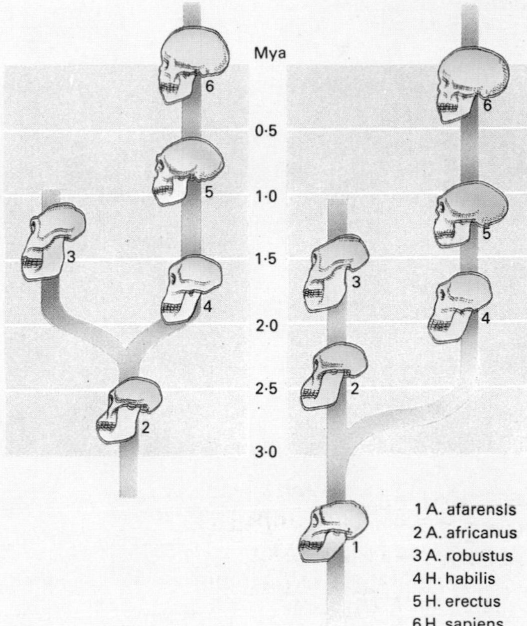

▼ **Australopithecus afarensis** ***was first discovered in 1974. It lived 3-4 million years ago and so is one of the oldest hominids. Some biologists maintain that*** **A. afarensis** ***still spent a great deal of time in the trees, as well as on the ground, and that its lifestyle was therefore between that of apes and humans.***

▲ ***Before Don Johanson found*** **A. afarensis** ***most anthropologists accepted that*** **A. africanus** ***was the first hominid, and had given rise both to*** **A. robustus** ***and*** **Homo.** ***But Johanson suggests that*** **A. afarensis** ***is really the first hominid ancestor and that*** **A. africanus** ***was merely an early form of*** **A. robustus.**

Biologists estimate that there may be 10 million species of organism in the world. Many are microscopic, others as large as a row of houses. Some exist in countless billions and a few are rare almost to the point of disappearance. Species like the elephant of the African grasslands have an enormous ecological impact (➧ page 238), while others go almost unnoticed. Some are extreme specialists, dependent on the continued existence of another organism or some quirk of topography or climate. Others are versatile, able to carve a niche in almost any environment (➧ page 90).

There is one species, however, whose impact is of a quite different order of magnitude to that of any other. This animal transcends the rules of ecology, bends the rules of evolution, and is having as great an effect upon its fellow species as the catastrophes of climate that wiped out 90 percent of all marine species at the end of the Permian period 225 million years ago. The creature that has wrought these changes calls itself *Homo sapiens*.

We see our achievements through human eyes. We like to point out that we are the only creatures that can properly be said to have developed arts, sciences and formal sports. In practice, it is not easy to distinguish such achievements from those of other animals.

Apart from a few bone fragments found in 5-4 million-year-old deposits, the earliest hominid fossils are dated to just over 3 million years ago

◄ **Homo erectus,** ***who lived 1.5-0.5Mya, was the first human to use fire. It acted as a tremendously powerful tool, assisting in hunting, enabling our ancestors to eat tough meat, and, much later, helping humans to clear wild vegetation to make way for agriculture.***

► ***With modern machines humans can transform entire landscapes in a very short time. But we cannot control or even predict the ecological consequences of our engineering projects. Dams, for example, are intended to bring wealth but often merely bring soil erosion and disease.***

▼► ***A spider's web is no less remarkable a piece of engineering than the Eiffel Tower. But spiders can only build webs, while we can build anything we choose, using any materials. It is this versatility that gives us such power over our environment.***

Our ecological peculiarities

The first and most obvious of these is that we are by far the most numerous large animal that has ever lived, with a population now approaching five billion. As for our size, among land animals, only bears, big cats, and a few hoofed creatures are bigger, and there is a rule of ecology: large animals are rare. Thus the rise of the human population (➧ page 134) is unprecedented in all of natural history.

The reason we are so numerous lies in our second ecological peculiarity. Working individually or, as is more usual, in groups, we exert far more control over our environment than any other creature. We are not the only creatures that build, but our cities, harbors, airfields and motorways fragment and obliterate natural environments to an extent that greatly exceeds the impact of other animals.

More significant, however, is the fact that we manipulate the natural world so that it will provide us with more food than is available from "unimproved" nature (➧ page 132). But how can one creature be so dominant? What makes us so different?

The evolution of an all-powerful species

When we consider such people as Leonardo da Vinci, Einstein or Brunel we have no cause to wonder at human dominance. With such depth of understanding and breadth of imagination, coupled with manual skill and not inconsiderable strength, there could be no doubt that we would dominate. But the human species did not begin with Leonardo or Brunel. They are latecomers. It began, so modern evidence suggests, with a creature about a meter high who lived in the dwindling forest and plains of Africa and was called *Australopithecus*. This creature had neither the physical nor the mental equipment to design flying machines or suspension bridges. Indeed it had a small brain, did not speak, and did not even make stone tools.

Regarding the evolution of the genus *Homo*, here a few general

points must be considered. First, evolution does not have direction. Any one organism has to be able to compete with others of its own kind and of different species, and has therefore to be able to function as a complete, finished creature. No animal is simply a prototype for some later creature. We speak; that is one of the crucial differences between ourselves and other animals (➧ page 137). At some stage our human-like – hominid – ancestors must have made meaningful noises, from which speech evolved. But those "noises" functioned as a useful means of communication.

Second, evolution is opportunist. An organ may evolve in response to one set of selective pressures and then be turned to other purposes. Alternatively, it may be made redundant because of some change of evolutionary direction and simply become vestigial, or be acted upon by new selective pressures and evolve into something new. Thus, both human beings and ostriches evolved as bipedal animals, able to run quickly over the plains and see long distances. The ostriches' upper limbs became vestigial. But when our ancestors learned to walk on two legs and freed the upper limbs from the burden of locomotion, natural selection operated in such a way that the forepaws were further developed into agile hands and fingers.

A third point concerns game theory and evolutionary advantage (➧ page 14). If one organism has even a very small advantage over another then it will tend to obliterate the other. The reason is statistical. Evolution operates over many generations. If one animal has only a one percent reproductive advantage over another then, repeated in each generation, this soon results in total annihilation.

In summary, our first ancestors were creatures that were competing with all others and were just successful enough to survive. They were subjected to selective pressures quite different from those that now afflict us. That we are their descendants is just the result of evolutionary chance and opportunitism.

Improvements in sanitation and medicine have allowed an exponential world population-growth curve to develop, with massive ensuing problems of health, food and work

The birth of farming

The most significant act ever committed by a human being probably occurred in West Africa, about 30,000 years ago. There, according at least to anthropologist John Yellen of the National Science Foundation in Washington, a man, woman, or child first consciously decided to thrust a twig or stick into the ground in the hope and expectation that it would grow into a new plant. This was the first act of cultivation.

By cultivating their favorite plants, our predecessors increased the amount of food available and ensured a steady supply. But why did they develop large-scale agriculture, which began in the Middle East about 10,000 years ago? It used to be thought that farming was much easier than hunter-gathering, and that once people realized that plants could be grown and animals herded, then they gave up their earlier ways of life. Modern evidence shows the reverse is true.

Indeed, the true reason for the spread of agriculture seems to be that people were forced into it. Cultivation produced more food than was provided by unaided nature. This in turn allowed the population to grow. Once the population grew beyond a certain point, cultivation had to continue or people starved. In addition, because cultivation was hard work, the first farmers needed large families to spread the load. Hence began a vicious spiral of population growth.

◄ !Kung bushmen return to their camp with the day's kill. Professor Richard Lee of Toronto University has shown that the !Kung bushmen need to work for only about 6 hours a day, on 3 days a week, to provide enough food; and the children and old people do not work at all. Yet the !Kung live in one of the harshest environments on Earth. In areas such as the Middle East, where there was abundant fish, game, and wild fruit and grains, the hunter-gathering life must have been extremely easy. By contrast, farming is extremely hard unless aided by modern machinery and pesticides. The bushmen are generally well aware of the skills of cultivation, but prefer to hunt and gather, simply because it is so much easier given the local soil and climatic conditions.

▼ 20,000 km-long rice terraces in the Philippines. These have remained unchanged for almost 2,000 years and represent one of the earliest examples of large-scale agriculture.

World population

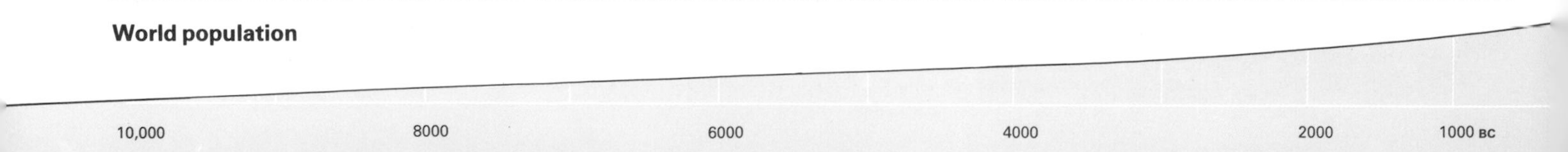

10,000 8000 6000 4000 2000 1000 BC

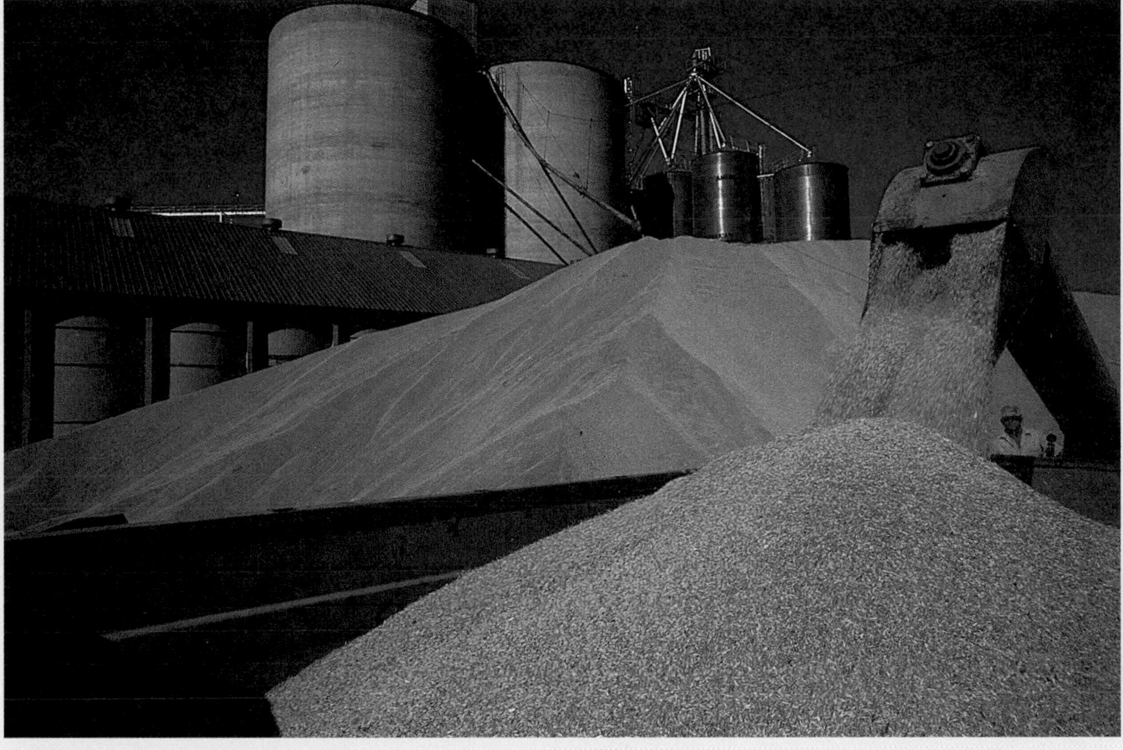

▲ ***The United States and Europe have huge surpluses of grain that are uneconomic to sell off***

The mass of humanity

Large animals tend to be rarer than small ones. They need more space and food, and they tend to reproduce more slowly. Predators are inevitably less numerous than the animals on which they prey (◆ page 82).

Human beings are large. Among land animals, only some of the hoofed animals and the largest carnivores – bears, lions and tigers – are bigger. We are also to some extent predatory. Yet there are almost five billion of us, and our population is still rising rapidly. We are the commonest of mammals, surpassed if at all only by rats and mice.

The reasons for the tremendous growth of the human population are two-fold. First, the development of agriculture and then of industry have made it possible for vast numbers to exist. Second, as people change from one way of life to another – from hunter-gathering eventually to an urban, industrial life – so their attitude to child-rearing changes.

The development of industry and of science has greatly increased productivity over the last 200 years, and particularly over the past few decades. Thus primitive farmers might obtain a tonne of wheat per hectare. The British farmers of AD 1900 obtained about 2.5 tonnes. Modern European and North American farmers average 8 tonnes. In addition, farmland has spread such that arable land now occupies 10 percent of the entire land surface and permanent pasture occupies another 23 percent (◆ page 210). For such reasons, the world population has increased three-fold since 1850.

But change of attitude to population is at least as important as change in possibility. Thus, hunter-gathering peoples in general want to limit the size of their families. Farming people need large families because working the land is labor-intensive. When people became industrialized their families at first tended to stay large, as in early 19th-century Britain. In general, poor people in under-developed countries tend to favor large families because children are a form of "insurance". In the absence of a welfare state and a state pension people need children to look after them in their old age. Hence, 80 percent of today's population growth is in poor countries in Africa, Latin America and parts of Asia (◆ page 135).

Whether or not there are too many people in the world, or in any particular country, is a matter of deep and sometimes bitter controversy. We could easily feed the present world population of around five billion, and probably about three times this number. It seems to follow, then, that the present famines are not caused directly by overpopulation, but by an inability or reluctance to distribute food to the people who need it.

Indeed, the spectre of "overpopulation" has often been used as an excuse for not helping people. In 1798 the English cleric Thomas Malthus published his "Essay on the Principle of Population", in which he pointed out that the food supply can be increased only arithmetically, whereas the population was bound to increase exponentially. Overpopulation was inevitable, he claimed, and the excess numbers would be curbed by disease or famine. Accordingly, the widespread poverty in early 19th-century Britain was regarded as the consequence of natural law. Yet at that time Britain's population was less than 10 million. It can support 60 million. Early 19th-century Britain was merely technically backward.

4,000
3,000
2,000
1,000
Population (millions)
AD 1000
2000

Inexorable growth

The present population of South America is around 300 million (➧ page 135). If numbers continue to increase at the present rate, in 500 years people will stand shoulder to shoulder over the entire continent. Clearly, people would die of famine or disease long before this. If this situation is to be avoided, people must make a conscious decision to have fewer children.

The governments of countries such as India and China have already decided that their population growth must be curbed: India has around 750 million people and China around one billion. But other countries have yet to make such a decision. In recent decades, for example, Tanzania has argued that it has too few people to work the land properly, and Argentina has argued that it has too few for national defence. Elsewhere, as in much of South America, there are religous objections to practicing birth control.

In addition, population control can be technically difficult. For example, contraceptive methods that are reliable and safe are also expensive, and are not readily available in the world's villages and shanty towns where population is growing most rapidly. Hence, even in India numbers are still increasing rapidly and will top 1,000 million by the end of this century.

But even if everyone in the world agreed to have only two children, and contraception was universally available, the present world population would be bound to increase by at least 50 percent, and would probably double, simply because people nowadays tend to live longer, and for the next few decades at least the birth rate is bound to exceed the death rate. The human population is set to exceed six billion by the end of this century.

...or a gradual slowing-down?

In general in the animal kingdom, the rate of population growth tends to diminish as the population grows. But with humans, the rate of growth has increased as numbers have grown.

Several long-term scenarios are possible. One is that the human population will rise to levels that simply cannot be sustained and that there will be mass famines. The overall result will be a boom-crash cycle of population growth and decline. This would spell disaster not only for the human species but for most other species on Earth as well, as we would strip the world bare in our efforts to feed ourselves. Another possibility is that efforts to curb growth will succeed, in which case the human population could level out some time in the middle of the next century, at around 7 to 10 billion. Such a population could be sustained indefinitely.

Many would argue that 7 to 10 billion is too many. The world would be crowded. Other species would be packed into parks and gardens, and wilderness would have disappeared. But if the human population is to diminish – other than through war, famine or disease – then people will have consciously to decide to have only one child. A theoretical snag is that there would inevitably be a time at the start of the decline when the number of old people in the world greatly exceeded the number of youngsters. This could be economically embarrassing as the young people tend to do most of the work and generate most of the wealth. Thus there is built-in resistance to population decline. However, one conservationist, Dr Michael Soule, has suggested that if this hurdle were overcome then the ideal number of human beings would be around 100 million – less than twice the population of present-day Britain.

▲▲ *A shanty town in Ecuador. In South America, where the birth rate continues to soar, most major cities are unable to accommodate the bulging population. For many people, home is a crude wooden or corrugated iron shack or a hut built on the edge of a refuse tip.*

▲ *A family planning poster in Nigeria, a country where in the 1960s a sudden great burst of population growth occurred following several years of high production, medical aid from developed nations and imported agricultural technology.*

Population profiles

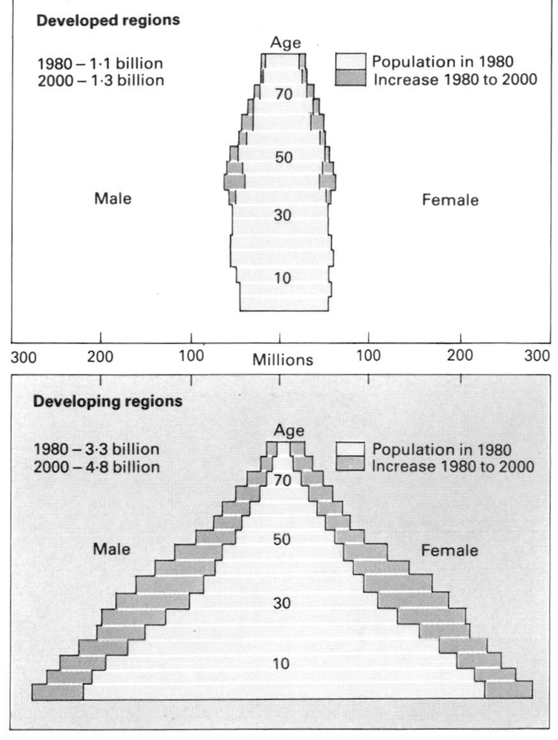

◄▼ *The very high birth rates in the Third World give a surplus of young people in those countries, whereas the stable populations and low birth rates of the developed countries have led to a possible excess of middle-aged and old people. Age-structure diagrams, or populations profiles, help scientists not only to determine the current growth status of a population but also to predict what is likely to happen in the future. A broad base as in the lower profile, indicates that a large number of persons will soon be entering the reproductive age and a "baby boom" is imminent. In contrast, a column-shaped profile, in which the prereproductive (0-14 year-old) age group and the reproductive (15-44 years) are almost equal in size, describes a population that is not expected to grow much in the years to come.*

Natural birth control

Biologists have long wondered why populations of hunter-gathering peoples do not seem to grow beyond the carrying capacity of their environment. Numbers do not generally grow to high, unsustainable levels and then crash, as is the pattern with lemmings. In practice, disease and inter-tribal war undoubtedly play their part. However, studies of the !Kung bushmen of the Kalahari desert in Southwest Africa show that the most important factor is the most obvious: the women produce children only at about 4-year intervals, which means that they rarely produce more than five children each throughout their child-bearing years. As infant mortality is inevitably high, such a number is just sufficient to keep the population size constant.

!Kung women prefer to have few children because they are semi-nomadic, and walk more than 4,000km per year. They must carry their infants, and could not carry more than one at a time. So they cannot afford to have more than one child under 4 years old. In addition, soft food suitable for infants is hard to come by in the desert.

*But !Kung women do not use contraceptives. So how are they able to restrict childbirth? There are two reasons. First, men and women often refrain from intercourse after childbirth, sometimes for 2 years or more. In addition, women remain infertile after childbrith, sometimes for up to 3 years. Part of the reason for this is nutritional. Milk production requires enormous amounts of energy – a lactating woman requires at least 2,700 kilocalories a day – and !Kung women are on a marginal diet. Menstruation is suppressed when the body is under such nutritional stress (**◆ page 59**).*

Furthermore, however, the mothers carry their infants constantly, and the infants suck at the breast whenever they feel hungry, all through the day and night. This continual sucking maintains output of the hormone prolactin from the anterior pituitary gland at a high level. This in turn suppresses production of gonadotrophic hormones from the pituitary, which suppresses ovulation.

The loss of sexual cycles during lactation is termed lactational amenorrhea. This natural mechanism of birth control breaks down when women are encouraged to feed their babies at 4- to 6-hour intervals, rather than feeding them continually. Bottle-feeding has helped to produce an enormous burst in population worldwide partly because it removes the stimulation of the breast and partly because mothers are freed from the nutritional burden of lactation. Aboriginal women under "modern" conditions, as in Australia, may nowadays produce babies every year, beginning at age 15 and remaining fertile for at least 30 years.

◄ ***A !Kung bushwoman breastfeeds her offspring. Throughout her reproductive life, a !Kung woman has only about 48 reproductive cycles. She is pregnant for about 4 years of this period, and experiences lactational amenorrhea (see above) for about 15. By contrast, women in modern societies have around 420 cycles during their reproductive life. Breastfeeding, as well as preventing conception also provides the newborn with protective antibodies – colostrum, a secretion from the breasts for the first few days after birth, is rich in these proteins – which bottle-feeding fails to do.***

The upper limit to human brain size appears to be determined by the amount of food the mother can supply while the fetus is in the womb

A 3 million-year transformation

The changes that have been brought about from *Australopithecus* to *Homo sapiens* are many and varied. We are taller and have lost most of our hair. We have developed a different distribution of body fat. The hips of our females are proportionately wider. However, the most critical physical difference is the size of the brain. Our brain is about 1400ml capacity, whereas that of *Australopithecus* was around 480ml. To some extent brain size is simply related to body size, but we are not three times as heavy as Australopithecines were. Relative to body size, their brains were about halfway between ours and a chimpanzee's. The most important point, though, is not that we have a large brain, but that we can perform remarkable things with it.

Anthropologists, in less philosophical mood than Descartes (◆ page 19), have tried to define the crucial difference between humans and the rest of the animal world in terms of what we do rather than what our brain is capable of. Many animals use tools: some finches use thorns to probe for insects and thrushes and sea otters use stones as anvils for smashing snails or shellfish. But, it has been argued, only humans make tools. However, primatologist Jane Goodall showed in the 1970s that chimpanzees remove the leaves from twigs so as to probe for termites, and screw up leaves to make sponges to gather water from crevices (◆ page 27). That is tool making, of a sort.

In fact, there is only one faculty of human beings that can truly be said to be different from that of other animals. That is the ability to speak; not simply to mimic sounds, as a parrot seems to do, but to convey a potentially infinite number of different ideas through the medium of articulated sound. The advantages of speech are immense. For instance, sound, but not light, can travel around corners. Thus animals that can communicate by sound can attract each others' attention and convey ideas even without making eye-to-eye contact. This is a great advantage in group living, and not least in group hunting (◆ page 56).

The ability to articulate sound – to produce the range of consonants and vowels that can be superimposed over changes in pitch, tone, and volume – creates a speed and versatility of communication that is without compare. Descartes was correct in pointing out that we do now think to a very large extent in words – unspoken "sounds" within our heads that give our thoughts a degree of precision that is difficult to envisage being produced in other ways.

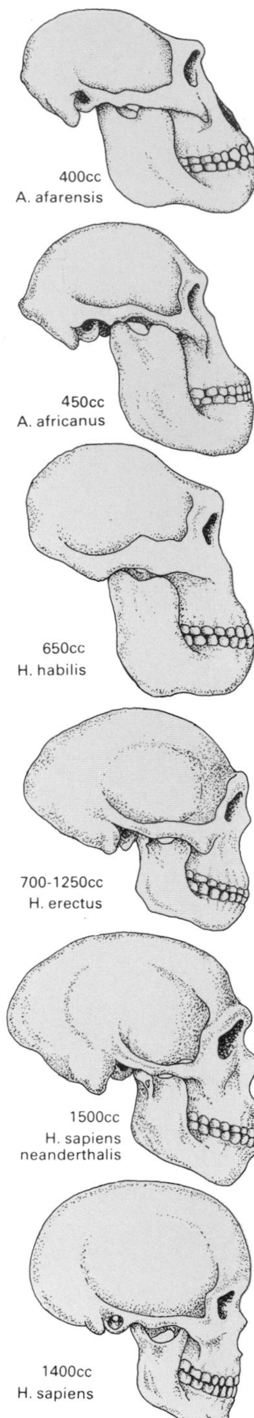

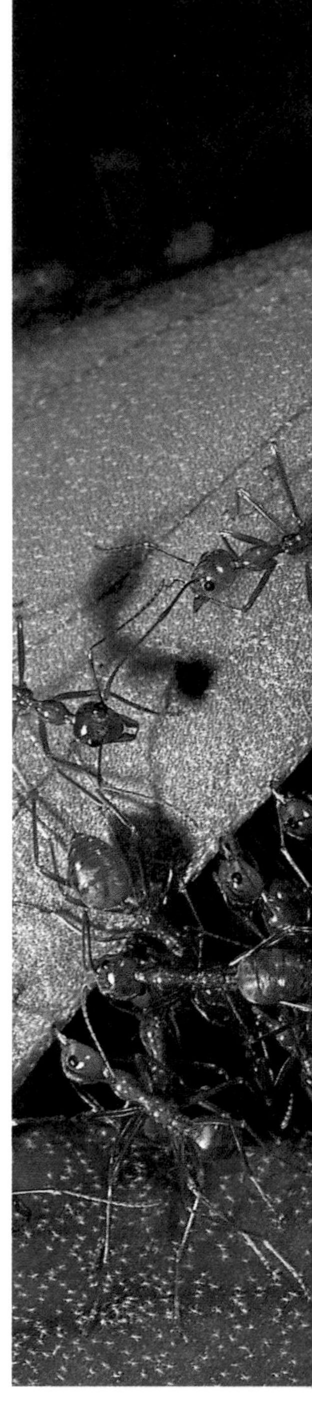

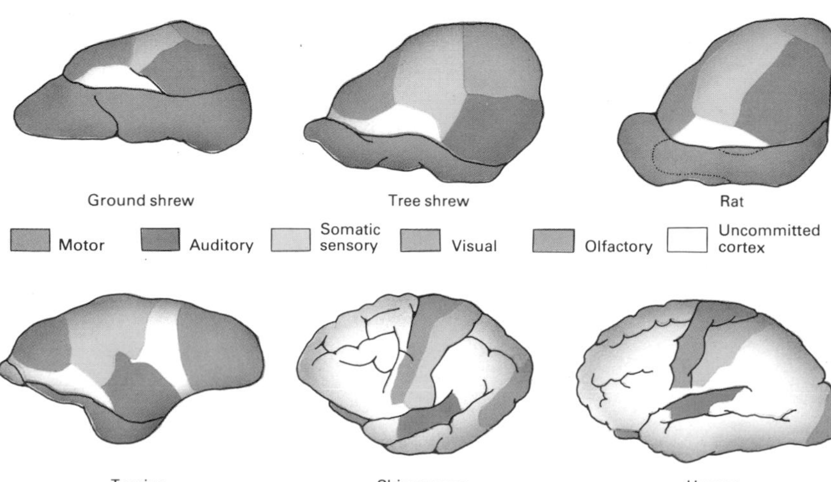

▲ Brains are not preserved as fossils but skulls are, and they show the size and shape of the brain that once occupied the cranium. The expansion of the human brain since Australopithecine times reflects increase in body size and rising intelligence. (A gorilla's brain is 500ml.)

◄ The evolutionary line that led from lower mammals to primates to humans has greatly changed the proportions of the different parts of the brain. The human brain is dominated, indeed enveloped, by the huge cerebrum – the part that does the thinking.

▲ To cooperate in bending leaves for nest-building, weaver ants must communicate precisely and efficiently. But insect communication tends to be highly stereotyped; each species can convey only a few "ideas". Human language is effectively infinite.

► Chimpanzees have been taught to use sign language. But many biologists argue that such trained chimps are not able to use the signs they have learned to express new ideas. Humans can express as many ideas through sign language as they can through speech.

The evolution of speech

No one doubts that many animals can make a wide variety of sounds and gestures that each relate to specific objects and moods, and together constitute a vocabulary (♦ page 32). What is in doubt is whether even the brightest animals, such as chimpanzees, whales and gorillas, are truly able to manipulate those sounds and gestures so as to generate new meanings – whether, in fact, they can generate true "sentences" out of their various "words".

Philip Tobias, Professor of Anatomy at the University of the Witwatersrand in South Africa and a world authority on human evolution, suggests that there are two main clues as to the timing of human speech. The first is cultural. He argues that human beings could not have developed the skills necessary to make complex stone tools unless they had been able to convey complex ideas to one another. The first creator of such complex tools was the first true human being, Homo habilis, who first appeared in Africa around 2 million years ago (♦ page 129). According to Professor Tobias's theory, these creatures must have had some rudimentary speech.

In addition, speech in modern humans is controlled by two centers in the brain – Broca's and Wernicke's – that other animals do not possess. According to Professor Tobias, the shape of the inner surface of the cranium of H. habilis shows that its brain included those areas.

Modifications to the brain are not sufficient, however. Speech also requires a highly controlled method of producing sound. American anatomist Dr Geoffrey Laitman has now shown that the structure, and more particularly the position, of the human voice box, or larynx, is unique.

All mammals except human beings that are past infancy are able to breathe and swallow at the same time. (Human infants can do this too: they breathe as they suck at the breast.) They can do this because as they swallow the back of their throat rises and completely separates the passage to the lungs (the trachea) from the passage to the gut (the esophagus). The necessary mechanism involves the larynx, which in nonhuman mammals and human infants is positioned high in the throat.

As human infants develop, however, the larynx moves to a lower position in the throat. This brings about one enormous advantage. A space is created above the larynx in the throat and this space, the pharynx, acts as a sound box and echo chamber. This gives humans enormous control over the volume and pitch of the sound they produce. This, combined with the extra articulation afforded by the complex musculature of the tongue and lips, enables humans to produce a far greater variety of controlled sounds than any other animal.

Exactly why the larynx changed its position during human evolution is not known. Perhaps it was simply because the human skull changed shape, possibly in response to changes in diet. If this were the case, then the ability to speak, which gives human beings such sophisticated means of communication and one far superior to that of other animals, would originally have been an accident.

► *The bones of whales, the largest of all mammals, left to rot in Antarctica. Because we have an advanced capacity to communicate with each other and hence to work together, we have become extremely successful predators, able to defeat animals far larger than ourselves. But perhaps we are too successful. We have already hunted many animals to extinction, though whales, mercifully, still survive.*

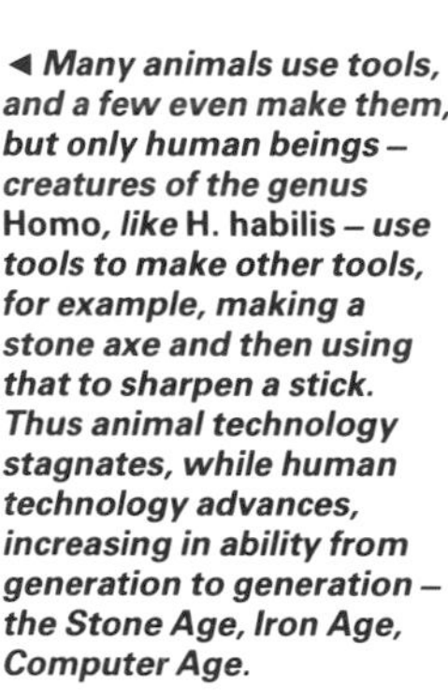

◄ *Many animals use tools, and a few even make them, but only human beings – creatures of the genus* Homo, *like* H. habilis *– use tools to make other tools, for example, making a stone axe and then using that to sharpen a stick. Thus animal technology stagnates, while human technology advances, increasing in ability from generation to generation – the Stone Age, Iron Age, Computer Age.*

***The specialist* Australopithecus robustus**
A. robustus *had enormous jaws and teeth, and apparently endeavored to live on the harsh plains of Africa by chewing its way through vast quantities of the coarse unpromising vegetation.* Robustus *first appeared more than 2 million years ago and existed for over 1 million years, so he was not an unsuccessful creature. However, evolution over the past billion years has shown that specialists are more prone to extinction than generalists, partly because they tend to be fewer in number, but mainly as they cannot easily adapt to change.*

The basic strategy of human survival – to control the environment rather than to bow before it, and to be generalist – was present in the hominid lineage from the beginning. Except for the diversionary robustus, the australopithecines (♦ page 129) undoubtedly were generalists. The abilities that humans subsequently developed were evolved in response to the pressures imposed by those strategies, and in turn reinforced those strategies. As humans became more intelligent and dextrous, the more readily they could overcome many of the forces of natural selection and the more varied were their methods of exploitation of the environment. Generalist controllers that we are, we now shape entire landscapes and dictate the form, as well as the survival, of our fellow species. But there is a logical progression in this, from the first upright, man-like ape that sought to stay comfortable and eat a varied diet.

Human ability and human survival

Different organisms have evolved different survival strategies. Some animals, such as bison, stand and freeze in the winter snows, while others, like beavers, build nests to keep out the cold (♦ page 112). Again, some organisms survive by specializing, and some by being generalists. Thus a koala bear is an accomplished digester of eucalyptus, but eats nothing else, while a rat will eat anything at all.

The survival strategy of human beings is, on the whole, active. Admittedly, Australian aborigines tend to sleep naked on the open plain, enduring the freezing temperatures of the night, and Africans may walk hatless beneath the noonday Sun. By and large, however, humans survive by manipulating the environment. Rather than endure the vagaries of weather, they build shelters, and rather than search for food they grow it. Furthermore, human beings are not specialists. In fact they are the supreme generalists. Some humans, such as Eskimos and Masai, are almost entirely carnivorous, and others, like the aboriginals of Japan, are practically vegan. Most are somewhere in between, and eat whatever is at hand.

Evolution, culture, and responsibility

Human beings have stepped beyond the normal bounds of animal intelligence and communication; and as a result of their ability to control the environment, they have bolstered their population to a level many times greater than could ever be achieved by any other animal of comparable size. They have also broken one further boundary. Natural selection made the human species what it is, just as it shaped the

form and survival strategies of all other animals. But human evolution is no longer determined by natural selection. For example, at least in organized technological societies, individuals that are physically weak do not necessarily die, and ambitious or clever individuals do not as a matter of course achieve a reproductive advantage. Thus there is no obvious reason why the total gene pool of the human species should change significantly from one generation to the next, or one century to the next (➧ page 202). Unless there is a shift in the frequency of genes, then evolution in the simple biological sense cannot take place.

There is, however, another form of evolution that runs parallel with physical, or genetic, evolution. This is the gradual change of culture. The more intelligent animals do pass on at least some of their experience from one generation to the next. Many songbirds learn their songs from their parents, or at least other adult birds (➧ page 10). But human beings, aided by articulate speech, and in the last few thousand years by writing that puts that speech in visible form, have taken this process much further. We can effectively pool the information in each of our prodigious brains with that of all other human beings. And, unlike other animals, we have access to information not only of our contemporaries but also of many preceding generations. And we can record all our knowledge for use by our offspring.

This accumulation of knowledge, ideas and technique from generation to generation is in essence Lamarckian. Lamarck's principle of evolution does not apply to the living world, but it does apply to the culture of human beings. The mode of our evolution is hence qualitatively different from that of other animals.

Human being in nature

It is often argued that people living in "primitive" societies live in harmony with nature and that modern technology has alienated us so that we now behave in extremely destructive ways.

*Yet attitudes towards other species of animal vary tremendously from society to society. In northern Europe, laws forbidding cruelty date only from the 19th century. Before that, animals were regarded simply as material objects that could be beaten, starved or tortured according to the whim of the owner. Such attitudes still prevail over much of the world today. In primitive societies lactating women may suckle puppies or piglets, but those same animals are eaten when convenient (**➧ page 170**).*

*Whether the capricious attitudes of human beings make any significant impact on their fellow creatures is a matter of power. Thus, there have been people in North America for about 40,000 years. Throughout that time they have competed to some extent with the wolf. But it was not until the 19th century that modern fire-arms and rapid transport enabled them to "defeat" the wolf. However, there is strong evidence that early human beings did wipe out huge numbers of species from North America (**➧ page 198**). The people who effected these extinctions – known as the Pleistocene overkill – were precisely those primitive people who are said to live in harmony with nature.*

See also
The Limits of Animals 25-34
Groups and Troops 47-62
The Search for Better Foods 141-152
The Shadow of Extinction 195-208
Threats to the Environment 209-224

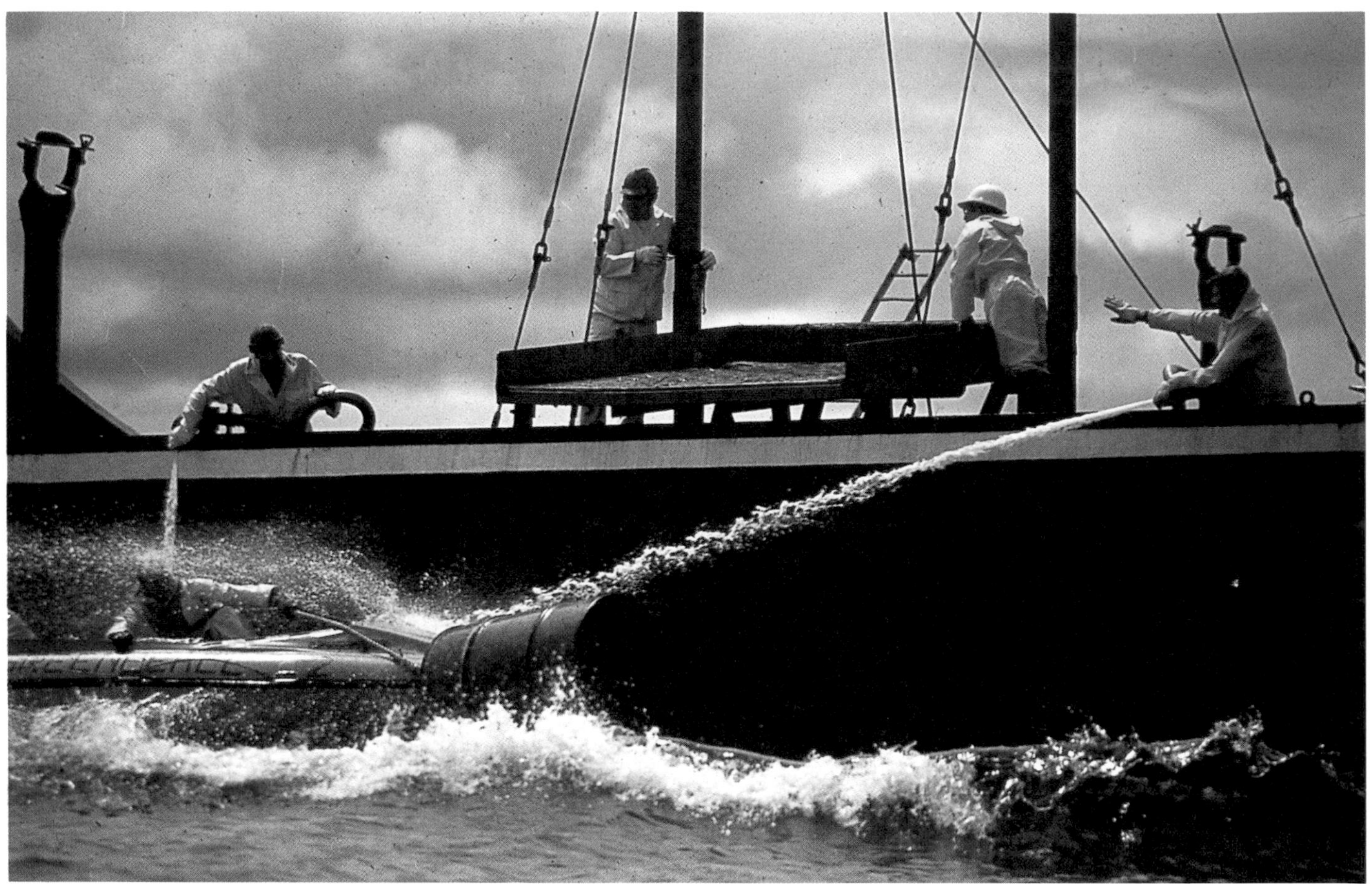

▲ ***Members of* Greenpeace *take direct action to prevent dumping of nuclear waste in the sea. After years of careless or deliberate destruction of their fellow creatures, human beings are at last waking up to the need for conservation. But governments often move too slowly, and ineffectively.* Greenpeace *has unquestionably helped to bring conservation issues to public attention.***

Shall they live or die

The continent where human beings seemed to have least impact on the native fauna – until the last century or so – is Africa. Perhaps this is because human beings evolved in Africa, and during the course of that long evolution all other animals were able to evolve in parallel, and adapt to human presence. In other continents human beings arrived as invaders, and took the native animals by surprise (▶ page 200). Invasions of one species into new territories often result in extinctions among the native fauna, even when the invader is less destructive than human beings.

Clearly, the last few hundred years have seen an enormous acceleration in the rate of extinctions of other species (▶ page 198). But this is not because technology has alienated us, but because it has made us far more powerful, and taken us in large numbers into territories that previously contained few or no human beings. However, we also now have the power to use our technological skill to preserve our fellow species. We have an innate tendency to destroy because we are big and powerful, but if we choose we could still save much of what remains (▶ page 235).

The future for humankind

There is no doubt that the life strategies of organisms are shaped by natural selection. Whether an animal chooses one mate or fights to acquire a dozen depends on Darwinian forces of evolution. Monogamy and polygamy are alternative reproductive strategies that suit different animals in different circumstances (◀ page 42). Or again, there may be selective advantage in altruism in nature. An animal may lay down its life for its own offspring, or even for its sibling's offspring, but if it did so for a non-relative it would simply become extinct (◀ page 61).

Human beings, however, have the mental ability, if they choose, to contravene biological strategies. Perhaps our ancestors were polygamous and we have inherited genes that give us a natural predilection to take more than one mate. Whatever the truth, we have the mental ability to judge whether any action is "right" or "wrong" or better or worse than another and to choose what we do. This again effectively makes us unique in nature.

Thus we find ourselves in charge of the world. The accumulation of small advantages in thought and communication gave us a physical significance and ecological dominance that is quite unprecedented. That same ability enables us to choose what we do with this enormous power. Destruction of the rest of the world would be relatively easy. To maintain the world in a state of stability, to look after our own species and yet leave room at least for a large proportion of our fellow species is a much greater challenge. In many parts of the world this challenge is being met by the creation of nature reserves and wildlife parks (▶ page 225).

The Search for Better Foods

12

Human nutrient requirements...World crop and livestock yields...Changes in agricultural practices... Future trends and a rational approach...PERSPECTIVE... Advances in plant breeding...Intensive and extensive farming...Plant stores...Manipulating plant genes... Combatting crop diseases

Worldwide, we obtain only about 5 percent of our food from the wild. The human population is now approaching five billion. Without farming, we could support only a fraction of this number.

The chief components of food are carbohydrates and fats, which supply us with most of our energy, and protein, which provides much of the body structure and the enzymes that control metabolism. We also need to consume small but significant quantities of essential fats, which are also involved in forming the structure of body cells, plus vitamins and minerals (➧ page 157).

In practice, each human being needs to obtain around 2,000 Kcal of energy per day in order to stay alive, and this means consuming around 0.5kg of carbohydrate – sugar or starch – or about 250g of fat. Nutritionists disagree about the amount of protein we need, but it is roughly in the order of 50g per day. The crops and livestock that supply the bulk of our energy and protein also supply most of the essential fats, vitamins, and minerals as well. In rich countries and in countries that specialize in raising livestock, people obtain about a third of their energy from animal products, and nearly three-quarters of their protein. In poorer countries animal products supply less than a tenth of the daily energy, and perhaps only an eighth of the protein.

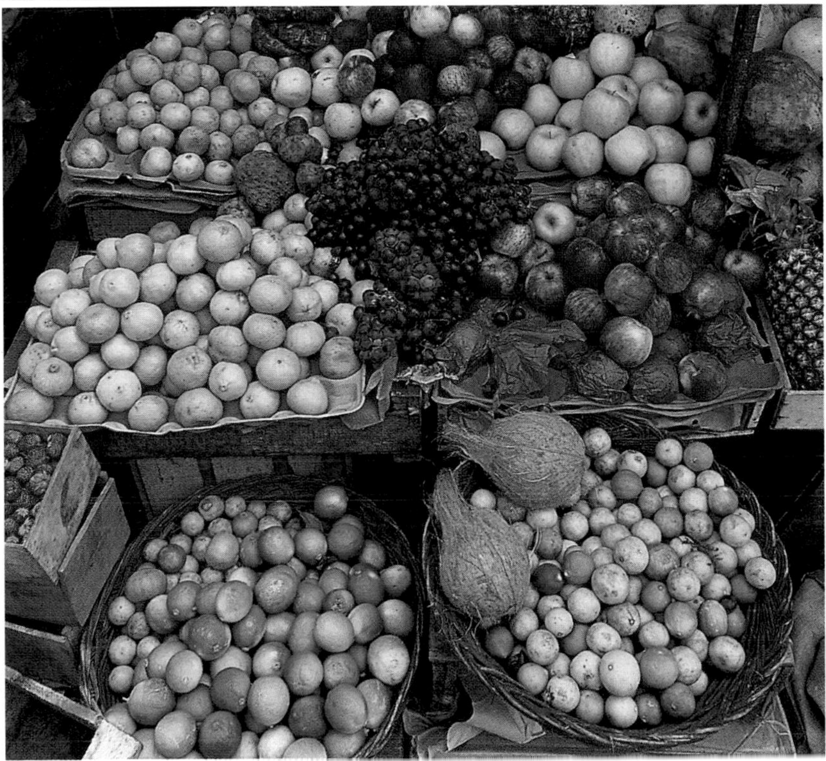

The improvement of crops

However good or well-bred a crop may be, it can always be improved. Sometimes the aim is to improve quality, for example the amount of protein in barley. Always breeders try to increase resistance to various stresses, such as drought or salty soil, and always include resistance to disease (➧ page 152). Thirdly, they seek to increase yield (➧ page 153).

Sometimes breeders can make many of the improvements they require simply by selecting and crossing the best of the crop. For example, millet is a very ancient crop grown in poor countries, which has not been well bred in the past. But some individual plants have big heads with big grains. Others are resistant to mildrew or are particularly drought resistant, and so on. Breeders at the International Crops Research Institute for the Semi-Arid Tropics (ICRISAT) in Hyderabad are producing excellent new varieties simply by selecting the best individuals from among the ancient varieties, and then crossing them to combine their good qualities.

Breeding is simplest, though, when the required genes can be found in other cultivated varieties. Then the improvement is made by crossing the two plants. If no other cultivated variety contains the appropriate genes, the breeder may have to search among wild relatives of the plant. The disadvantage here is that wild plants contain a great many genes that are not required. These are eliminated by back-crossing; the offspring are re-crossed with the original cultivated parent, so as to dilute the genetic contribution of the wild one, and this is repeated through five or six generations until only the desirable genes from the wild plant are left.

If there are no wild close relatives, then the breeder seeks genes among more distant relatives. Often the offspring of such wide crosses are sexually infertile, or partially so (➧ page 206).

When two plants can be crossed and the offspring are fully fertile, the two plants are said to belong to the same primary gene pool, or GP1. When they are more distantly related and some of the offspring are infertile, the plants are said to be of the same secondary gene pool, or GP2.

Plants such as wheat and rye which can be crossed, but whose offspring are infertile unless specially treated, are of the same tertiary gene pool, or GP3. As we have seen, breeders prefer to find the genes they require within a plant's GP1. They spread the net wider only when forced.

In addition to the conventional methods of crossing and selection, breeders now have two additional, powerful techniques for crop improvement. The first is tissue culture (➧ page 151) and the second, genetic engineering (➧ page 146). By genetic engineering, it is theoretically possible to acquire genes from any other plant or indeed from animals or bacteria.

◀ Worldwide, plants provide us with about 80 percent of our daily energy and two-thirds of our protein. Some fruits, notably the date in the Middle East, are significant sources of energy, and most fruits and vegetable provide useful amounts of minerals and vitamins. In poor countries, dark green leaves such as spinach, and yellow fruits like papaya, are vital sources of vitamin A.

The semi-desert of much of Australia is too dry for growing conventional crops, but enough plants grow to support vast herds of sheep and cattle

Why raise livestock?

At first glance, livestock farmers are hopelessly inefficient producers of both carbohydrates and protein. In Britain, for example, farmers expect to produce around 6 tonnes of wheat per hectare. (In North America, where there is more space, yields per hectare are only about half this). Six tonnes of wheat provide about 20 million kcal of energy. By contrast, the best farmers might expect to maintain two cows on one hectare, producing around 2,000 gallons (9,000kg) of milk between them. This is equivalent to about 6.5 million kcals. Yet high-grade dairy farming is the most efficient form of livestock husbandry, in terms of food per unit area.

High energy crops outstrip the best livestock even more spectacularly. Oil palm yields around 5 tonnes of oil per hectare – equivalent to 45 million kcals. Neither, at first glance, do livestock compare well as producers of protein. Meat is rich in protein, and the protein is of high quality. But animals acquire their protein by eating plants, and a great deal is wasted in the passage from plant to animal (◀ page 85). Similarly, within the plant, seeds will probably contain more protein per unit weight than leaves. But the raw material of protein – amino acids – is produced in the leaves, and the amount of protein contained in the leaves when the plant is young is greater than in the old plant. The old plant bears the protein-rich seeds but its leaves are dead.

Thus a hectare of cabbage should easily supply about 2 tonnes of protein, and a hectare of potatoes or of wheat around 0.5 tonne. But the high-yielding dairy cows would give only about 300 kgs (0.3 tonne) at most, while sheep would yield less than 100 kg per hectare. Yields of all crops and livestock are immensely variable, worldwide. But good crops grown in good conditions are bound to out-yield good livestock raised in good conditions.

But despite these figures, there remain several extremely powerful reasons for raising livestock, and for continuing to conduct research into livestock. The first is nutritional. Though animals provide relatively little protein per hectare (compared to the best crops), what they do supply is of very high quality. Second, animal products contain nutrients that are not easily obtained in adequate amounts from plants. These include some essential fats. They also include vitamins such as B12, which is not produced in plants at all, and results in pernicious anemia if deficient. Milk has many nutritional shortcomings (80 percent of the calories it provides are in the form of highly saturated fat) but is it also one of the few sources of vitamin B2 (riboflavin) and of calcium. Third, though highly concentrated food is a disadvantage in the West, where calorie and protein intakes are too high, they can be of great benefit in Third World countries, where the predominantly vegetable food is often too bulky. Children and nursing mothers in Africa may be unable to take in enough calories in a day simply because they cannot physically consume enough maize, sorghum or cassava (▶ page 154). A small amount of meat or milk can make all the difference.

There are agricultural advantages in producing livestock, too. In many areas it is virtually impossible to grow good crops at all, but cattle, sheep, or goats, can survive. Similarly, livestock can thrive on material that human beings simply do not want to eat, and would otherwise be wasted. Pigs and poultry can eat anything from household scraps to substandard corn (▶ page 170). Cattle worldwide can be raised largely on straw, a by-product of cereals that is often simply burned or composted.

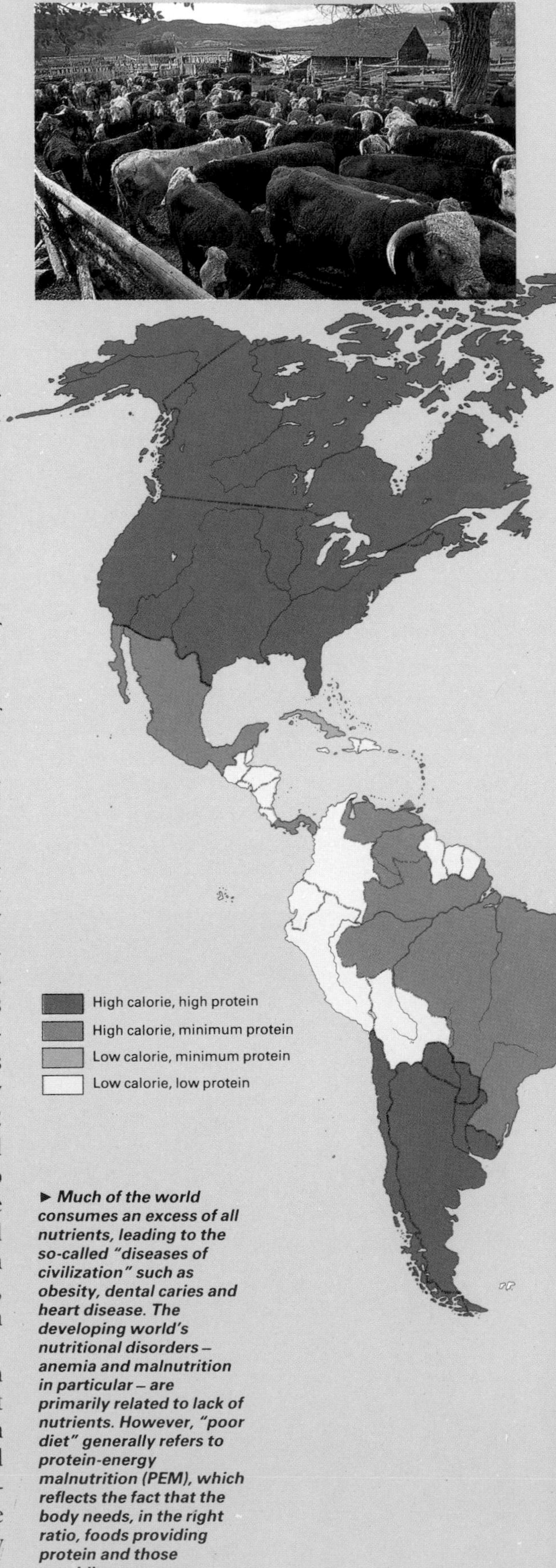

► Much of the world consumes an excess of all nutrients, leading to the so-called "diseases of civilization" such as obesity, dental caries and heart disease. The developing world's nutritional disorders – anemia and malnutrition in particular – are primarily related to lack of nutrients. However, "poor diet" generally refers to protein-energy malnutrition (PEM), which reflects the fact that the body needs, in the right ratio, foods providing protein and those providing energy.

◄ On North American ranches as many as 10,000 beef cattle at once can be reared by one man in automated feed lots. The animals are fed on rich rations of grain, alfalfa and corn as well as hay. The animals are ready for market at 11 months.

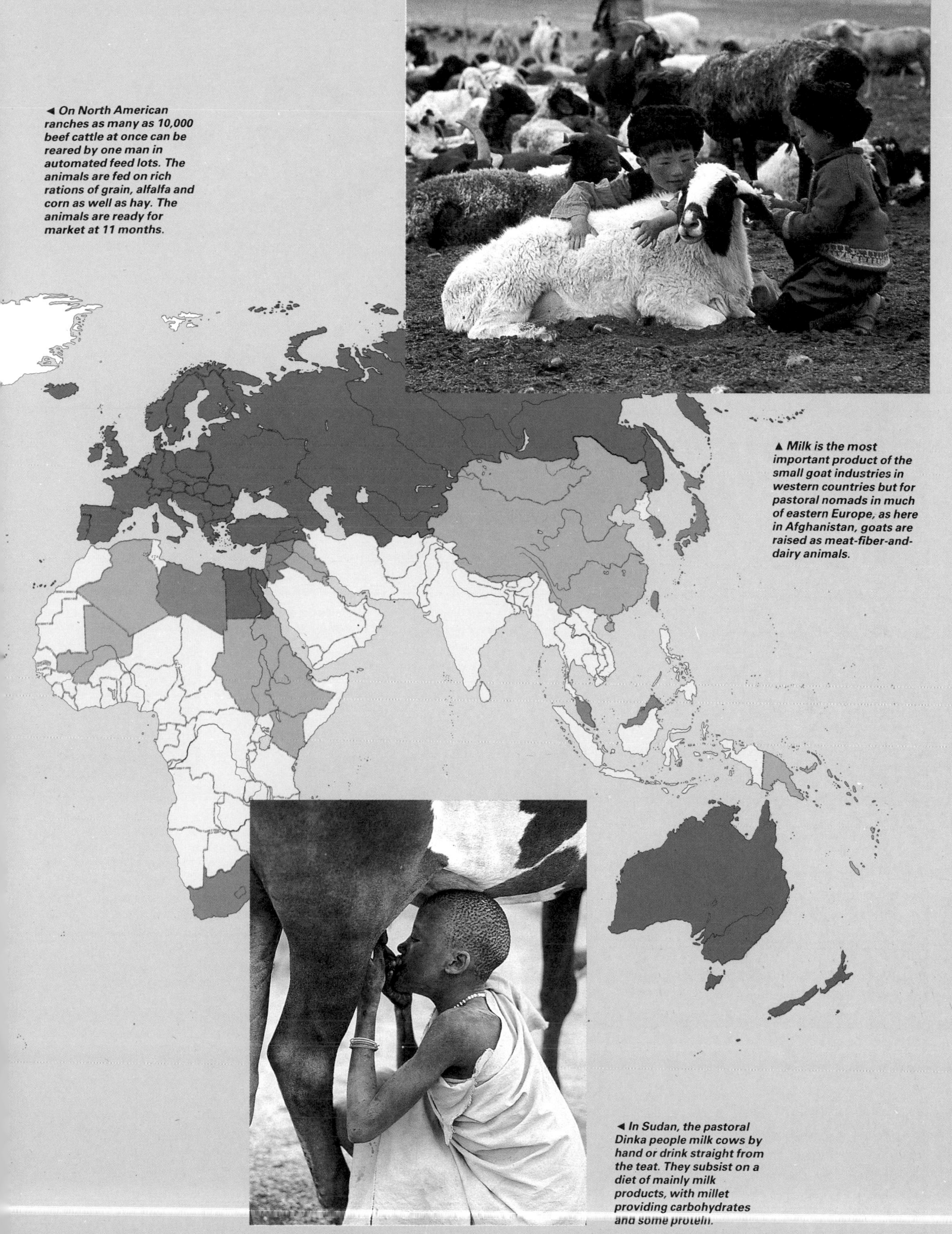

▲ Milk is the most important product of the small goat industries in western countries but for pastoral nomads in much of eastern Europe, as here in Afghanistan, goats are raised as meat-fiber-and-dairy animals.

◄ In Sudan, the pastoral Dinka people milk cows by hand or drink straight from the teat. They subsist on a diet of mainly milk products, with millet providing carbohydrates and some protein.

In western Europe wheat fields can average 13 tonnes per hectare per year, while in semi-arid regions of the world crop yields rarely exceed a tenth of this figure

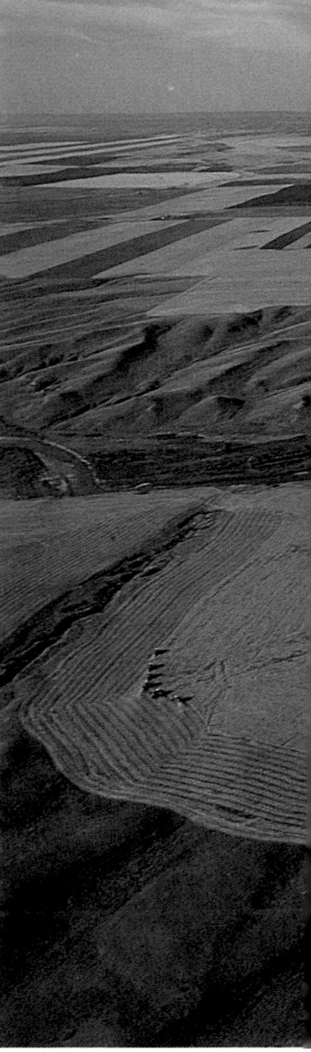

▲ **A Dervish herd of sheep and goats in North Africa. The sheep eat mainly grass and the goats prefer bushes. Between the two they make excellent use of very unpromising vegetation. Extensive farming is justified where land is cheap or not suitable for irrigation but the vegetation is sparse.**

In many countries, livestock serves other functions besides providing food. Worldwide, animals provide leather, wool and dung to fertilize the fields. In India, cow dung is an important fuel (➧ page 164). In much of the world, and notably in India, cattle, including water buffalo, are the principal form of transport. They would need to be kept even if they provided no food at all.

The real objection to animals is that in many regions people keep too many of them. Instead of feeding waste food to pigs and poultry, and simply raising cattle and sheep on highlands and grasslands, countries such as Britain and the United States give at least half of the cereals and pulses that they grow to their livestock. Thus the animals may compete with humans, as well as serving them.

Where is farming going?

Twenty years ago, it seemed that farming was destined simply to become more and more intensive, increasingly "technological". All crops, it seemed, would be high-yielding varieties, heavily fertilized, irrigated where necessary, and protected from pests, weeds and diseases by insecticides, herbicides and fungicides. Beef cattle would be fed on "feedlots", standing shoulder to shoulder and feeding continuously on silage or grain (➧ page 162). Dairy cattle would be kept in extremely large herds, up to 100 animals, and milked in mechanical milking parlors. Pigs and poultry would be kept exclusively in "factories" (➧ page 171). Third World countries, which inevitably raised crops and livestock by more traditional labor-intensive methods, would move more and more towards the highly mechanized systems of Europe and North America. Nonetheless, the frenetic dash toward intensiveness and factory methods is now widely acknowledged to have many disadvantages and drawbacks, and there are encouraging signs that the systems of the future might be far more convivial than was previously envisaged.

Two main types of farming

When farmers go all out to produce the maximum yield of crops, meat, eggs or milk from the smallest possible area, then the system is said to be intensive. When they are content with a small output per unit area, the system is extensive.

Intensive systems include modern arable farming, raising high-quality crops in greenhouses, grazing of cattle on heavily fertilized, high-yielding grass; and "factory farms" in which poultry, pigs, and sometimes cattle are confined throughout their lives and given feed that will promote maximum growth or output of eggs or milk.

Extensive systems include hill farming, in which sheep or cattle are allowed to graze the wild grasses and heathers, or the very extensive systems of the world's pastoral nomads, who may range with a wide variety of livestock over huge areas. A very good example are the Ngisonyoka people of Kenya who graze camels, cattle, sheep, goats and donkeys.

The main advantage of intensive farming is that it produces an enormous amount of food in a small area, and of extensive farming that it need not upset the balance of nature.

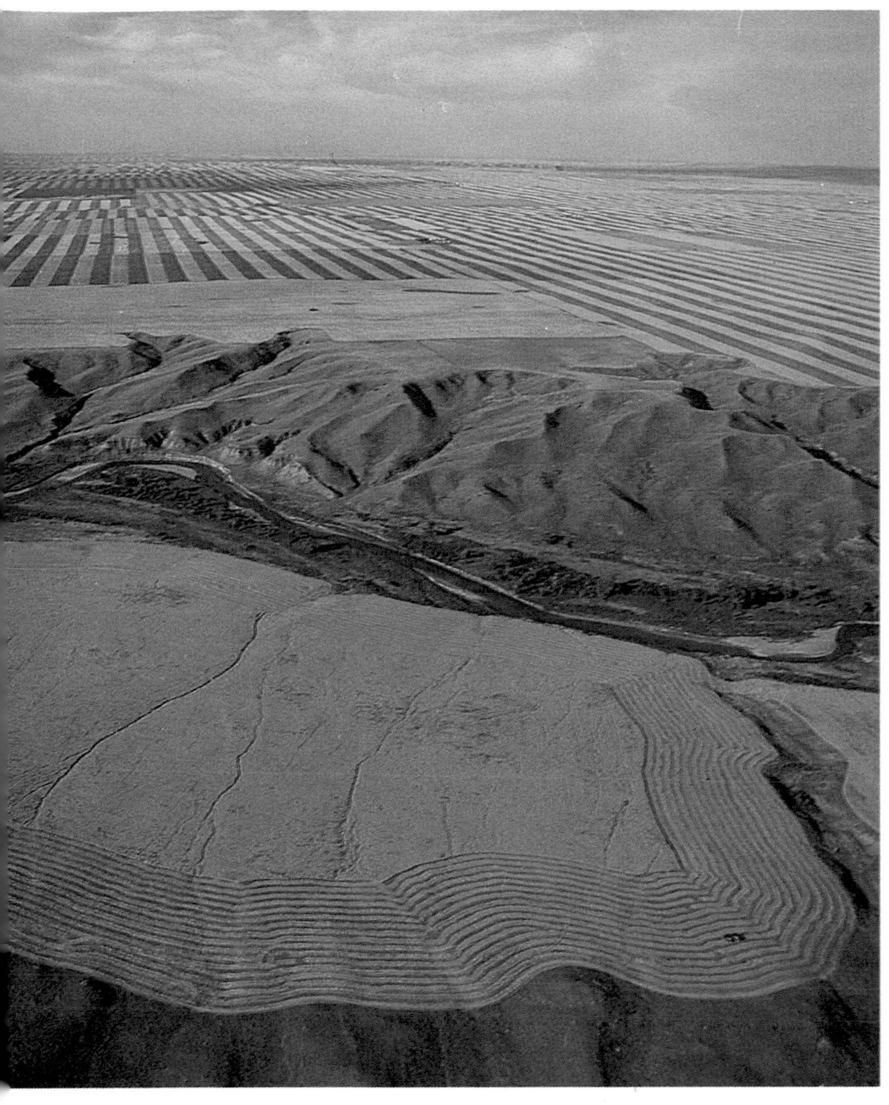

The trouble with being too intensive

The drawbacks of too much intensiveness are many and varied. For example, people in Third World countries have often rejected new strains of rice because they simply were not palatable. If they are substandard, meat and other luxuries must be added. What is gained by extra yield is lost through the need for supplements.

Second, traditional crops are generally extremely varied genetically. The modern, highly bred crops are far more uniform. The more uniform the crop, the more liable it is to be wiped out by a single pest or disease.

*The heavy use of fertilizers has two main drawbacks: first, artificial fertilizers are expensive, and, second, if such chemicals are not used carefully they can run off fields and pollute waterways (**➧ page 215**).*

Over-reliance on pesticide raises similar problems, only worse. In the Third World especially, controls on use are often non-existent or ignored and over-use of DDT in agriculture has encouraged DDT resistance among malaria-carrying mosquitoes, which are now spreading.

It is now clear, too, irrigation cannot be extended to all the world's drylands, and when it is laid on, it raises problems of its own such as increase in water-borne parasites, notably the schistosoma worm that causes bilharzia.

◀ When land is fertile and the climate is good, farming is intensive. Inputs of fertilizer and pesticide are high, but the expense is justified because yields are also high. However, there is very little room for wildlife on intensive farms and intensive livestock production is at times cruel and not viable in semi-arid regions.

▼ In much of the world, lack of water is the chief restraint on crop yields. At Hebes, on the west bank of the Nile, ox-power pumps water from underground. Research is also under way to reduce evaporation of surface water, and to control the sudden bursts of water in "flash floods", which at present is largely wasted.

ICRISAT has developed sorghum strains that yield a tonne per hectare even in the worst drought years in the Sahel when given only minimal fertilizer

Gene banks

Breeders obtain new genes for the improvement of crops mainly from primitive varieties, or landraces, of existing crops, or from wild relatives. To ensure that these sources are constantly available, and to prevent them becoming extinct, specimens of the world's principal crops are now kept in a series of gene or germplasm banks. For example, the International Rice Research Institute (IRRI) in the Philippines stores 10,000 varieties of rice and hopes to have all 12,000 by 1990. The International Potato Center (CIP) in Peru has 12,000 kinds of potato landraces. ICRISAT (◆ page 141) stores the world's main collections of sorghum, millet, chickpea and pigeon pea, and has an extensive collection of groundnut varieties and relatives. The Royal Botanic Gardens, Kew, in England keeps 5,000 species of wild plants in store.

Normally plants are kept in store in the form of seeds (◆ page 236). The general rule is to store these dry and cold. Thus, an onion seed that contains 4 percent moisture will last a thousand times as long as one with 14 percent. Below 4 percent the fats it contains begin to degrade. "Cold" means anything between +4 and −196°C. However, the seeds of some tropical trees are killed if they are dried, and so they cannot be stored. One possibility for these is to store them in the form of frozen growing tissue, or meristem, and then regenerate whole plants from it by tissue culture (◆ page 151). Or such plants might be stored as frozen embryos or seedlings. Cassava meristem has already been frozen to −186° and then regenerated.

Most seeds can be brought "back to life" after storage simply by restoring warmth and moisture. Some, however, enter a distinct physiological state of dormancy and must receive some additional special treatment before they will germinate. Some, for example, need to be exposed to light.

Soil management and appropriate technology

Many people, including most governments, still believe that high-tech alone represents the future for agriculture. But there are other lines of development.

One highly encouraging trend is in the improvement of crops that previously were neglected, notably sorghum and pigeon pea, grown in the semi-arid tropics. The emphasis is not on achieving very high yields, but on establishing steady, acceptable yields even in bad years. The method is to develop strains that are more drought-resistant and more resistant to pests and diseases (◆ page 141).

Pesticides will probably never be abandoned altogether, but increasingly emphasis is placed on integrated control. The chemicals are used sparingly, and are accurately targeted on the pests they are intended to destroy. Techniques for biological control are used alongside, and the crops themselves are bred to be as pest-resistant as possible.

Engineers are looking at ways of improving soil management without investing in expensive dams that have wide-reaching ecological effects (◆ page 220). A technique developed for fruit orchards in California in response to serious droughts in 1976 is to utilize groundwater supplies for a trickle irrigation system. Small amounts of water are metered out to plants from buried pipes. Another technique is to

build minute earth barriers, only a few centimeters high, on land that is only very gently sloping. This reduces the speed of run-off, and again can greatly increase cropping.

Native animals are also being improved. In India, surplus cows, which eat a great deal but produce nothing, are being sterilized. In Bangladesh, the oxen (castrated males) that pull carts are slowly being replaced by strong cows, which give milk as well as draught power. The carts, too, are being improved, first, by fitting them with rubber tyres, second, by putting bearings on the axles, and third, by replacing the uncomfortable traditional yokes with modern harnesses that enable the animals to get better purchase. Easy-running carts mean the cattle need less food, as they do not have to work so hard. Also, used with a good harness, they are kinder to the animals.

Finally, as oil becomes more expensive, it is likely that agriculture will be increasingly called upon to produce materials that at present are supplied by industry using oil as the raw material (➧ page 158). These new uses may compete with food production but to some extent they will complement it, utilizing surpluses. They should also supply valuable income for Third World countries that have a great deal of sunshine and reasonable rainfall, and can produce huge crop yields, but have little income.

Genetic engineering

In the early 1970s molecular biologists in the United States showed that DNA, the genetic material of all living organisms except a few viruses, could be broken up into individual genes by means of enzymes, and then transferred from one organism to another. This is genetic engineering.

In principle, genetic engineering can be used to introduce functional genes from any organism into any other, be it animal, plant or bacteria. So far, though, it has been put to commerical use only in a few species of bacteria.

Genetic engineering for the improvement of crop plants is at an early stage. There are numerous problems. First, biologists do not yet know which specific pieces of plant DNA actually form particular genes, and until this is known useful genes cannot be transferred.

Second, methods of transferring genes reliably have yet to be perfected. For example, the best way to introduce plant genes into other plants is to make use of a bacterial parasite of plants Agrobacterium tumefaciens. *This organism invades plants by squirting some of its own DNA into the host. This DNA then takes over the plant's cells. Genetic engineers endeavor to attach the required plant gene to* Agrobacterium *DNA, which is then piggy-backed into the new host. But* Agrobacterium *has limitations as a gene vector. Notably, it does not invade cereals, the most important crop plants that breeders would particularly like to improve.*

A third problem is that of gene expression. A gene may be introduced successfully into an unrelated plant, but then may fail to function. Alternatively, it could in theory function too well. A gene introduced into a cereal to improve the protein in the seeds might begin to produce that same protein in the roots, where it would simply be unproductive.

Finally, genes newly introduced into foreign plants may disrupt the normal workings of the cell.

Such problems will be overcome, however. Within a few years, crops like sorghum may be given genes from wheat to improve the quality of their seed proteins. Wheat protein is rich in gluten, which produces flour suitable for making leavened bread. Sorghum flour is low in gluten and can be used only for unleavened bread. Eventually, cereals might be given genes from leguminous plants or from bacteria to enable them to fix nitrogen from the atmosphere.

However, genetic engineering will not replace conventional plant breeding, and will not produce the kind of instant improvements that are sometimes suggested. After a foreign gene is introduced, breeding will have to continue to find the strains in which the gene functions the best, and in which its products are not degraded. The two processes must be used together.

◄◄ The Royal Botanic Gardens, Kew, both exhibits unusual plants and acts as a safeguard for endangered species or the exhaustion of the potential of cultivated varieties.

◄ In the Negev in Israel, shallow depressions are dug around plants to provide a catchment area for the minute amount of rain that does fall (a few centimeters a year).

Agricultural Efficiency

The meaning of "efficiency"

In all walks of life, "efficiency" means the amount got out divided by the amount put in. The immediate question is – "amount of what?"

In most contexts the answer is easy. For example, steam engines perform a certain amount of work when plied with a known quantity of energy in the form of coal or oil. Efficiency, here, is the amount of useful work got out, divided by the amount of fuel energy put in. But in agriculture, "amount" may mean many different things. Thus, crops require a certain area of land, plus water, fertilizer, sunlight, warmth and energy for cultivation, which may be provided by human labor or by machine. These are the chief inputs. In return, the crops may produce fibre, fuel, drugs or food. Food, in turn, may take the form of a high-energy food, such as oil or sugar, or have good balance of energy to protein, as in cereals. It may have little or low nutritional value and yet be highly flavorsome, as with spices; or the crop may serve not as food but as feed, for livestock. Because the inputs and the outputs in agriculture are both so variable, here the word "efficiency" tends to be used to mean many different things.

▲ Though organic farming is becoming fashionable and produces high-quality food, it cannot alone produce enough food for all the people in the world.

▲ Agriculture needs energy but can also produce it. China has tried to solve its energy problems by building village-scale biogas plants.

▼ The increasing use of energy for agriculture over the past 30 years partly reflects increase in output, and partly a change to the use of modern machinery.

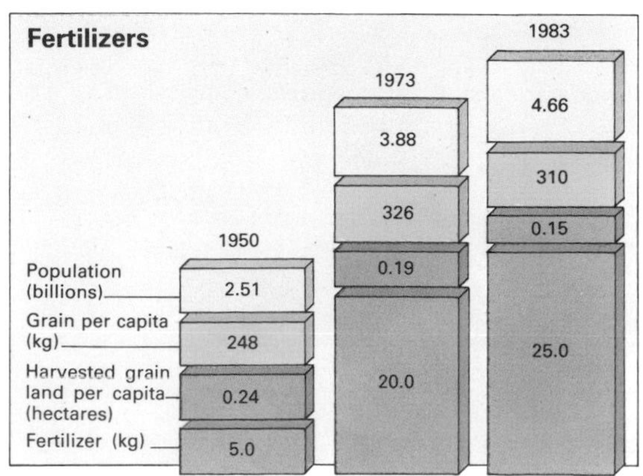

Fertilizers and nutritional value

In Britain, for example, the average yield of wheat per hectare – about 8 tonnes – is roughly twice what it is in North America. Clearly, if the input we are considering is land, then British wheat farms are about twice as efficient as North American. But British farmers need to apply twice as much nitrogen fertilizer to achieve those high yields. In addition, most British wheat is sown in autumn and harvested late the following summer. Because their winters tend to be so cold, American wheat is sown in spring. Thus the growing season for British wheat can be almost twice that of North America. Relative to the inputs of fertilizer and sunlight, then, North American wheat grows just as efficiently as British (➧ page 154).

Modern varieties of wheat, on the other hand, whether British or American, unquestionably utilize fertilizer far more efficiently than those of 50 or even 20 years ago. Modern varieties convert a high proportion of the nitrogen fertilizer they are given into grain, but when old-fashioned varieties were given extra fertilizer they simply tended to produce bigger leaves and stems – that is, more straw.

We might also measure efficiency in terms of the nutritional value of the food produced relative to the inputs. For example, a hectare of land would produce five or six times as much protein if it were used to raise wheat than it would if it were sown with grass which was then used to produce beef. Thus wheat is in general a much more "efficient" protein producer than beef (➧ page 165). However, a hectare of cabbage provides at least twice as much protein as a hectare of wheat. But cabbage is extremely watery and fibrous, so the protein it contains is extremely diluted and human beings cannot eat enough of it to satisfy their needs. So although cabbage produces protein very efficiently, the protein is not in a useful form, and cabbage, unlike wheat, is not an efficient provider of human staple food.

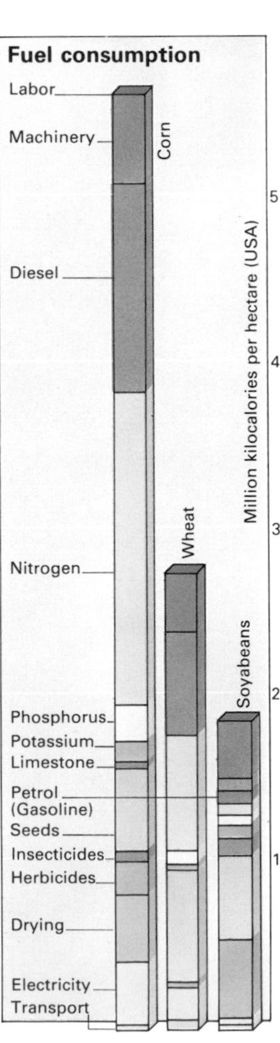

▲ Modern farming – livestock or crops – is high input, high output. Even the inputs need energy: phosphorus requires energy to mine it.

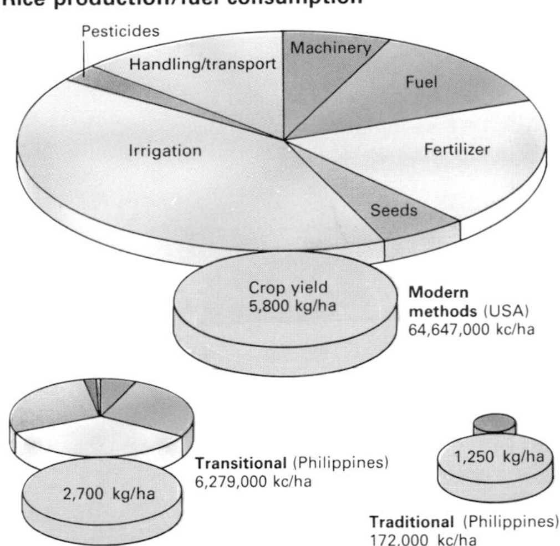

The arguments become even more convoluted when we consider the energy input into agriculture. For example, for every calorie's worth of energy a peasant farmer uses to cultivate the land, he may expect to grow 10 calories-worth of food. By contrast, modern heavy tractors and combine harvesters may use a calorie of energy in the form of oil for every calorie of food produced. In energy terms, then, modern machines have only one tenth of the efficiency of human labor. But a single person working without machines or animal power cannot cultivate more than a hectare or two. Aided by tractors and combines, he can easily plough and harvest 100 hectares. In terms of human labor required per unit of land, mechanized agriculture is far more efficient. Furthermore, tractors can run on alcohol from fermented agricultural wastes; highly mechanized agriculture is feasible without using fossil fuel.

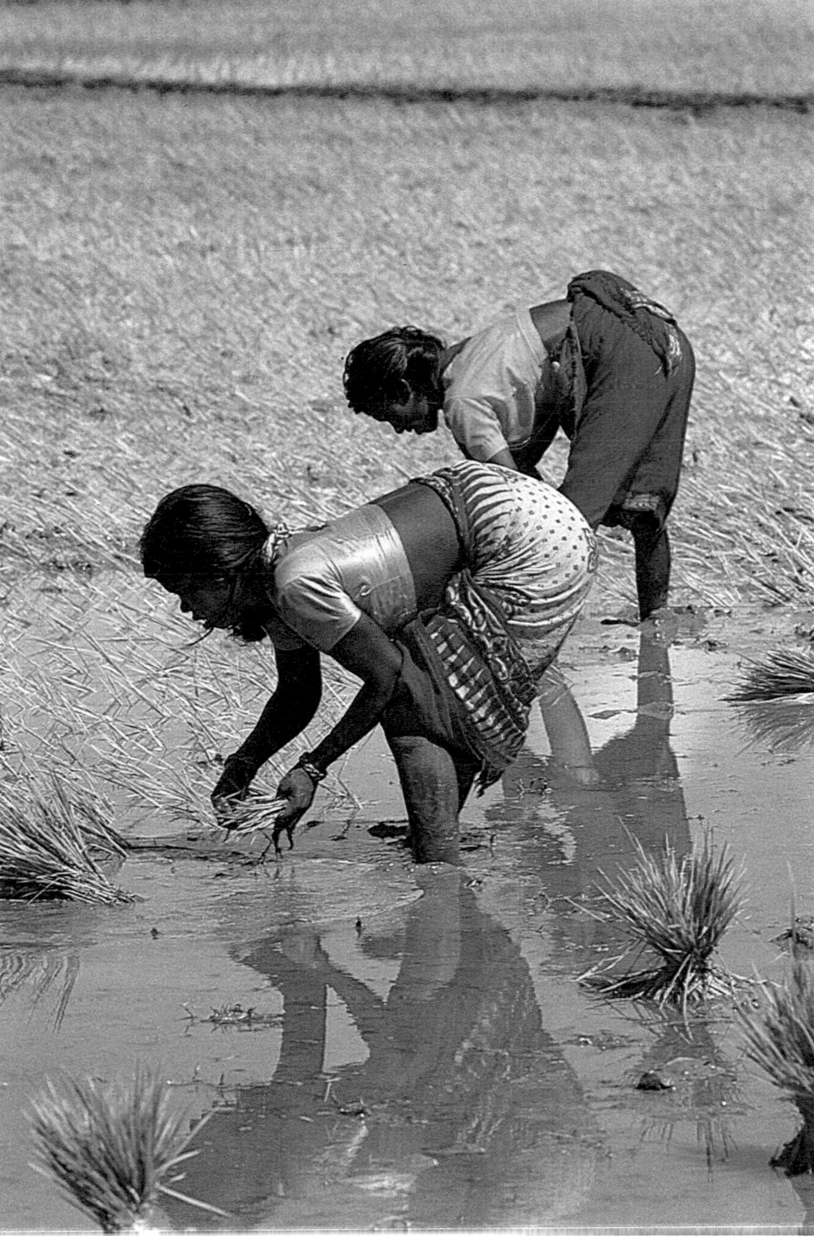

▲ Modernization of farming leads to higher yields. But the world cannot sustain the inputs needed for this, and may have to reinvoke traditional practices.

◀ Traditional farming requires little or no fuel energy but often makes intolerable demands upon human labor. A compromise is needed.

Cash value

In the end, people try to solve the problems of measuring efficiency in agriculture simply by giving every input and output a cash value. Thus if land is cheap and labor is dear, highly mechanized agriculture producing low yields (as in the North American wheat belt) may indeed be efficient in cash terms. If land is dear and labor is cheap, then it might be worth employing hundreds of people to produce the highest possible yields of high-value crops such as coffee.

However, the cash-register approach does have some severe shortcomings. For example, labor may be cheap because people volunteer to work for nothing – as on an Israeli kibbutz – or because they are enslaved – as once was commonly the case in the cotton and sugar plantations. In the latter case, high efficiency in cash terms conceals grave social injustice. Nevertheless, it may be socially desirable to employ lots of people on the land, even at low wages, to avoid unemployment.

Then again, systems of agriculture that generate only low yields may well be efficient in cash terms if land is cheap. But land has often been undervalued. For example, farmers in much of Europe may find it profitable to plough up cheap heathland to produce wheat. But if the conservation value of the land were taken into account (▶ page 226), then that heathland would not be so cheap, and the poor yields that are inevitably obtained on it would not be profitable.

In short, "efficiency" is a necessary concept in agriculture but it is used in many different ways to mean many different things. Sometimes "efficiency" is a viable and useful concept: modern wheat varieties really are more efficient plants than old-fashioned varieties. More often, though, the word merely relates to the way the figures have been juggled. It is important, when you see the word applied to agriculture, to try to work out what is implied by it.

At the Food Research Institute in Norwich, England, cells from the Cinchona *tree are being cultured to produce quinine, which is used as an antimalarial drug and as a flavoring, notably in tonic water*

Triticale

In the 19th century the Scottish botanist Stephen Wilson attempted to cross wheat *Tritium* with rye *Secale*. Wheat flour, in general, is of better quality and more versatile, but rye is extremely hardy. A hybrid of the two would, he thought, be very useful. However, his hybrids were infertile (➧ page 206).

The reason is now clear. Wheat plants either have 28 chromosomes (in the case of *durum*, or pasta wheats) or 42 (in the case of bread wheats). Rye has only 14 chromosomes. Thus a hybrid between the two has either 21 or 28 chromosomes – 14 or 21 inherited from wheat, and 7 inherited from rye. When a hybrid between the two tries to form sex cells (gametes) of its own it cannot do so because gametes cannot be formed unless the chromosomes inherited from each parent first line up and exchange some of the genetic information they contain. But if, as in this case, the chromosomes from each parent are different, then this preliminary pairing off cannot occur.

However, in 1937 Pierre Givaudon of France discovered an alkaloid chemical called colchicine, within the corm of the autumn crocus, *Colchicum autumnale*. Colchicine has the remarkable property of causing the chromosomes in a cell to double, without subsequent cell division. A cell with the usual two sets of chromosomes is said to be diploid (and with just one set, as in gametes, haploid). A diploid cell in which all the chromosomes have doubled to give four sets is said to be tetraploid. Any cell with more than two sets of chromosomes is said to be polyploid.

If colchicine is applied to the 21-chromosome hybrid of wheat and rye, then the chromosomes double to give 42. If colchicine is applied to a 28-chromosome hybrid then they all double to give cells with 56 chromosomes. In either case, each chromosome thus acquires a partner with which to pair during gamete formation. Thus, such treated hybrids are fertile. In fact, the first fertile hybrid of wheat and rye was produced in the mid-1950s by Dr F.G. O'Mara of Iowa State University. The hybrids are called *Triticale* – a name which is itself a hybrid of the scientific names of the parent plants.

The first fertile triticale plants were not terribly successful. Though they were theoretically fertile, in practice they often failed to set seed. Now the teething troubles have been overcome and triticale is now a successful heavy-yield grain crop both in the developed world (including Canada and Eastern and Southern Europe) and in the Third World (including Argentina and Mexico). The Soviet Union has the greatest area planted with triticale.

Several other crops are polyploid interspecific hybrids, including the swede, formed as a hybrid of turnip and cabbage, and the loganberry, a hybrid of blackberry and raspberry. But both those fertile hybrids occurred spontaneously. Triticale is the first man-made interspecific polyploid hybrid. It is in fact a new man-made genus – the first new grain crop to be brought into cultivation for at least 2000 years.

In agriculture today there is one major irony. High-technology is not necessarily the answer to the world's agricultural problems. But this does not mean that we can or should abandon science. For example, the development of high-yielding or pest-resistant crops requires extremely sophisticated research and development techniques (➧ page 151). However, the finished crop is easy to grow by traditional techniques. The new crops are indeed better than the old, but the old ways of life do not have to be abandoned in order to accommodate them.

Scenarios for the future

By examining present trends and projecting them into the future, we can arrive at several quite different visions of how life will turn out.

We could muddle on as we are, with the high-technology high-input agricultures of the west producing over-fed people who die young from "diseases of affluence", such as coronary heart disease, and with frequent famines in much of the Third World.

Or matters could get even worse. The present high-tech factory farming of the west could become even more intensive, and western systems continue to be imported into Third World countries in which they are often not only inappropriate and uneconomic, but also damaging to the fragile ecosystem.

▼ Triticale, a hybrid between wheat and rye, is a manmade species that has become the first new cereal to be introduced into agriculture for several thousand years. Unlike traditional cereals it would appear to have a role in both rich countries and the Third World.

◄◄ Nubian tribesmen threshing sorghum in the Sudan. Sorghum and millet are the principal cereal crops of the world's drylands. Breeders are now trying to produce strains of these plants that yield more heavily and are even more resistant to drought and heat.

▼ Plants of many species can now be grown from single cells in a test tube. This technique of tissue culture enables breeders to produce clones. The method is valuable in species that reproduce slowly, and cannot be raised from cuttings, such as coconut and oil palm.

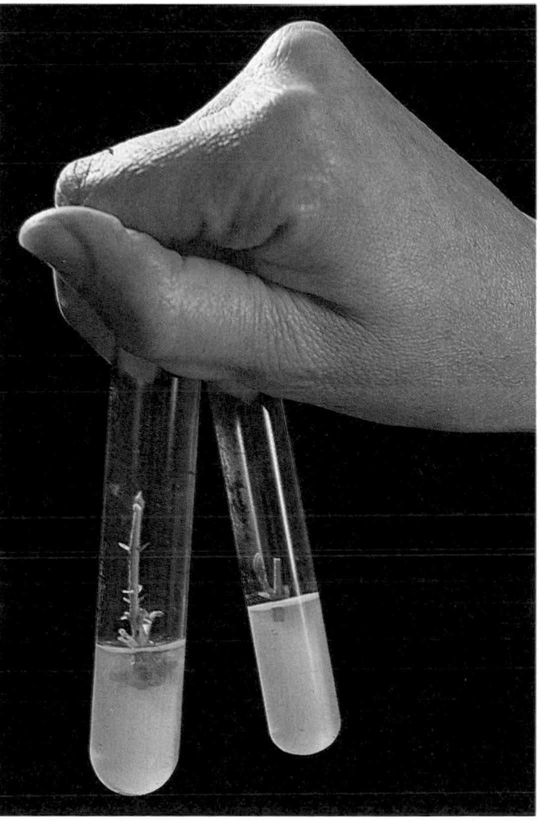

Tissue culture

In plants many cells retain the quality of totipotency – the ability to grow into a complete individual – throughout life, even many differentiated cells. Biologists can in theory take millions of individual cells from a plant and generate each of them into new individuals. This is done by putting the tissue into culture and feeding it with sugars, minerals and vitamins, and treating it with hormones. This technique of "tissue culture" is now an extremely important part of plant breeding. It has four main advantages.

First, many crops are chronically infected with virus, which is passed from generation to generation. But not every cell in the plant is necessarily infected. Breeders can develop disease-free stock by culturing disease-free cells. This has been done with potatoes in Britain, and is now being carried out on cassava in California.

Second, some crops are very difficult to breed. For example, some coconuts yield 400 nuts per season and some yield only 30. However, it takes many decades to breed new coconut varieties because, as trees, there is 10 years between generations. Furthermore, coconuts do not produce twigs or suckers that can be used as cuttings. There is no conventional way to multiply them up. So breeders at Hindustan Lever in Bombay and at Wye College in Kent are developing methods of culturing coconut tissue so that each individual cell may grow into a complete new tree.

The third use of tissue culture makes use of a very peculiar fact. The plants produced by regeneration from different cells of the same plant are not all identical to their parent. Yet all cells in the same plant contain the same genes. The differences may have several causes, including changes in the arrangement of genes brought about by culturing. This kind of alteration is called somaclonal variation. It is an embarrassment if the aim is simply to produce facsimiles of the original plant, but is also a potential source of new varieties. Scientists in Hawaii have produced somaclonal variants of sugar cane that are resistant to the highly destructive fungus disease red rot. Breeders in Cambridge have now produced somaclonal variants of Britain's leading potato, Maris Piper, with increased resitance to the scab fungus.

Fourth, for 20 years scientists have dreamed of creating hybrids between cultured cells of different species that cannot be crossed sexually. This is called somatic hybridization. The technique is to remove the walls from the cultured cells by means of enzymes, to produce naked protoplasts. If naked protoplasts from different species are brought together and treated electrically or chemically (so as to excite their outer membranes) then they may fuse, and develop a new cell wall, to form a hybrid.

The original hope was to produce very wide crosses by such means but so far somatic hybrids have been produced only between plants that are closely related, such as tobacco and petunia (both of which are related to the potato and tomato).

Finally, many drugs and other valuable agents are produced by plants. Some of these could be produced simply by culturing the individual cells that produce the valuable agents.

See also
An Ape Called Man 129-140
The World's Crops 153-162
The World's Livestock 163-172
The Shadow of Extinction 195-208
Saving Habitats 225-234

Integrated pest control

Usually a plant's mechanism of pest resistance is unknown, but sometimes it is clear. For example, the American wild potato Solanum bertholdii *is resistant to attack by aphids because its leaves are hairy, which deters the insects. This potato is now being crossed with European varieties to breed aphid-resistant domestic potatoes.*

The second way in which plants resist disease is simply by being variable. If all the plants in a crop differ genetically, then some may die in any one year, but some will survive. Primitive varieties of crops tend to be highly variable, so although they are always attacked by disease, the farmer rarely loses all his crop. Modern varieties are particularly vulnerable because they are bred to be uniform.

Third, crops may be protected by spraying with pesticide or fungicide. Modern pesticides are extremely effective but have several drawbacks. They are expensive, and therefore not available to poor farmers. Many are potentially dangerous pollutants (▶ page 216). Thus organochlorine pesticides, which include DDT and dieldrin, are known to kill birds of prey as they accumulate in the animals' fat. In addition, inaccurately applied pesticide may kill all insects in the surrounding areas. Most pesticides are undiscriminating, killing predatory insects such as ladybirds as well as the pest insects. Finally, insects and fungi eventually build up resistance to chemical controls; many insects are now unaffected by DDT.

The fourth method of pest control is by biological control. For example, ladybirds may be released to kill aphids. There are problems, however. In the 1870s, the Indian mongoose was released in the West Indies to control rats in sugar cane, and after it had killed the rats it also destroyed much of the native wildlife. On a different tack, work at Southampton University in England now shows that the insects that do most to control aphids are not ladybirds, as previously believed, but the much less conspicuous ground beetles. Ladybirds appear on the scene only after the aphid numbers have built up to substantial levels. Ground beetles take a steady toll all through the summer.

However, the watchword for pest control these days is integrated control. All methods are used together. Thus crops should be bred that have several different lines of defense against all the main pests. Where fungicides and pesticides are used, they should as far as possible be specific and kill the disease-causing organisms but not the animals that feed on them, or useful insects such as bees. For example, several insect controls are now based on pheromones – hormones that insects produce to attract other insects at mating time (▶ page 38). Pea moths can be lured to their doom by traps containing insecticide but exuding the scent of pea moth mating pheromone. (A similar technique has been successful in combatting tsetse flies, which cause disease in cattle and humans.)

Finally, scientists now monitor the populations of major pests and warn farmers just before a particular pest is liable to strike their fields. The pesticide can then be applied precisely when it is needed so that damage to resident insects (such as ground beetles) is reduced.

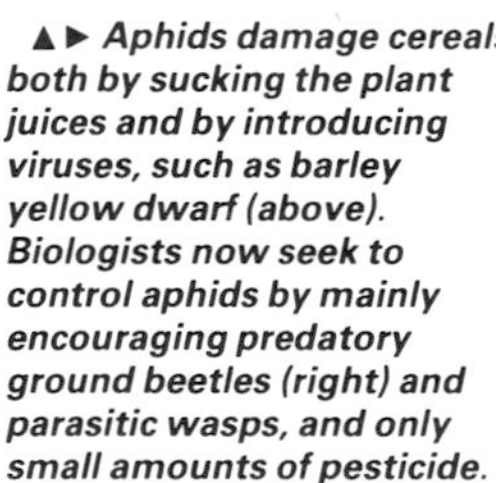

▲▶ Aphids damage cereals both by sucking the plant juices and by introducing viruses, such as barley yellow dwarf (above). Biologists now seek to control aphids by mainly encouraging predatory ground beetles (right) and parasitic wasps, and only small amounts of pesticide.

What if we adopt "rational agriculture"?

Diets worldwide would be primarily grain-based, as indeed is already the case with most traditional cuisines (◀ page 142), from India and China to the Middle East and the Mediterranean. Livestock would be kept only in those situations in which the animals did not compete with human beings for food: cattle and sheep on the grasslands, where crops often cannot be grown, and pigs and poultry used primarily as scavengers and cultivators (▶ page 170). Traditional systems of agriculture would be improved from within; the farmer being helped to produce bigger crops of higher quality with the aid of improved crops and subtle but significant changes in cultivation.

By the middle of the next century the human population may be 10 billions. People could be hungry, harassed and in a world deprived of most other species of animals, and of wild plants. Alternatively, with good research astutely applied to rational policy, it could be better than has been experienced for many years.

The World's Crops

13

Food plants and their relative nutritive values...Plant oils and sugar crops as food, fuels and feedstocks... Improving food flavor and palatability...Animal feeds... PERSPECTIVE...Biomass and harvest index...Major crop-producing areas of the world...Advances in wheat and rice production...Exploiting as yet uncultivated plants...Natural fertilizers...Plant growth factors... Utilizing organic wastes

In the course of our history we are known to have consumed around 3,000 different species of food plant, and cultivated about 150 on a significant scale. These 150 can be divided into five subgroups.

The most important of these are the staples, which supply both energy and protein in large amounts (♦ page 142). Second in importance are fruit and vegetables, which are valuable sources of flavor, serving to make the staples more palatable, and also supply some energy and protein, vitamins and minerals. Third are the crops that supply oil and sugar – highly significant sources of flavor and energy worldwide. Fourth are the spices and herbs. These supply little or no nutrient, but they are vital sources of flavor, again increasing palatability, and are useful as preservatives. In cases such as peppermint, pepper, endive and sage, herbs and spices act as aids to digestion. Finally, there are the fodder crops, sometimes grown specifically for animals, as with clover, lucerne, or fodder grasses, and sometimes as a by-product of food crops (sugar-beet tops, straw and so on).

Staple crops

These, invariably, are storage organs; specialist structures that plants have evolved as food stores for the next generation of plants. The storage organs that serve as staples include seeds – cereals, pulses, non-cereal grains such as buckwheat, and the coconut – tubers, notably the potato and cassava – and, for some Africans, the banana.

The chief of all the staples are the cereals, which comprise large seeds of the grass family, Gramineae.

Yield, biomass and harvest index

In general, it seems best to produce the highest possible yields of crops. But there are three caveats.

First, farmers may produce too much. Europe during the 1980s has harbored a "grain mountain" – millions of tonnes of unwanted wheat which eventually are generally sold very cheaply to the Soviet Union. In Third World countries, subsistence farmers often have no incentive to produce more grain than they can eat themselves because local markets cannot absorb the surplus.

The obvious answers to this drawback are, in Europe, to take land out of production and turn it over to conservation (♦ page 226) and, in the Third World, to increase wealth so apparent surpluses can be absorbed.

Second, it can be dangerous to try to achieve high yields. In the Third World farmers who attempt to grow high-yielding strains of grain by applying fertilizers sometimes go bankrupt simply because the rains fail. And when there is no water, all other inputs are wasted. In temperate countries, farmers may attempt to produce extra high yields of grass by growing varieties that continue to grow in winter. But winter-growing varieties are less hardy than those that are dormant in winter, and are killed by heavy frost.

Third, yield should not be achieved at the expense of quality. For example, varieties with a very high protein content inevitably yield less than those with low protein, because a plant needs to use more energy to produce a gram of protein than it does to produce a gram of starch. So high protein must be balanced against high yield.

In practice, there are two ways to increase yield. The first is simply to increase biomass – the total size or number of plants produced per unit area in a given time. The second is to increase harvest index – the proportion of the crop that is actually useful. Wheat yields in Britain have risen more than threefold since 1900, but modern wheat plants are actually smaller. All that is increased is the size of the heads relative to the straw.

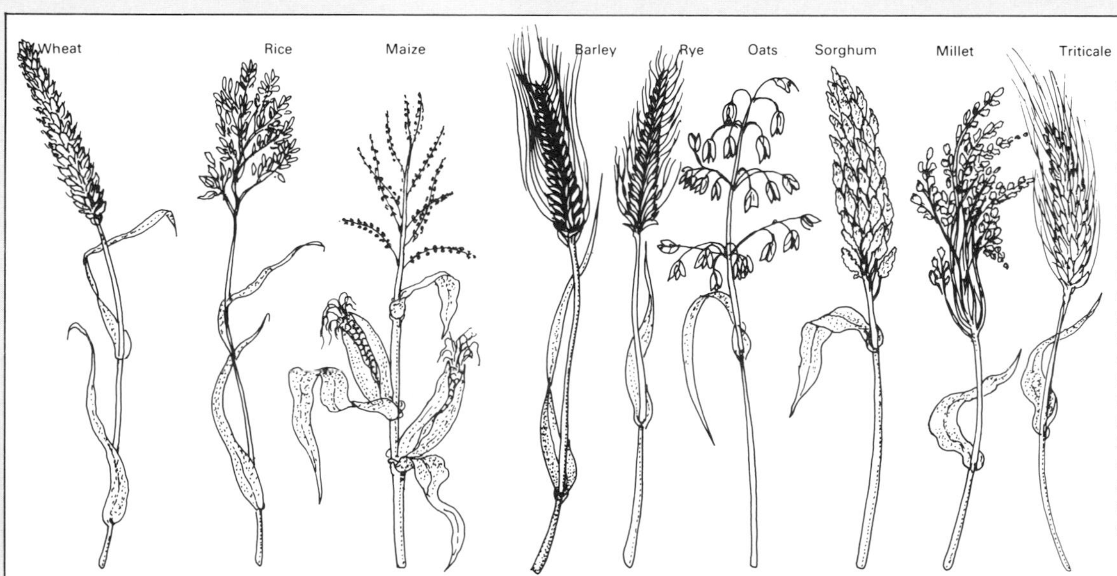

◄ There are nine basic kinds of cereal: wheat, rice, maize (in North America known as corn), barley, rye, oats, sorghum, and the various plants known as millet, plus a hybrid of wheat and rye called triticale (♦ page 150). The straw as well as the seeds, or ear, are valuable. In poor countries in particular, straw serves as fodder for animals. In rich countries, technologies are being developed to process straw into paper or building materials, to pack it into bricks for fuel, or to ferment or otherwise process its cellulose to produce alcohol for fuel or feedstock for the chemical industry.

Worldwide, the most important tuberous food plant is the potato, which is related to the tomato, tobacco and sweet pepper

Analysis of food production statistics

Among staple crops, wheat is grown in by far the greatest amounts (both total tonnage and area), with rice and maize also very high, and the rest seeming almost minor by comparison. Such figures can be deceptive, however. About half the wheat grown in Europe is fed to livestock (◀ page 142). By contrast, almost all the rice grown in the world is eaten directly by humans. In addition, the rich people of Europe and North America who eat wheat also eat a great deal of potato, vegetable and animal products besides. For most of the world's rice-eaters, rice is the principal source of energy and protein. Thus more people eat rice than wheat, and they rely more heavily upon it.

Sorghum and the millets are the chief cereal for the 750 million people who live in the semi-arid tropics – lands around the equator where water comes only from a brief annual rainy season. As with rice, people who rely upon sorghum and millets do so to a very high degree. Yet the world grows twice as much barley as millet and sorghum. But barley is used mainly for beer, whisky and animal feed. Only a few people in the southern Soviet Union and some mountain regions of Asia rely upon it as staple.

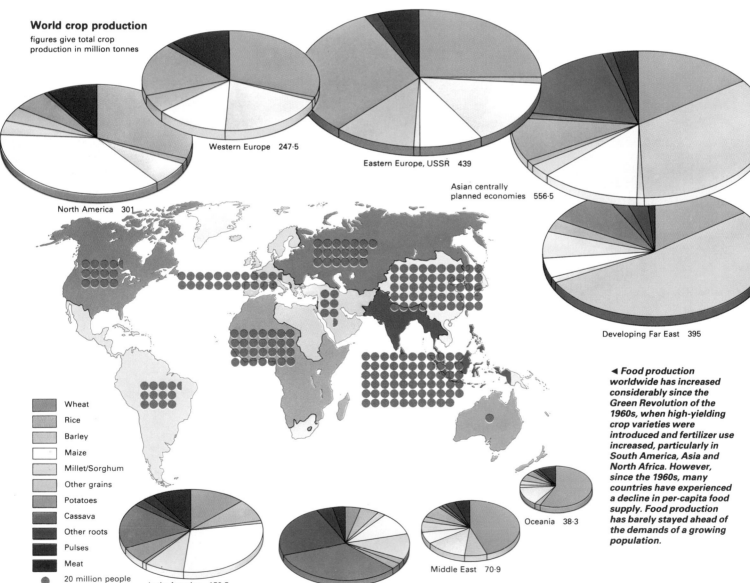

World crop production

figures give total crop production in million tonnes

North America 301

Western Europe 247·5

Eastern Europe, USSR 439

Asian centrally planned economies 556·5

Developing Far East 395

Oceania 38·3

Middle East 70·9

Africa 138

Latin America 159·5

Wheat
Rice
Barley
Maize
Millet/Sorghum
Other grains
Potatoes
Cassava
Other roots
Pulses
Meat
● 20 million people

◀ Food production worldwide has increased considerably since the Green Revolution of the 1960s, when high-yielding crop varieties were introduced and fertilizer use increased, particularly in South America, Asia and North Africa. However, since the 1960s, many countries have experienced a decline in per-capita food supply. Food production has barely stayed ahead of the demands of a growing population.

◀ Packing cassava roots in baskets ready to take to market. Also known as manioc or tapioca, cassava is the world's sixth most important food crop after wheat, rice, maize, potatoes and barley. It belongs to the Euphorbiacae, an extraordinarily varied family of plants that includes the common spurges, the rubber tree, Hevea, and the cactus-like candelabra tree of Africa. Cassava originates from South America (Brazil is still the world's largest single producer) but is grown widely also in Africa and Asia. It is a good source of energy, high-quality protein and vitamin C. The plant is a shrub growing to 2m tall. It has woody stems, sections of which are used to propagate the crop, and swollen tuberous roots.

Dwarf varieties and the Green Revolution

Farmers began to use artificial nitrogen fertilizers on a very large scale soon after World War II. However, when they applied these to cereals – wheat or rice – the results were sometimes undesirable. The plants grew tall but then lodged – they fell over because they outgrew their strength.

In the mid-1940s breeders discovered a variety of wheat in Japan known as Norin 10, which had very short stems. Dr Orville Vogel of the US Department of Agriculture introduced this plant to the United States in 1947, and from it bred a series of semi-dwarf varieties, that do not grow tall and lodge even when they are given high doses of nitrogen. Varieties containing the Norin 10 dwarfing genes are now grown throughout the world.

In the early 1950s rice breeders found similar dwarfing genes in varieties such as Fee-geo-woo-gen, in China. The International Rice Research Institute in the Philippines, founded in 1960, subsequently produced a series of semi-dwarf rices. The first famous one was IR 8. Now there are nearly 30 such varieties. These dwarf rice and wheat varieties were the foundations of the Green Revolution.

*Thus, modern temperate wheats are around 75 to 90cm tall, whereas those of the turn of the century were around 130cm. The modern dwarfs yield more heavily because a much greater proportion of nutrient is processed into grain rather than straw. The harvest index of modern wheat is around 50 percent, whereas in old-fashioned varieties it was less than 40 percent (**◀ page 153**). The story is the same for rice varieties.*

Plant physiologists calculate that the harvest index of wheat cannot be increased beyond 60 percent or the leaves will be too small to provide enough energy for the grain. When this limit is reached, the only way to effect further increases will be by increasing the biomass again. But the stems must be bred thicker, to avoid lodging. The wheat of the 22nd century could be as tall as a man, with stems like pokers and heads like foxes' tails.

Cereals for all

Cereals (◀ page 153) are excellent food. They provide energy in the form of starch and sometimes of fat, several vitamins, and contain roughly 8-12 percent protein. Until recently, nutritionists believed that human beings needed to eat large amounts of protein, and that the protein in cereals was inadequate both in amount and in quality. It is now held that the human protein requirement is not especially great and that the protein in cereals is perfectly adequate. Thus, although there are many more people in the world today than there were 20 years ago (◀ page 134), the task of feeding everybody in some ways seems easier. The chief requirement is simply to produce enough cereal. Appropriately, most of the agricultural research in the world is devoted to cereals, and crops that formerly were neglected, notably sorghum and the millets, are now receiving a considerable amount of attention (◀ page 141).

The importance of pulses as staple crops

The pulses include beans, peas, lentils and chickpeas. Again, the amounts that are grown worldwide (see map opposite) do not properly reflect their importance. Pulses are of particular value for three main reasons. First, they are members of the family Leguminosae, the chief exponents of nitrogen fixation (▶ page 158). With the aid of their root bacteria, legumes make their own fertilizer. In lands where artificial fertilizer may be prohibitively expensive, they are the prime source of fertility.

Second, the protein of legumes perfectly complements that of cereals. All proteins consist of chains of amino acids, of which there are about 20 different kinds. From a nutritional point of view, it is important that each protein contains all the various amino acids, in the right proportions. But cereal protein contains relatively small amounts of the essential amino acid lysine, which is why cereal protein eaten on its own is not as beneficial as the same weight of protein from, say, an egg. But the protein of legume seeds is particularly rich in lysine. So pulses and cereal eaten together provide an excellent protein balance. All the world's popular cuisines include cereal-pulse dishes: chapatis and dhall in India; frijoles and tortillas in Mexico; beans on toast in Britain. Finally, some pulses are particularly valuable because of their great versatility. For example, pigeon peas in India are a valuable source of firewood (▶ page 177) and the remarkable winged bean is now grown in 80 countries as a source of vegetable as well as of green or dried beans.

◀ Harvesting soya or soybean. This pulse is believed to be native to southwestern Asia, where it has been grown since ancient times. It is the most important food-legume in China, Japan and Malaysia. The seeds yield an edible oil which, when refined, is used for cooking, in salads, and for making margarine. Soya flour is rich in protein and low in carbohydrate and, with wheat flour, is now much used in bakery.

Of the 250,000 or more flowering plants known to exist, only about 3,000 have been cultivated, and of these only 150 on a large scale

Vegetables and fruits

"Vegetables" are leaves, stalks, tubers, bulbs, and roots; staples such as potatoes and beans are more commonly referred to as such. "Fruits" comprise predominantly apples, oranges, dates, cherries and their relatives, as well as bananas, gourds, and other parts of plants such as the stalks, as in rhubarb. The chief contributions they make to diet are, in the case of vegetables, protein, carbohydrates and flavor, and for fruits, vitamins and minerals.

Leafy vegetables include cabbage, spinach and lettuce. They provide more protein per hectare of land than any other crop, and far more than any livestock. But the protein they contain is dilute; the chief component of leaves is water, and they are also very fibrous. A British scientist, N.W. Pirie, of Rothamsted Experimental Station in southern England, has long advocated extracting the protein-rich pulp from leaves by machinery, leaving the fibrous residue for cattle. The theoretical snag is that if it is possible to produce a good crop of leaves suitable for pulping – of lucerne (alfalfa), for example – then it should also be possible to grow cereals or pulses that need no such processing. However, some dark-green leaves such as spinach may provide significant amounts of protein even without processing.

Leaves are not generally particularly rich in energy. However, recent research suggests that the fiber – that part of the plant that escapes digestion in the small intestine – is broken down in the large intestine, the colon, to form volatile fatty acids, or VFAs. These can be absorbed through the wall of the colon, and for people on a high-leaf, high-fiber diet may be a significant source of calories. VFAs are the chief source of energy for ruminant animals, primarily cattle and sheep. In these animals they are produced by bacteria in a compartment of the stomach, the rumen.

New crops

Many crops that are not now cultivated could be cultivated in areas where present crops languish, or have special nutritional or gastronomic possibilities that could be exploited.

For example, even the most drought resistant of accepted crops, such as millet and sorghum, are killed by complete drought, but in Central Australia a grass called Echinochloa turnerana *produces a good crop of grain even if it receives only one good watering the course of its whole life.*

Another dryland plant needing little attention is the gemsbokboontjiee of South Africa, Bauhinia esculenta. *This is a legume – it fixes its own nitrogen (▶ page 158) – and grows as a vine up to 6m in length. Its tuberous root contains 37 percent protein and 33 percent fat, and weighs up to 12kg.*

The chaya of Mexico and Honduras, Cnidoscolus *species, is a bush up to 2m high, with excellent spinach-like leaves. It dies off in extreme drought, but rapidly recovers when the rains return.*

Finally, the wax gourd, Benincasa hispida, *has melon-like fruit and is grown throughout Asia, though not elsewhere. Its fruits can be stored without refrigeration for a year.*

▼ A typical vegetable stall in a village market in Thailand. Asia is the continent where many of the world's food plants grown today first originated. For this reason, it has had less need than any other region to supplement its food sources with foreign plants, and to this day has done so to the least degree. Yet those few it has imported, such as red pepper – **Capsicum frutescens** ***– from South America, it now cultivates on a large scale. Thailand is the world's second largest exporter of capsicum fruits.***

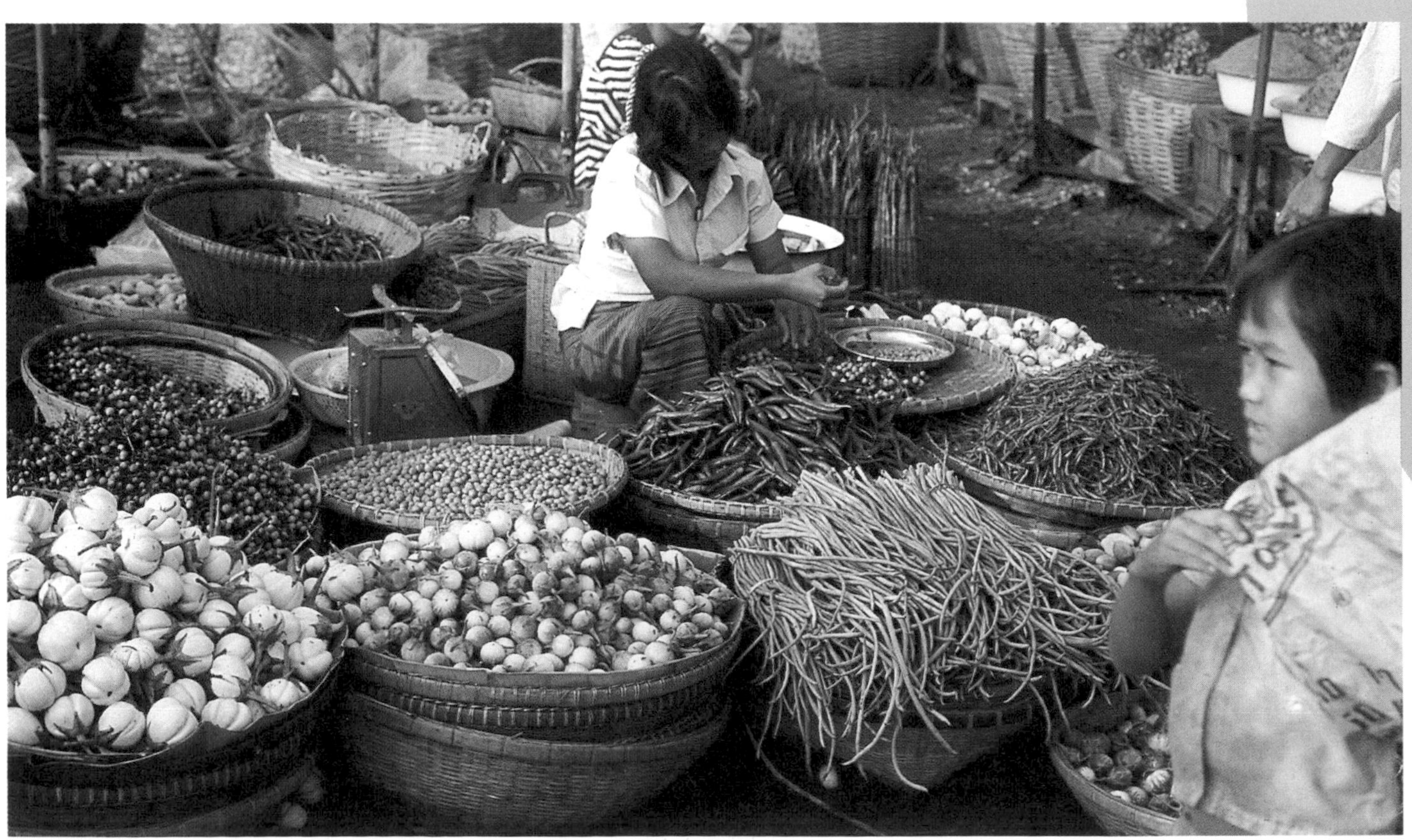

▲ *The fruits of many trees have a tough, fibrous, rather non-porous skin or a hard shell that permits long-term storage without spoilage. "Sapoti" fruit of the sapodilla tree* Achras sapota *has a brown fruit filled with a luscious pulp in the middle of which are embedded the black shiny seeds, which are not eaten. The sapodilla tree also yields another edible product, chicle gum, which is the basis for chewing-gum. The tree is widely grown throughout the American tropics but could do well in other hot, moist climates.*

▲ *Few modern crops are resistant to salt, but much of the world's land is extremely saline. The remarkable tamarugo bush from northern Chile,* Prosopis tamarugo, *above, can grow through salt several meters thick. It is an evergreen legume. There are plans to plant 40,000 hectares of tamarugo in South America. Various saltbushes of the genus* Atriplex *(➧ page 160) are able to excrete surplus salt from hairs on their leaves, and make good fodder crops. Salinity is a common problem associated with irrigation – as the water evaporates there is a buildup of salts.*

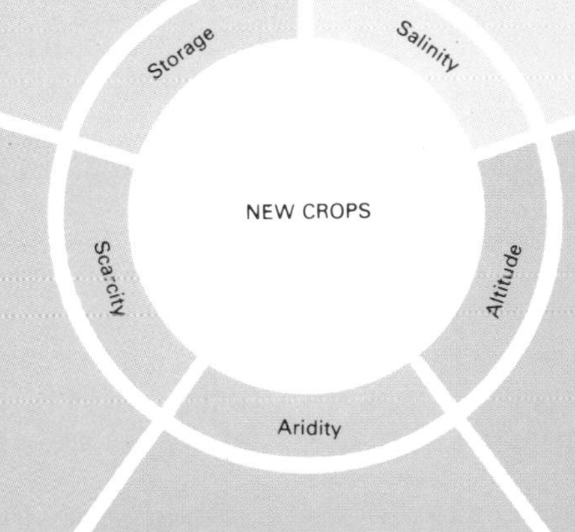

▲ *The jojoba,* Simmondsia chinensis, *grows in arid country in northern Mexico and the southwest United States. Its seeds contain a liquid "wax" which is of great value for specialist purposes in industry. The jojoba is now being widely cultivated. The only other source of comparable material is oil of the sperm whale. There is now no excuse to hunt this animal.*

► *This small-cobbed variety of maize,* Zea mays, *grows particularly well in dry highland areas of South America, where maize meal is often cooked into cakes known as tortillas.*

▲ *Two grain crops that do extremely well in mountainous areas are the amaranth,* Amaranthus *spp, above, and the quinoa,* Chenopodium quinoa. *Amaranth is related to the garden plant love-lies-bleeding and its highly nutritious seeds were a staple of the Aztec people. Quinoa is related to the wayside herb Good King Henry and was cultivated by the Incas in Peru and by Indians in the eastern United States. The Spanish conquistadores largely replaced those ancient crops with imported barley, which is less nutritious than either. However, both plants are now the subject of improvement programs and could profitably be grown in the highlands of Asia and possibly Africa, as well as South America*

The world's principal sugar crops are sugar cane, a grass grown in the tropics, and sugar beet, a temperate plant that is a relative of the garden beetroot

Free-living N-fixing organisms

Two major groups of organism possess enzymes that are able to convert atmospheric nitrogen into soluble nitrogen salts, i.e. to "fix" nitrogen. Between them, they provide 60 percent of the nitrogen made available to crops (*◆ page 88**).*

First are the Cyanobacteria. These are sometimes called blue-green algae. They live on wet surfaces, including the surface of the soil. Cyanobacteria of the genus Anabaena *are particularly important. They live within leaves of the tiny aquatic fern* Azolla, *that floats in the water of rice paddy fields. Together with manure (including human excrement), the* Anabaena-Azolla *duo provides Asian rice fields with most of their nitrogen.*

The most important nitrogen-fixing organisms worldwide are species of Eubacteria – "ordinary" bacteria. Most live free in the soil, and of these some live in the region of the roots of certain crop plants, notably millet. With these plants they have a symbiotic – mutually beneficial – relationship. The plant exudes carbohydrate from its roots, which the bacteria require as a source of carbon, and the bacteria provide the plant with nitrogen.

Rhizobium *bacteria and leguminous plants*

Plants of the Leguminosae family have evolved an especially specialized relationship with nitrogen-fixing bacteria. They include herbs such as clover, lucerne (alfalfa), vetches and medics; the big-seeded pulse crops, namely peas, beans, lentils, chickpeas and groundnuts; shrubs such as gorse and broom; and trees, that include the extremely important acacias of Africa. All these plants harbor bacteria of the genus Rhizobium *in their roots.*

When leguminous plants are given heavy doses of nitrogen fertilizer, natural bacterial nitrogen fixation is inhibited by the soluble salts that the microbes themselves produce. So plant breeders have not tried hard to increase the nitrogen-fixing abilities of commercial pulses such as soya or runner bean. However, forage crops such as clover and vetches make an enormous contribution to the fertility of pastureland. And pulses such as chickpea and pigeon pea are particularly important in poor countries whose farmers cannot afford nitrogen fertilizers.

Breeders nowadays are therefore trying to breed plants that are extremely hospitable to rhizobia, and to develop strains of rhizobia that are particularly good at fixing nitrogen and are not inhibited by additional nitrogen salts already present in the soil.

Since the 1970s, biologists have dreamed of using the techniques of genetic engineering (*◆ page 147**) to develop varieties of wheat that can fix nitrogen. One approach is to introduce the necessary genes from legumes into wheat plants so that they too are able to form root nodules to hold rhizobia.*

The second possible approach is to transfer the nitrogen-fixing genes from rhizobia into wheat, so that the wheat can fix its own nitrogen. Then the plant will not need to form a symbiotic relationship with rhizobia. The problem is that bacterial genes probably would not be properly expressed inside the plant cell's nucleus.

▲ Roots of leguminous plants such as beans are covered with nodules of nitrogen-fixing bacteria. With some crops, seeds are coated with the bacteria prior to sowing.

◀ A sugar factory in Barbados, where sugar cane is processed into refined sugar, paper products (from the fiber residues) and molasses (the uncrystallized syrup), which are used for animal feeds.

Sugar and oil

Some crops are grown either for the fat-rich oils in their seeds or fruits, or for the sugar in their stems or roots. They are excellent sources of large amounts of energy. But do we really need "pure" energy foods? Westerners eat too many calories and in addition obtain about 40 percent of their energy from fat (◆ page 142). Such a high fat intake is believed to contribute to coronary heart disease (heart attack) and to various common forms of cancer, including cancer of the bowel and of the breast. Their need for more fat in the form of plant oil seems equivocal.

However, plant oils tend to be less saturated than animal fats, and so preferable. Already, lard and dripping have been largely replaced with sunflower, corn and safflower oil.

On the other hand, about a billion of the world's people are undernourished, or dangerously close to malnutrition, and the chief reason for this is lack of calories. Superimposed on this, many people in Asia and Africa now derive less than 10 percent of their calories from fat, and could double their intake without any significant risk of heart disease or fat-related cancers. For many Indians and people of the Sahel, fat is a luxury. A spoonful of peanut oil or palm oil will turn a bowl of cereal and pulse into a pleasant dish – and also double the calorie content. Sugar, too, can usefully raise Third World calorie intake and increase palatability.

Both oil and sugar crops can be used for many things besides food. They can serve as industrial feedstocks. Oil can be burnt as fuel (and can indeed be used to run diesel engines) while sugar is easily fermented into alcohol, also a prime fuel. Brazil in particular, where massive harvests can be produced all year round, is growing millions of hectares of sugar cane for precisely this purpose. As time goes on and fossil fuel supplies diminish, such crops' industrial value may vie with their food value.

▼ **Spraying crops with fertilizer, pesticides or fungicides using light aircraft is fast and energy-efficient, but can introduce environmental hazards if the chemicals are non-selective and are spread by the wind beyond agricultural areas. In Third World countries, with a large cheap labor force, treating crops with artificial fertilizers and some other chemicals is still done largely by hand.**

Organic farming

Farmers utilizing "organic" methods of cultivation feed their crops exclusively on natural fertilizers in the form of animal or human excrement or compost (➧ page 162), or by nitrogen supplied by biological fixation by clovers and other legumes. In addition, they do not apply chemical pesticides (➧ page 152).

Organic methods have advantages and disadvantages. Advantages include the fact that some organically grown produce is of extremely high quality, with excellent flavor, and does not contain residues of chemicals normally applied to crops. As pesticides and artificial fertilizers are not applied, there are toxic run-offs from fields to cause water pollution (➧ page 214). Third, organic material, by definition, is high in carbon-based materials, notably cellulose, and this tends to increase the texture of the soil, producing a tilth that provides a good balance of oxygen and water. Soils low in organic material tend to be too free-draining, like sand, or too close-packed and cold, like clay.

However, some organically grown plants do suffer badly from disease, and toxins from potentially edible fungi are liable to be far more dangerous than those from fungicides. In addition, flavor is to a large extent a quality of the variety, rather than the method of cultivation, and organic farmers often concentrate on well-flavored rather than on high-yielding varieties.

A larger disadvantage is that yields tend to be smaller than if chemical pesticides or fertilizers are applied. The fertility of organically fed fields can be very high, but only if many tonnes of manure or compost are applied. This is expensive, however. In addition, one small field may absorb organic material from a great many animals or human habitations all around. Thus, organic methods cannot be applied universally because there is not enough organic material to go round.

Finally, organic farming in practice can produce its own problems of disease and toxicity. Thus, where human excrement is used as fertilizer, as in much of Southeast Asia, people suffer greatly from water-borne parasites, notably hookworm. Farm manure or sewage may contain toxic heavy metals. For example, pig manure may be rich in copper, which is used in pigs as a growth promoter, and sewage tends to be polluted by industrial effluent, which may have a high metal component.

Thus organic farming has a place, but cannot provide the answer to everyone's food problems. To date its most effective use has been in China, where "ecological agriculture" has ensured that the world's largest population now experiences little or no malnutrition or starvation. In the West, it is still more of a fashion than an established practice.

Some specialist plants have a need for elements such as nickel, sodium, silicon and aluminum that are toxic to many plants and almost all animals

The principal oil crops of the tropics are oil palm, coconut and groundnut or peanut. The chief problem with the palms is to provide large numbers of uniform, high-yielding plants. The plants, being trees, take a long time to mature, so conventional plant breeding as carried out on cereals is far too slow. However, unlike fruit trees such as apples and pears, they produce no sideshoots, so the grower cannot simply multiply the best, or elite, trees by cuttings. These problems are now being overcome by tissue culture (◀ page 151).

In more temperate countries the principal oil crops for human food are soya (a pulse), maize (a cereal), sunflower and safflower (of the daisy family, Compositae) and olive (related to the ash and privet). Olive is particularly valued for its gastronomic qualities. Sunflower and safflower are the most unsaturated of oils, now finding their way into specialist margarines and cooking oils. Finally, two plants related to the cabbage are of growing importance worldwide as oil seeds. These are mustard and rape.

Herbs and spices

Every country produces its own range of herbs and spices. Some may aid digestion, perhaps because they are appetizing and stimulate the flow of digestive juices. Several apparently have antibiotic qualities, and are useful in storage. The ability to provide the kind of aromatic chemicals which give herb and spices their flavor tends to run in plant families. For example, the tropical family Zingiberaceae provides ginger, turmeric and cardomoms, and the Umbelliferae – carrot family – gives us coriander, fennel and angelica. In addition, families that produce spices may also produce other chemicals of interest. The Leguminosae, the pulse family, for example, provides fenugreek and liquorice, plus the dye indigo and a range of novel insecticides.

Plants for fodder

Plants grown for animal feed range from the wild lichens of Northern Europe, which provide the feed for the Lapplanders' reindeer (◀ page 80), the vast areas of semi-wild but heavily grazed "permanent pasture", that accounts for a quarter of the world's land surface, to the highly bred, heavily fertilized, tightly managed lush green pastures of parts of Europe and North America.

The question of what should be done with or for fodder plants is not so easy as with food crops. In much of Africa and India the permanent pasture is wild country, with a wide variety of native plants and attendant animals. The people who graze cattle or goats are often nomads, and are effectively part of the ecosystem. Often their activities lead to overgrazing, which in turn encourages the spread of desert. But should the nomads give up their way of life and settle on permanent ranches? This is an insult to ancient ways of life, and it may also be counterproductive; nomads, in general, move around because the areas they inhabit are too dry to sustain enough vegetation for permanent settlement.

Should the grazing lands themselves be improved, perhaps by planting more productive species and/or adding fertilizer? This raises the problem of expense – permanent pasturelands in the tropics tend to be vast and the income they provide is low – and of land ownership. Solutions involve land reforms so that people who use the area pay their way and are responsible for its maintenance, and research to discover exactly which plants can grow in arid regions and what are their optimum nutritional requirements.

◀ The Australian spongy saltbush Atriplex spongiosa is a potential feed crop. Such wild plant species may prove useful as fodder for livestock reared in difficult conditions.

► High-cost food production is justified only by high-value crops. Expensive polythene tunnels at Barcelona, Spain, provide out-of-season strawberries.

Saltbushes and clover as fodder

Around the world, groups of biologists are searching for new fodder plants or, rather, for ancient ones that have been neglected, to plant in difficult areas. For example, a US National Academy of Sciences special committee advocates various saltbushes of the genus Atriplex, *which grow well and produce good fodder even when the soil is highly saline, as it tends to be in many dry regions. These saltbushes come from Australia, but might be grown with advantage in India, Africa and South America. They also recommend the remarkable Tamarugo tree, a legume from Chile, which will grow through salt a meter thick and still produce excellent forage. Scientists at the International Livestock Center for Africa at Addis Ababa have recently shown that several Ethiopian clovers will give spectacular yields if only they are given just a little extra phosphorus.*

◄ ***In the desert of the Somali Republic nomads follow herds of animals to wherever there is grazing. Such systems can be very efficient, making use of seasonal herbage.***

Fertility and limiting factors

The rate at which a plant will grow depends on three factors: what species it is, the prevailing physical conditions – temperature and light (♦ page 97) – and the availability of the various materials it needs for growth. When plants are raised in greenhouses, every condition of growth can be controlled. Most food plants, however, must be grown outdoors. The farmer can control only the supply of essential materials.

Plants use growth materials either to form cells and tissues or for chemical reactions, when they are known as micronutrients or trace elements.

Water is needed in greatest amounts. It forms much of the physical structure of the plant and is split by photosynthesis to provide hydrogen – an essential element in all organic materials. Water is mainly obtained from the soil through roots, though some desert plants and forest plants may capture it from the air.

Carbon is the principal element in all organic compounds. Plants obtain it through their leaves in the form of carbon dioxide gas (♦ page 88).

Nitrogen is a key component of some organic compounds, including proteins and nucleic acids. Plants absorb nitrogen from the soil in the form of soluble inorganic radicals, chiefly nitrates. They can also absorb nitrogen through the leaves.

Phosphorus and potassium are needed in relatively large amounts, though usually far less than nitrogen. Compounds containing the three elements nitrogen (N), phosphorus (P) and potassium (K) are the principal components of artificial fertilizers.

Plants also require small but significant amounts of sulfur, calcium and magnesium, and minute quantities of iron, manganese, zinc, copper, boron, chlorine and molybdenum. Plants that fix nitrogen (♦ page 159) need cobalt.

If any essential component is missing, growth ceases. When this happens, the relevant component is called the limiting factor. In high latitudes, in winter, temperature is limiting. In greenhouses, carbon may be limiting and growers may enrich the air with extra carbon dioxide. However, when the sun is shining, providing light and warmth, and plants are adequately watered, the limiting factor may be lack of nutrients in the soil. Sometimes the nutrients may simply be unavailable. In acid soils, for example, many plants are unable to obtain the nitrogen that may be present. Liming increases fertility by reducing acidity, which increases the release of the nutrient from the soil.

More often, though, the nutrient supply is deficient because there simply is not enough in the ground. Most soils in wild areas contain very little nitrogen, and the productivity of these regions is low by agricultural standards even though the vegetation may at times appear luxuriant.

See also
Interactions 81-96
An Ape Called Man 129-140
The Search for Better Foods 141-152
The World's Livestock 163-172
Threats to the Environment 209-224
Saving Habitats 225-234

On-farm fermentation – compost and silage

When organic material is piled up, it rots under the action of fungi and bacteria. This may happen in one of two ways: aerobically, if air is allowed to circulate, or anaerobically, if air is excluded.

Anaerobic fermentation is slow and does not generate heat. Breakdown of organic materials is only partial, so that many highly complex organic products are produced, many of which are extremely smelly. If well controlled, however, it effectively is a form of pickling. Organic acids are produced, such as butyric acid, which soon arrests the action of bacteria and acts as a preservative. Thus farmers employ anaerobic fermentation when conserving grass, or other vegetation, in the form of silage for food for cattle or sheep.

Compost requires aerobic fermentation. The heap is made open, incorporating strong material such as straw which will maintain an open structure. Aerobic bacteria invade and produce a rapid and thorough breakdown of material, reducing many nitrogen-containing compounds to ammonia. Aerobic decomposition creates enormous temperatures, which destroy most parasites if the heap is well made.

Compost heaps must be big enough to retain heat, but not so large that they are crushed by their own weight and become anaerobic. They must be damp, but not so wet that they remain cold. Above all, a good balance must be struck between materials high in carbon and low in nitrogen, such as straw, and those high in nitrogen but low in carbon, such as animal excrement. This is because the decomposing bacteria need both nitrogen and carbon to construct their own cell material and need them in the right ratio.

Compost should have particular relevance to Third World countries, where crop residues are often allowed merely to become pollutants (due to anaerobic fermentation) but where soil fertility is often low.

▼ In western countries grass is often grown as an intensive crop, heavily fertilized and high yielding. Often it is then served to livestock in preserved form as silage.

► Old-fashioned meadows can be maintained only by grazing animals or frequent mowing. They are not highly productive but are a valuable source of wild varieties of plants.

Developments in rich countries

Farmers in Britain in particular increasingly regard grass not as permanent pasture but as a short-term crop, to be cultivated as part of a rotation. Such crops, grown for one, two or three years are known as leys. In general, the farmers begin with a highly bred variety of rye grass, which is then very heavily fertilized to produce the maximum yields of the most nutritious foliage. Only with further developments in breeding or genetic engineering, to increase the efficiency of photosynthesis, can the yields of present grass varieties be increased.

Yet fields of just a single species are quite different from the traditional meadows, which were a form of permanent pasture, with a general level of fertility and a wide variety of species. People in Europe do not need to consume large amounts of milk or meat, but they do increasingly need leisure. There is much to be said for maintaining the old-style meadows, which are not particularly productive but are of great conservational value (➧ page 227).

The World's Livestock

14

The wide variety of animals that we eat...Breeds of cattle and their uses...The economics of raising livestock ...The origins of present-day domestic animals...Sheep, goats and poultry as inexpensive alternatives to cattle ...PERSPECTIVE...Possibilities in cattle breeding... Improving sheep-farming as a commercial concern... The advantages and disadvantages of free-range and factory farming...Domesticating various wild animals

Animals raised for food are of two main kinds: the herbivores and the omnivores. The herbivores live virtually entirely on vegetation. They include cattle, sheep, goats, deer, horses and rabbits. They are often in practice fed on high-grade plant material, including cereals (mainly wheat, barley and maize) and pulses such as soya and field beans (relatives of broad beans) (◆ page 154). But their great merit is their ability to thrive on low-grade fodder, which comprises mainly grass in all its manifestations. This ability depends upon the bacteria that live in their guts. These break down the fiber – cellulose – in the plant material to form volatile fatty acids, which the animal then utilizes as the chief source of energy.

The chief of the omnivorous livestock are pigs and poultry – primarily chickens and ducks. These have digestive systems that in basic design are comparable to humans'. They cannot subsist on cellulose, and indeed cannot survive on food that could not, in theory, sustain a human. In practice, in modern intensive systems, they are fed primarily on cereal grain, augmented by pulses and perhaps by fish meal. However, their theoretical advantage is their low standard of gastronomy. They can thrive on waste food and are happy to spend hours scratching for grain, worms, roots, or whatever is available.

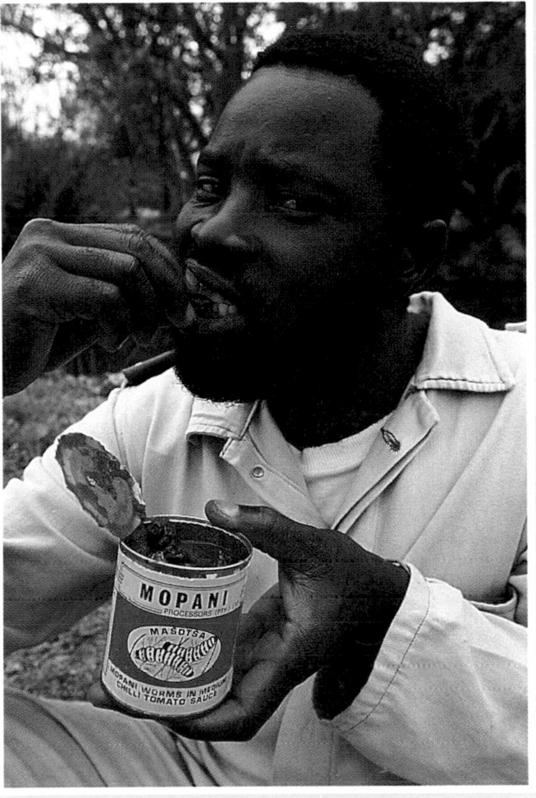

▲ In South Africa, "Mopani" – larvae of a species of emperor moth – are one of scores insects that are eaten, often with a staple food. They are rich in minerals and vitamins.

▼ Yanomani Indians in the Amazon skin a squirrel to eat. Westerners eat only a few forms of livestock but most animals are good sources of protein and energy (in the form of fat).

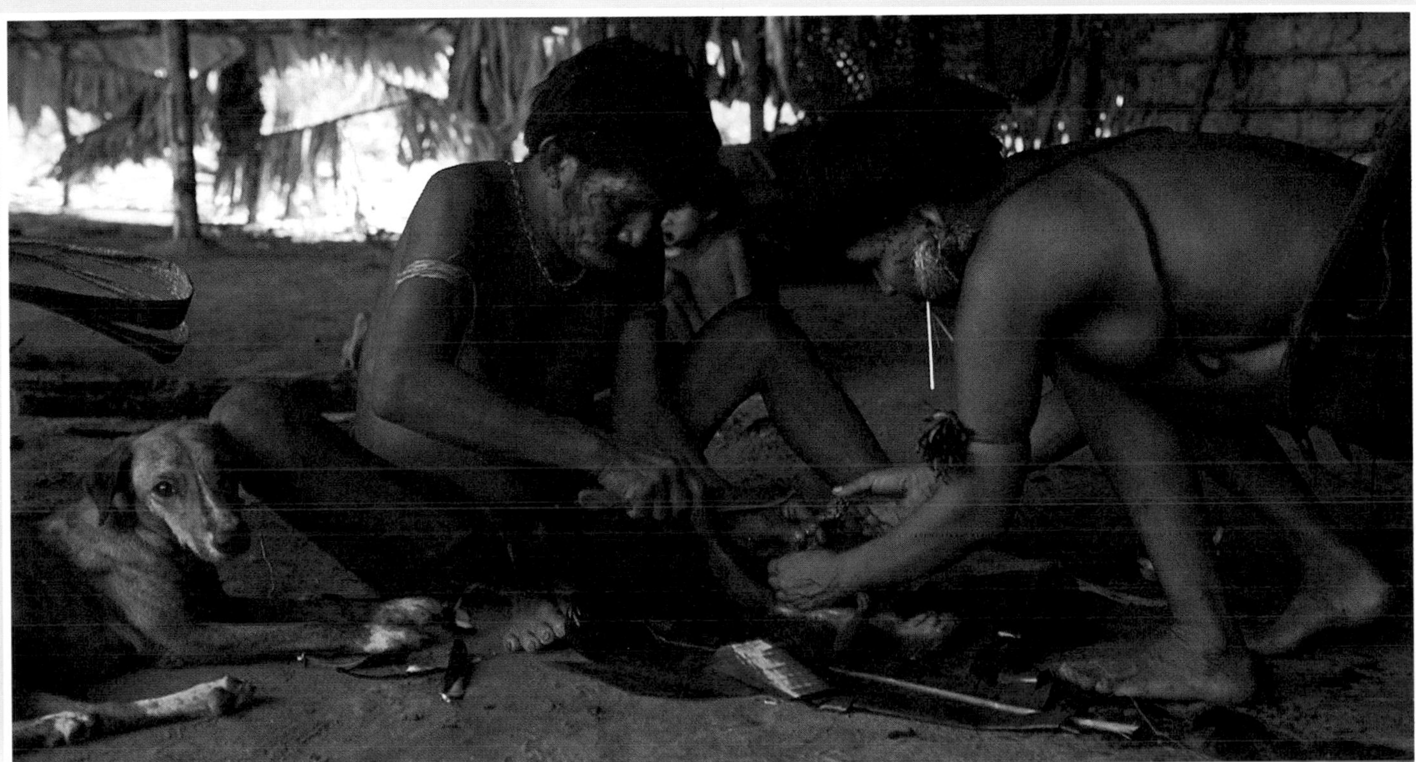

Cattle

In practice, cattle are kept under an enormous variety of circumstances, from near desert to swamp, living on food that varies from cereal straw, to lush grass, to cereal grains, including barley, wheat and maize. They also serve many different purposes: as providers of dung, fertilizer and fuel; leather, for shoes, clothes, shields, boats and tepees; milk; meat; and draft power.

When breeding cattle, it is essential to ensure that the right kind of animals are matched to the right conditions. If this is not done, the result can be disastrous. For example, many of the cattle of Africa and the Indian subcontinent are skinny and shockingly unproductive by European standards. Attempts have sometimes been made to improve them, therefore, by crossing with European breeds. But the tropical cattle are adapted to high temperatures and drought, in which

◄ Smithfield meat market, London. Beef cattle in Europe were traditionally bred and fed to be fat. But it takes more energy to produce a kilo of fat than a kilo of lean. Fat is also bad for the consumers – too much may lead to heart disease and some cancers.

► In Syria, large herds of cattle are being watered. Cattle in dry tropical countries are often unproductive by European standards. But they are generally well adapted to the harsh local conditions, and thrive where European cattle would perish.

▼ Vegetarians sometimes argue that livestock farming is wasteful. But animals fulfill many different and vital functions, particularly in poor countries. Even the dung is useful. Here, in Turkey, cowdung "bricks" are drying in the Sun, for use as fuel.

European animals often languish. Even more serious is the fact that tropical breeds in general can thrive on extremely low-grade fodder, such as coarse grass and rice straw. The reason they have this ability is that they have an enormous stomach, which occupies a third of their body. They can therefore eat a great deal, relative to their body size. The stomach of European cattle, by contrast, occupies only a fifth of the body. When the food is too low-grade, they simply cannot eat enough and lose weight rapidly.

Dual-purpose breeds

In Europe, cattle breeds were traditionally divided into muscular breeds, raised for beef, and more gracile creatures, that were kept for milk (➧ page 166). Cows cannot produce milk unless they first have a calf. Some female dairy calves are kept, eventually to become part of the dairy herd themselves, while the surplus calves, including most of the males, are raised for meat. However, the calves of some breeds were sometimes so skinny that it was hardly economic to raise them for meat, and at one time many of them were simply killed and thrown away.

Times changed, however, such that it was no longer economic either to keep cattle purely for beef (in which case the mothers' milk went to the calves, and produced no income) or to keep dairy cows whose calves were too feeble to be raised for beef. So the specialist breeds have largely given way to the dual-purpose Friesian. Friesians give enormous milk yields and are also large and muscular. Pure-bred Friesian calves make good beef animals.

Objectives in raising cattle

There have also been more subtle changes, brought about partly by an improvement in the general standard of feeding, and partly in the increasing aversion, among consumers, for fat.

When cattle are very young, the main growth takes place in the skeleton. As they reach adolescence, the bone continues to grow but the emphasis switches to muscle. As the animals mature, they accumulate fat. In general, butchers and consumers like at least some fat on the carcass; adolescent meat is rangy and somewhat tasteless. A fat animal is said to be "finished" – ready for slaughter.

In feeding cattle, the farmer thus has two aims. The first is to achieve finish as quickly as possible. If he can produce a good beef carcass in one year and not two, he should make twice as much money. But he also must ensure that the animal is of sufficient size. The way to raise animals quickly is to feed them well: plenty of high-grade food. But if he feeds them too well then their growth pattern is concertinaed. They begin to put on muscle before their skeleton is fully grown, and to lay down fat before they are particularly muscular. Overfed animals are too small and too fat.

However, the quality of feed has increased over the past few decades through improvements in grassland management and because the diet of cattle is often now supplemented with cereal. The traditional beef breeds were originally bred to do well on low-grade fodder. On today's feed, they become too fat too soon. By contrast, it was often difficult in the past to produce well finished carcasses in a reasonable time from large breeds, because they simply continued to develop bone and muscle. But when they are given large quantities of high-grade feed they do produce large "finished" carcasses in a reasonable, economically viable timespan.

There are estimated to be 1.2 billion cattle in the world, more than half of which are in Europe, North America and Australasia, where humpless cattle predominate

Cattle breeding

In addition to the switch to the dual-purpose Friesian (⧫ page 165), recent decades have seen a vogue for very large breeds, for beef: the Devon from England, the Charolais and Simmental from France, and the enormous Chianina from Italy, which stands 2m at the shoulder. These breeds were originally developed as draft animals – and indeed they still pull carts and ploughs in continental Europe. Now they are finding a new role. They are being raised to produce the leaner meat that people now prefer – carcasses that are "unfinished" by past standards.

The lesson here is that old breeds of livestock, whether cattle or any other animal, should never be allowed to die out. As conditions change, their special qualities may again prove useful. Old fashioned "rare" breeds are now being conserved to maintain a rich gene pool (⧫ page 202).

In the past, cattle breeding was a slow business. Nowadays, it would be possible, if considered desirable and if the money was available, to transform all the cattle of the world in a few years. A single bull can now inseminate tens of thousands of cows, by means of artificial insemination (⧫ page 242). "Elite" cows can be induced to produce as many as 40 ova in a year. These can then be fertilized in a petrie dish (in vitro fertilization) and the resulting embryos can be raised in the wombs of lesser animals. If necessary, the embryos can be taken across continents, in a frozen state. To increase the numbers still further, the embryo could first be divided into four pieces, each of which can grow into a complete embryo. These and other techniques are mostly still experimental, but such manipulations of living matter have been successfully achieved with some animals including sheep and mice.

►** The aurochs (1) or giant wild ox,* **Bos primigenius,** ***is the ancestor both of the humpless domestic cattle of Europe and North America, and of the humped "zebu" breeds, such as Brahman (10) and Gir (9). Cattle serve many functions, some of which are reflected in the different shapes and sizes of the modern breeds. Thus zebu provide milk, meat and traction power on low-grade fodder. The Hereford (2) is a typical beef breed, very stocky, and short in the leg. The great size of the Simmental (3) and Blue Belgian (4) reflect attempts to produce the most meat possible. By contrast, Brown Swiss (5), Jersey (6), Normandy (7) and Finn (8) are specialist dairy animals: they are excellent milkers, but they are too skinny for good quality beef. Longhorns – English (11) and Texas (12) – now serve no useful purpose; farmers breed them mainly for their appearance.

11

12

10

9

8

6

5

7

Sheep

Sheep, like cattle, are ubiquitous, and are almost as versatile. They do not as a rule provide draught power but they do provide milk (used in yoghurts and cheeses such as the Greek fetta and the French Rocquefort), meat (known as mutton or lamb), wool, leather, and such by-products as tallow. In much of Europe, wool was once the prime reason for keeping sheep. In Australia, where the principal breed is the opulently fleecy Merino, originally from Spain, it is still extremely important.

Sheep have two principal drawbacks. The first is that they are extremely susceptible to pests and diseases. The second is that their productivity is low. Most breeds of sheep produce only one or two lambs per year. A few breeds, such as the Finnish Landrace or the Greek Chios, may have as many as five at a time. However, such litters tend to contain runts, too feeble to survive, and few ewes can look after more than two lambs at a time, or three at most.

The economics of sheep farming

When animals are raised for meat then the price obtained from the carcass has to cover not only the cost of raising the animal, but also the cost of keeping its mother. With sheep, the cost of keeping the ewe may be borne by a single offspring. Thus it is all too easy for sheep farmers to find themselves on a descending economic spiral. Because sheep do not as a rule generate much income, the temptation is to keep them cheaply, for example on open hillsides covered in wild grass and heather. But if conditions are too harsh, then productivity falls even further. A great many upland ewes fail to produce any lambs at all in difficult years.

In general, then, the prime task in sheep breeding is to increase productivity. There are three possible approaches. One is to induce small ewes to produce big lambs, so that the mother (being small) can be kept cheaply, while her fast-growing offspring fetch a good price. Two, persuade the ewe to produce more lambs per litter. Three, induce the ewe to produce more than one litter per year. Ideally, these approaches could be employed simultaneously.

1

2

Big lambs from small ewes
In Britain, hill farmers keep flocks of small, resourceful animals that can live on coarse feed and even survive the winter outside. The small ewes are mated, in autumn, with enormous rams from the lowlands. The lambs, when born, are not significantly larger than pure-bred lambs – the size at birth is determined largely by the mother – but their growth rate after birth, and their final weight, are roughly half-way between the ewes and the rams. Ewes and lambs spend the summer together on the hills. The lambs are taken off in the autumn, leaving the winter grazing to the ewes.

◀ Longwool breeds – Border Leicester (5) and Wensleydale (6) – live on the uplands, while Corriedale (4) and Blackface (3) are at home on lush lowlands. Merinos – Spanish (1), Tasmanian (2) and Arles (8) – and the Sudanese (7) tolerate hot, dry conditions and poor pasture, producing meat, milk and wool. The primitive, hardy Soay (10) survives the cold, wet wind of the Outer Hebrides, and the West African Dwarf (9) tolerates the humid forest zone of West Africa.

The first option is already widely practiced (see panel). The second approach – more lambs per litter – can be achieved in part by cross-breeding with highly prolific, or fecund, breeds, such as the Finnish Landrace. There are difficulties however. Finnish sheep are rather scraggy. Cross-bred animals may produce more lambs, but if quality suffers, nothing is gained. Breeding must proceed so as to retain the genes that bring fecundity, without the attendant scragginess.

The second problem is that ewes cannot, or will not, suckle more than two or three lambs at a time, and if they are kept on the hills, they would be hard-pressed to do so, because they could not eat enough to produce sufficient milk. Thus more and more farmers are employing systems for feeding artificially – generally consisting of a number of teats attached to a central reservoir of milk. The lambs need to be trained to use these devices. This is time-consuming, but not generally difficult.

The third approach is still experimental. There are several problems of which the most important is that most farmers prefer to raise lambs outdoors where they are very little trouble. This means the lambs should be born in spring. Autumn-born lambs are a nuisance. But sheep, unlike cattle, are seasonal breeders. The ewes come into heat in autumn in response to shortening day-length, and if they are not then inseminated, or covered, by the ram, they go out of season for another year. A cow that is not made pregnant will simply come back into heat a month later.

The basic task, then, if the farmer wants to produce lambs other than in spring, is to bring the ewe into heat. This can be done using hormones, by keeping sheep indoors where the day-length can be regulated artificially, and by using ewes that are not particularly light-sensitive.

If all the above techniques were applied, then future sheep could produce large litters of big lambs, at intervals of around 8 months. Such animals would be as expensive as cattle, however, and require the same level of husbandry. But present-day sheep that produced just a few more lambs, a little larger than at present, and more reliably, would be an advance.

In the Perigord region of France, pigs are still used to locate and dig up truffles, a prized underground fungus

Pigs

People have a love-hate relationship with pigs. Some societies, such as the Maring of New Guinea, are besotted with them. They look after vast herds of semi-domestic pigs in the forests as carefully as they look after their own children, until an appointed day when they decide the pigs have become too numerous, and they slaughter them by the score, in an orgy of pork. Yet to other societies, notably the Moslems and the Jews, pigs are taboo, though Israel does have a thriving large-scale pig industry.

Nowadays, in Europe and North America especially, the pig has become almost the ultimate factory animal. The old breeds have given way to modern hybrids, which grow faster and produce more offspring than pure breeds. In traditional systems, sows gave birth every 6½ months (two litters in 13 months). Pregnancy lasts 3½-4 to four months. The sow was allowed to suckle her young for 8 weeks, and after a 2-week rest she was returned to the boar, with which she mated immediately. Nowadays the piglets are commonly weaned at 5 or 6 weeks, so that the sow can fit in two litters in a year. In some very intensive systems the sow does not feed the young at all. They are put on to artificial milk at birth. The sow therefore needs less feed (because she needs most when she is lactating) and more piglets survive. When they are fed naturally, the sow often crushes one or several by lying on them, while the smallest ones in the litter, the runts, are sometimes prevented from feeding. Thus, whereas farmers until recently were very content to obtain 15 to 20 piglets per sow per year, 30 to 32 per year is now envisaged.

Pigs for every purpose

In general the husbandry of pigs covers the complete spectrum of intensiveness. At one end of the spectrum is the Maring method – the animals living natural lives, and feeding themselves in the wild. The traditional farmers of Europe kept pigs on mixed farms. They fed them swill, waste corn, and allowed them to graze. Pigs were, and are, also employed as cultivators. They would dig up a field with their snout and tusks in their search for roots and leave the soil well-fertilized and largely clear of weeds. Iron Age farmers may well have relied as much on pigs as on the plow. Ground prepared by pigs is known as pannage. In the past many breeds were kept for particular purposes. Berkshires were known for their ability to do well when kept outdoors on grass. Gloucester Old Spots were employed to eat windfall apples in the orchard.

The search for a better life

At the other end of the spectrum, the modern factory sow spends the whole of her pregnancy standing in an iron cage too small to turn round in, and the time when she is feeding the young (some suckling is still usual) is spent lying in a small "farrowing crate". But pigs are highly intelligent animals, and in a natural state live in intimate family groups. Scientists are now developing methods that are highly productive (providing at least 20 piglets per sow per year) but allow the pigs to live more natural lives.

▲ *"Battery" hens are kept for only one year, in which they lay about 300 eggs. The output is staggering but the system is cruel, and is already banned in parts of Europe.*

◀ *Pigs traditionally were bred to meet esthetic standards, the demands of the market or particular systems of husbandry. Durocs (2), Craons (5), Andalusians (4) and Berkshires (3) once were all bred to be hugely fat for pork and lard, and for their colors; Saddlebacks (6) did well outdoors; while the Landraces (1) of Scandinavia were bred lean.*

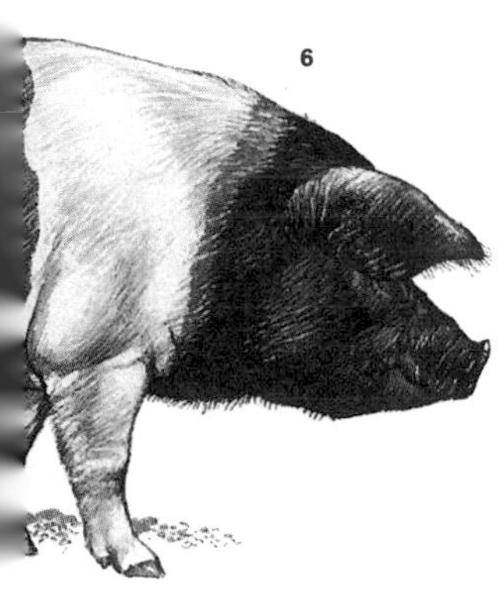

Poultry

"Poultry" for the most part means chickens (domestic descendants of the Indian jungle fowl) and ducks (descendants of the mallard). Turkeys and geese are minor by comparison.

Like cattle (◆ page 166), poultry are dual purpose: they are kept both for eggs and meat. Like pigs, they are essentially omnivores but also like to graze, and they were traditionally kept as scavengers and cultivators. They are still employed as such in gardens in Europe and Third World villages.

Hens and ducks are useful as egg-layers because they readily lay infertile eggs; they do not need first to be mated by the cock or the drake. In a natural state, the hen or duck tends to become "broody" after laying a dozen or so eggs (whether fertile or not). She stops laying, and sits on the eggs. But modern domestic breeds have largely had broodiness bred out of them. If the eggs are removed as they are laid (as is usual) they will continue to lay anything from 200 to 300 eggs in a year. Hens do not lay as many eggs in the second year as in the first, and whereas the backyarder may keep the flock for several years, the commercial farmer will start afresh each year.

As with pigs, however, both chickens and ducks have been coerced into highly intensive, factory systems of husbandry. The birds are crowded (hens in cages cannot stretch their wings) and they develop vices, such as feather-pecking, which are sometimes checked by removing the tips of the animals' beaks (known as "de-beaking"). Again, scientists are seeking to develop more humane systems and free-range poultry farming is gaining in popularity.

Factory farming

FOR

1. NUTRITION Nutritionists in the post-war years believed that humans needed large quantities of high-quality – animal – protein. The high demand for livestock could be met only by raising animals intensively.

2. EFFICIENCY Animals kept indoors in confined spaces need not waste energy keeping warm or in excessive exercise. They also escape many of the parasites that attack them outside. Hence they convert more of their food into growth, or milk or egg production, and so are more "efficient".

3. BREEDING When animals are not required to forage for themselves or look after their own young they can be bred exclusively for their productivity, without regard to other "irrelevant" qualities.

4. LABOR When animals are closely confined, a single worker can look after huge numbers – tens of thousands of birds, or up to 100 dairy cattle.

5. COST Because of increased efficiency, better breeding, and decreased labor, intensively raised meat, milk, or eggs are in general far cheaper. Thus, in real terms, poultry prices have fallen about 10 times since the 1940s.

6. WELFARE Though welfarists suggest that many modern farm practices are cruel, some of the intensive methods were in fact introduced to protect animals. Thus, animals outdoors traditionally suffered greatly from parasites; pigs froze in the mud; and modern hill sheep still die from starvation and exposure.

7. Q.E.D. Farmers argue that animals are not productive unless they are happy. As intensive livestock is extremely productive, the animals must be happy. They also argue that animals are effectively machines and it is absurd to ascribe human qualities to them, such as intelligence or misery. Close confinement does not therefore seem cruel as machines do not suffer. Until about the 1970s, animal psychologists shared this view.

AGAINST

1. NUTRITION Most nutritionists now feel that humans need far less animal protein than was previously believed, and also emphasize the dangers of consuming too much animal fat. Hence, they feel that we could get by, or even do better, if we produced far less livestock, and so do not need intensive systems.

2. QUALITY Rapidly grown animals fed on monotonous diets may be less flavorsome than those raised more slowly on semiwild vegetation. Small amounts of flavorsome meat are better than large amounts of bland meat.

3. WELFARE Welfarists point out that cruelty is not simply a matter of physical discomfort or of pain. Animals may suffer if they are bored or frustrated, just as humans may do. They often show this behavior through "stereotyped behavior": thus sows, confined in iron stalls during their pregnancy, in which they cannot turn round, spend much of their time chewing the bars. When kept in groups, bored animals may develop "vices": thus fattening pigs may bite each others' tails, and battery hens peck each other to death. Sometimes these vices are countered by surgery – tails are cut off or beaks are clipped – which seems to add yet more cruelty.

4. EFFICIENCY Many agriculturalists have now shown that although old-fashioned extensive husbandry was often extremely inefficient, it is possible to devise systems that allow animals to live fairly "natural" lives, and yet produce at an acceptable level.

5. COST Though extensively kept animals will always be more expensive than intensively kept, this is partly countered by the fact that we do not need to eat as much meat as once was thought. Besides, we have no right to expect animals to suffer just to give us cheap food.

6. Q.E.D. The argument that animals will not produce if they are suffering simply is not true. Overeating, leading to obesity, is a common sign of depression.

See also
An Ape Called Man 129-140
The Search for Better Foods 141-152
The World's Crops 153-162
The Shadow of Extinction 195-208
Saving Wildlife 235-248

New kinds of livestock

Many places in the world are too dry, cold, hot, steep, barren, or generally disease-ridden to support conventional livestock. In such situations, native animals may survive better and grow faster than the usual cattle, sheep or pigs. Some of these wild animals might reasonably be domesticated.

In dry, hot conditions (◆ page 160) several species of antelope do far better than cattle. The addax does not need to drink, and indeed is better adapted to the desert than the camel. The ancient Egyptians kept addax and in the 1970s scientists in the Negev in Israel crossed addax with goats to produce the goadex. The largest of the antelope, the eland, is also extremely tolerant of drought and heat. It can be domesticated and even milked like cattle. A herd has been maintained at Askaniya-Nova in the southern Ukraine since the turn of the century.

Many animals do better than cattle or sheep in rough, open country. The feeding strategy of red deer is to eat large amounts of poor fodder and to pass it through the gut quickly. Thus they get the best out of poor food, such as coarse grass or heather, and then quickly reject the residue. Sheep take longer to process their food, and so waste time trying to extract nutrient from forage that they would do better to avoid. The red deer is now a major meat animal in New Zealand, and ranching is minor industry in Scotland.

On the poor grazing of the mountains, llamas and their relatives, such as the vicuña (◆ page 205), may be the best meat animals. Horses, too, thrive better on the high, poor lands of the Massif Central than either cattle or sheep. Like red deer, horses and llamas process poor food extremely quickly.

In the heavy forests of the north, browsers are needed – animals able to live off trees rather than grass. The best is the moose, which can produce as much meat as a large bull. Moose are an important source of meat for Finns and Russians.

Many species can simply be allowed to scavenge, but may be more appropriate than conventional pigs and poultry. Thus pigeons and guinea fowl are able to range further afield than chickens. In South America, guinea pigs serve both as house-pets and as a significant meat supply.

In some cases, wild animals may mix well with domestic stock, and thus become semi-domestic. Thus many farmers in Africa maintain herds of antelope or giraffe on their farms. In general these animals prefer coarser forage than the cattle, and do not compete with them, while giraffe confine their feeding to the trees (◆ page 93). African buffalo cannot be allowed on to farms because they carry diseases such as foot-and-mouth.

Domestication raises several logistic problems, however. It is always tempting to try to keep such animals more intensively, but if this is done their advantage may be lost. Thus, huge guinea pigs are sometimes raised almost as battery animals in South America. In such conditions, they have no obvious advantage over poultry. If animals are culled rather than ranched or farmed, then there are problems of hygiene and of supply. Animals such as eland and red deer can be handled like cattle, but they are more skittish and swift. Both require expensive handling equipment if they are ranched. Tameness and reliability are very big factors if animals are to be handled. Bull elands are extremely dangerous. Musk ox can be lethal. On the other hand, if animals are kept semi-wild, then their environment has to be extremely well managed. This may not always be possible.

▲ Manatees, relatives of the seacows or dugongs, could serve as excellent dual-purpose livestock in the tropics: keeping the waterways clear of weed, and providing meat. Capybaras – giant rodents of South America – and hippopotamuses could serve in a similar capacity.

▼ Eland have long been promising candidates for domestication. These big antelope provide milk and are far more tolerant of disease and drought than are domestic cattle. They still present problems for the farmer, however; even when "tame", they can still be dangerous.

The Resource of Trees

15

What are trees?...Origins of present-day trees...Material resources...Food and medical resources...The impact of Man...Reafforestation...PERSPECTIVE...The communities of trees...The forest ecosystem...Tree-breeding... Agroforestry

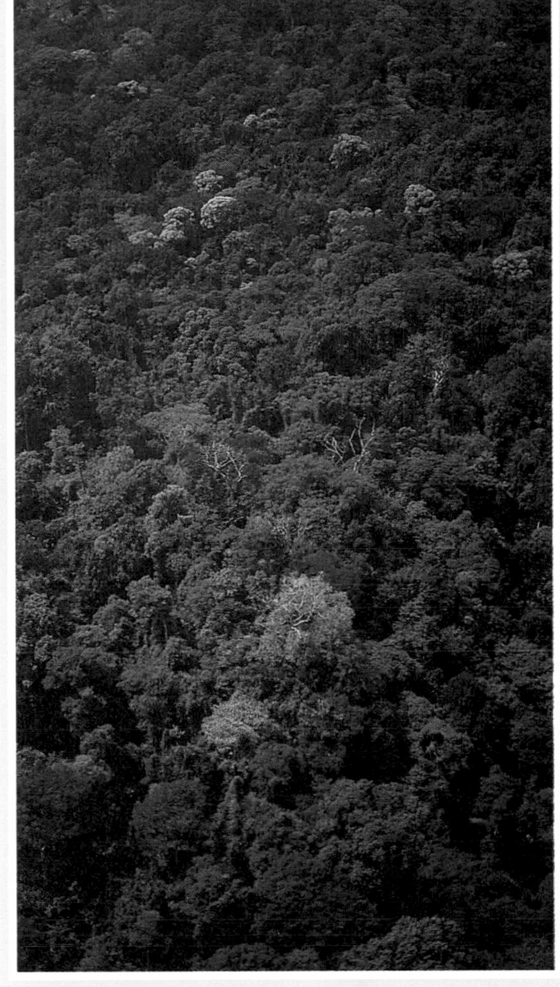

Most of us recognize plants as herbs, shrubs or trees. Shrubs and trees are woody, while herbs are at most only partially so. Typically, a tree will be usually at least 6m high and have a single stem (trunk) which is without branches for some distance above the ground, surmounted by the branched, leafy crown.

The origins of present-day trees

Forests are known as far back as 350 Mya years ago when, during the Carboniferous era, swampy lowlands were dominated by the extinct, treelike giant horsetails and early conifers. The conifers and their relatives (gymnosperms) subsequently rose in prominence but, during the Cretaceous system (c. 135 Mya), they were increasingly replaced by broadleaved trees of the flowering plants (angiosperms), which have dominated the world's vegetation ever since. By the early part of the Cenozoic era (70 Mya) forests of broadleaved evergreen trees and palms dominated large areas. More recently, the Pleistocene glaciations moving from and to the poles have caused latitudinal migrations of the forests, with concomitant extinction of many tree species, especially in Europe, and climatic changes over the past 16,000 years have led to the expansion and contraction of the forests.

▲ The species-richness of tropical rain forests has been attributed to their great age and ecological stability. But there is now evidence, especially from South America, that the Pleistocene glaciations at higher latitudes increased aridity in the tropics, with the forests retreating to refuges surrounded by grassland and scrub. Their expansion as the ice retreated led to much speciation to give their present diversity.

◄ The antiquity of forests is recalled by fossilized trees in the Petrified Forest of Arizona. These trees, forerunners of modern pines, were alive during the late Triassic Period, some 200 Mya. On dying they fell into a river that carried them to this location. Here they were covered by Jurassic sediments and impregnated with silica. Wind erosion, during dry eras such as that today, removed the overlying sediments.

Sclerophyllous forests in the Mediterranean basin and in California are, or were, dominated by evergreen oaks and, in southwest Australia, by gum trees

The communities of trees

The physiognomy (structure and appearance) of forests differs in various parts of the world in response to a number of factors, principally rainfall and temperature regimes (♦ page 97).

In the hot equatorial regions, with rainfall averaging over 2,000mm per year and no seasonal differences, evergreen tropical rain forests constitute the climax vegetation. These, comprising the most complex vegetation on Earth, normally have three or four levels of tree (♦ page 100).

In the tropics, as the annual rainfall falls below 2,000mm, a dry season becomes increasingly obvious and this leads to various kinds of seasonal tropical forest. Although, at first, only some of the emergent trees are deciduous, leaf-fall and flowering become increasingly correlated with seasonality and semi-evergreen seasonal forest gives way to deciduous seasonal forest, where the dry season lasts about 5 months. These forests are simpler in structure than the tropical rain forest, with fewer species.

Where the dry season extends beyond about 5 months, and the annual rainfall falls to around 500mm, the forest progressively becomes more open to give, eventually, savanna woodland. In this the frequently flat-topped trees are scattered over grassland or shrubland (♦ page 108).

In the western parts of most continents there are areas which have hot, dry summers and mild wet winters (rainfall 500-1,000 mm). The woodlands present in such regions of Mediterranean climate are composed of trees having sclerophyllous, leathery leaves which reduce water-loss. The rich shrub and herb layers contain many plants able to survive the fires which are common in these areas.

Warm temperate evergreen forests occur at higher latitudes than the sclerophyllous forest, and on the eastern sides of continents exposed to the monsoon or trade-winds. Rainfall is plentiful (up to 3,000mm per year) and distributed throughout the year. These forests are rich in species, with abundant lianas and epiphytes but, unlike the tropical rain forests, the trees lack plank buttresses. Most trees are broadleaved evergreens.

With the development of a marked cold season those parts of temperate Eurasia and North America having 700-1,500mm of rain per year can support summer deciduous forests as their climax vegetation (♦ page 102). These have a single canopy layer of trees which shed their leaves in autumn as an adaptation to the restricted or non-available water supply during the winter.

Encircling the Earth north of the summer deciduous forest is the boreal forest zone. Mainly lying between 45°N and 70°N, where the growing season is relatively short, the forests are dominated by conifers. Mostly evergreens, with conical crowns which prevent damage in areas of high snowfall, these trees can commence photosynthesis as soon as spring conditions permit. However, the northernmost parts of the forest, at 70°50′N in Siberia, comprise the deciduous larch. The deciduous broadleaved birches also reach the far north, so there is clearly no single group of adaptive features for trees surviving in these areas.

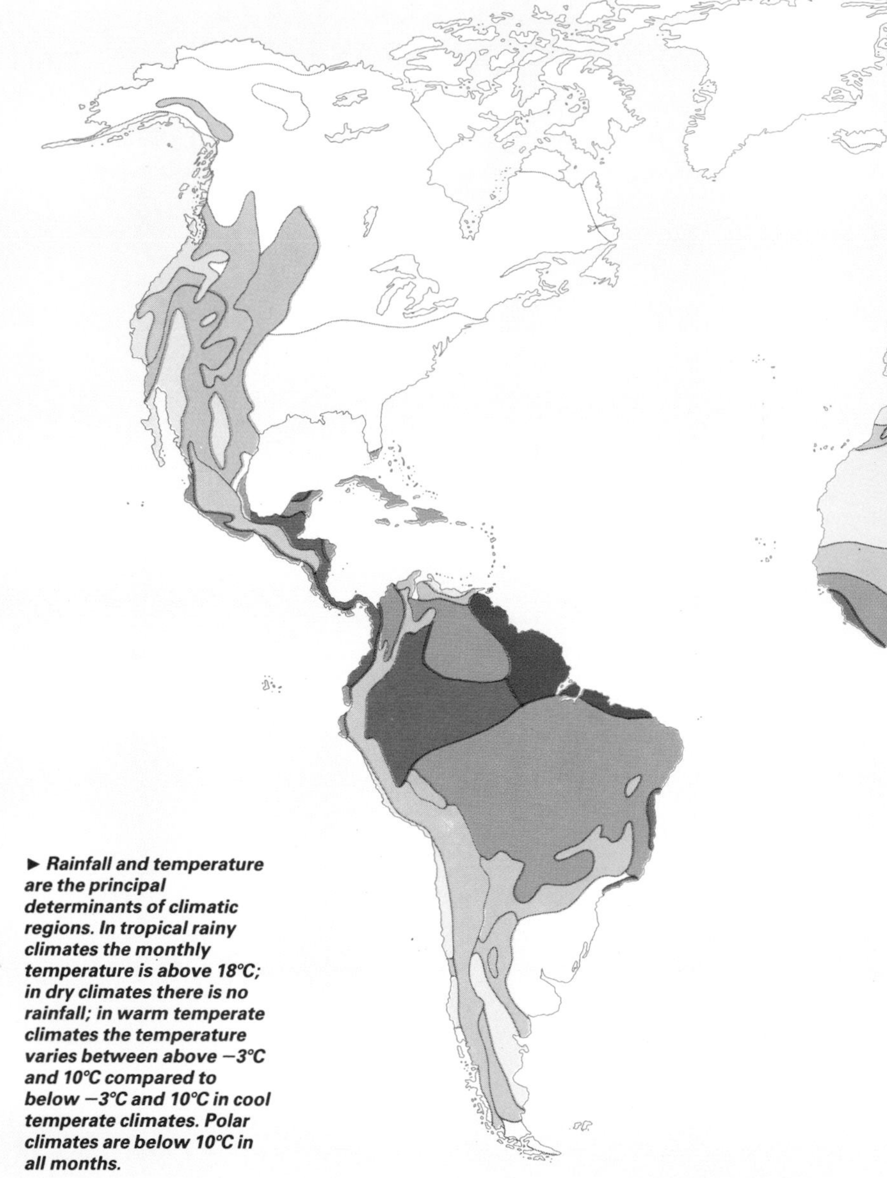

► Rainfall and temperature are the principal determinants of climatic regions. In tropical rainy climates the monthly temperature is above 18°C; in dry climates there is no rainfall; in warm temperate climates the temperature varies between above −3°C and 10°C compared to below −3°C and 10°C in cool temperate climates. Polar climates are below 10°C in all months.

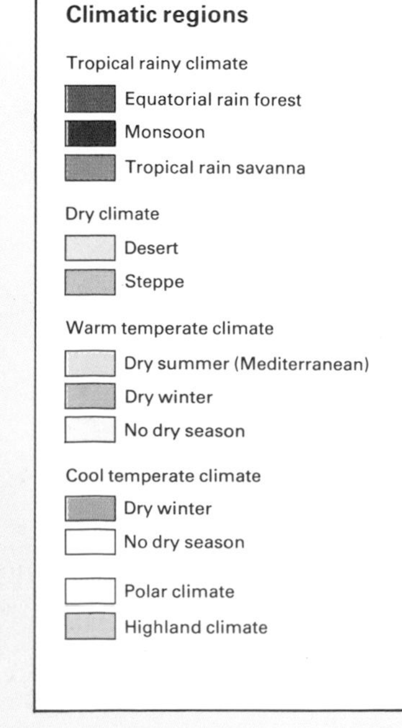

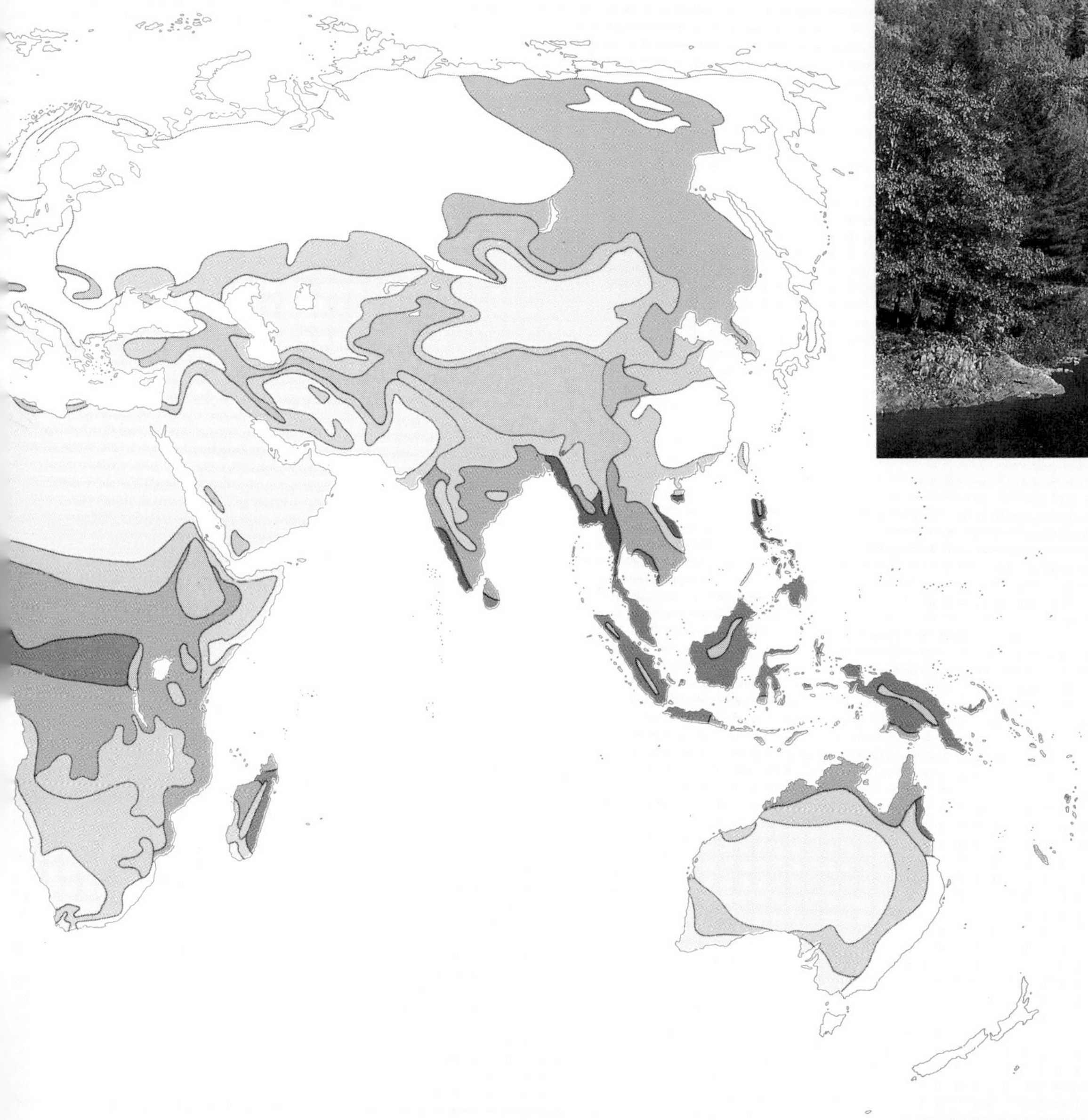

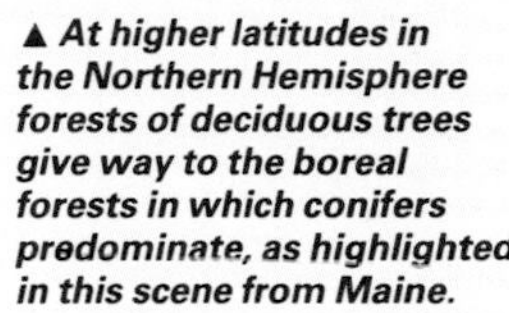

▲ ***At higher latitudes in the Northern Hemisphere forests of deciduous trees give way to the boreal forests in which conifers predominate, as highlighted in this scene from Maine.***

◄ ***Deciduous oak woodland, as exists today in, for example, the mountains of northern Spain, is typical of the original vegetation that covered much of western Europe prior to Man.***

▼ ***Beyond the rain forests large parts of the tropics carry savanna woodland as seen here along the Ruaha river in Tanzania. The grassland with trees and shrubs resembles a park.***

► ***Sequoia sempervirens, the coast redwood, seen here in Big Basin State Park, California, is a relict of an ancient line. This forest is an example of those found in coolish wet regions.***

▲ **Depletion of tropical rain forest by logging, Sabah.**

Material resources

About half of those trees felled each year are used as fuel, either as firewood or for preparing charcoal, on which some 90 percent of people in the developing world depend. Of the remainder, about two-thirds is turned into sawn timber, and much of this (c. 60 percent) is used in construction. Most of the timber used in modern building is softwood, derived from conifers, because of its straight grain and ease of manipulation. Hardwoods, derived from broadleaved temperate and tropical trees, are more usually employed in the production of furniture and for turning into such items as bowls, platters, ornaments and parquet floors; their thick walled cells give a denser wood which is very amenable to carving and hard wear. Most of the remaining timber is used in paper manufacture, with the rest being turned into plywood and other products, such as blockboard.

Many trees are "milked" for their valuable exudates. Rubber, *Hevea brasiliensis*, long tapped for its latex in the Brazilian forests, was introduced to southeast Asia where the greatest rubber production now centers. At least 36 species, including various figs (*Ficus*), produce an acceptable rubber substitute in times of shortage, as when the Japanese controlled southeast Asia during World War II. Turpentine is derived by tapping the oleo-resins of various pines, while oils of wintergreen and eucalyptus are produced by distilling the wood of, respectively, *Pernettya* and *Eucalyptus*. Cork is the bark of the evergreen oak, *Quercus subos*, of southern Europe.

Food and medical resources

From the apples and pears of cool temperate regions, through the oranges, lemons and grapefruit of warmer climes, to thc papayas, cocoa and mangos of the tropics, we are accustomed to the diversity of food provided by trees. Less obviously, the goats, deer and rabbits which browse the foliage and bark of trees are also part of our diet.

The forest ecosystem

Ranging from the tropics to high latitudes (about 70° in the Northern Hemisphere), and from sea level to over 3,500m, forests differ in their structure, constituent plant and animal species, and impact on the environment. However, despite their diversity, forest ecosystems have certain common characteristics. Climax vegetation is that which is the most complex that can develop under any set of environmental conditions, and forests are considered to be the most complex of all. Forest ecosystems normally develop in areas with a reasonably long growing season, mean temperatures generally over 10°C, a reliable water supply (at least 500-600mm annually) and protection from long exposure to drying winds. They are dominated by one to four strata of trees, with the development of shrub and ground (herbaceous) layers depending on the kind of tree canopy. Since the dense canopy absorbs or reflects up to 99 percent of the incident light, the development of the lower layers depends upon gaps in evergreen forest. In summer deciduous forest many of the herbs and shrubs pass through flowering and much leaf-growth before the leaves open on the trees. In northern deciduous forest the rich spring ground floras of bluebells, anemony, violets and aconites are characteristic.

Winds are greatly reduced by tree cover; wind speed at ground level in pine forest with trees 14m high is less than a quarter of that outside the canopy, and the reduction is proportionately greater in taller forest. Such factors, shade and protection, increase the humidity and ground moisture, with leaf-fall providing a gradual buildup in humus and modifying the parent soils. In this way forests develop their own microclimates. Within each a rich diversity of plants and animals have evolved and interacted to give the ecosystems in which 82 percent of land plant biomass is locked – from 40 percent of the land surface! *(▸ page 86)* *For this, if for no other reasons, remaining forests should be conserved.*

◀ **Down the ages many religions have considered that trees harbor spirits. From the oak of the ancient Britons and the fig tree revered by Romulus, to the sacred peepul of Indian Buddhists and the sausage tree of tropical Africa, we catch a glimpse of our ancestors living with the trees they feared, loved, exploited and worshipped. Totem poles, such as this carved by the Canadian Cespiok people, are a continuing reminder of the spiritual heritage given to us by trees.**

▲ *When wood is burned in the partial absence of air there remains a residue which is rich in carbon – charcoal. Charcoal has been prized by most civilizations as a very effective fuel for many purposes, from cooking to smelting. These charcoal furnaces in Afghanistan are and age-old symbol of this. Today, over 2,000 million people depend on wood, or on crop and animal wastes for fuel.*

▼ *We now consume about 3 billion cubic meters of wood a year. Although developing countries produce more wood most is used on fuel and charcoal of which only a fraction is exported. Developed countries produce the most processed wood and are the source of almost all paper and pulp. Cultivation of trees reflects these uses, including species suited as firewood crops.*

Main forest products

Developing countries

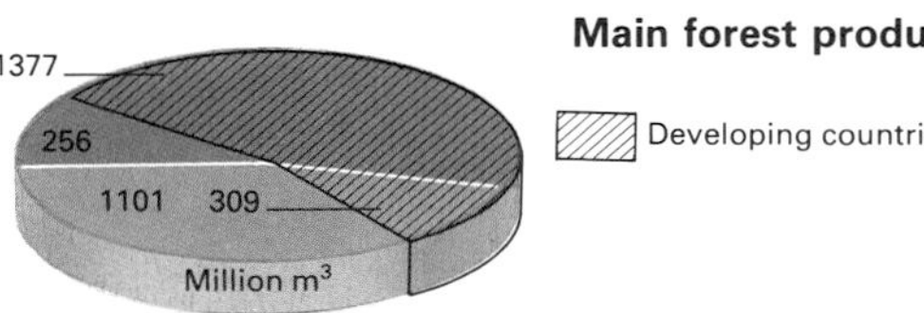

Roundwood

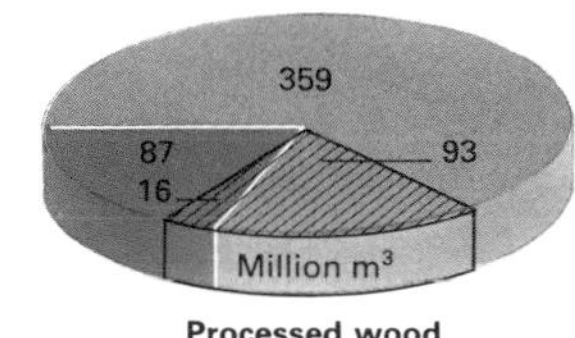

Processed wood

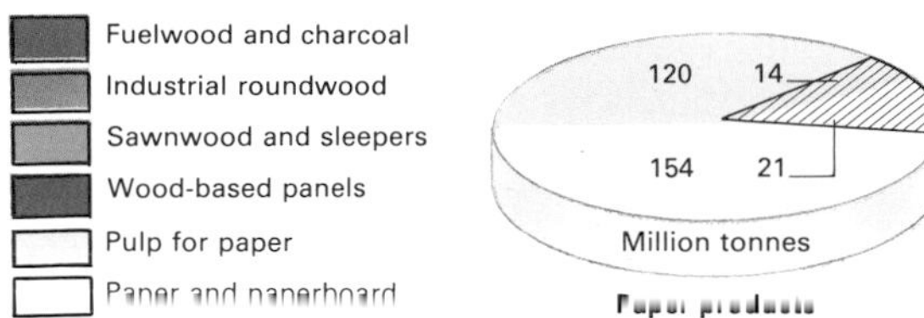

Paper products

Traditionally many tree-products have been used by Man as medicines. Most of these are now lost to modern cultures, though in the last 30 years or so there has been an attempt to rediscover and employ such products. Quinine, derived from the bark of various species of *Cinchona*, was long used as an antimalarial agent, the dried leaves of the Coca tree (*Erythrozylum*) are the source of cocaine, the oldest local anesthetic, while foxgloves (*Digitalis*) produce digitalin, which acts on the cardiovascular system. The role of the world's trees in generating the life-giving oxygen (◀ page 85) is increasingly prominent in discussions about the conservation of our forests.

Impact of Man

Before Man started the clearance of forests, forests and woodlands covered about 40 percent of the world's land surface. Today the figure is around 30 percent. In Britain and other parts of Western Europe virtually all the forest cover has been removed, as also around the Mediterranean and much of eastern North America. In southernmost South America a quarter of the forests have been cleared in just about a century, a story repeated elsewhere in the Southern Hemisphere. Of considerable current concern is the loss of the tropical forests. Tropical moist forest covered some 9-11 million km^2, or at least 6 percent of the Earth's land surface. Increasingly, it has been cleared for timber, ranching and agriculture; current estimates are that 100,000 to over 200,000km^2, about 2 percent of the remaining forests, are being lost each year. Most of the nutrients and energy in these forests are locked up in the trees, whose removal exposes the thin mineral-poor tropical soils to serious erosion (▶ page 214). The loss of numerous actually or potentially important trees, with the certainty that many are as yet unknown to science, and the threat to the oxygen-carbon dioxide balance of the atmosphere (▶ page 222), have contributed to concern over the depredations on tropical forests.

▲ *This stack of dipterocarp logs in Sabah, Malaysia, emphasizes the importance of the world's tropical rain forests (⧫ page 100) in providing a tremendous diversity of hardwoods. With its rich coloring and durability the wood is prized for furniture-making and other uses. The recent massive scale of extraction has caused serious degradation of this ecosystem and poses grave conservation problems.*

Reafforestation

The increasing knowledge of the world's plants from about the middle of the 17th century onwards was accompanied by the introduction of numerous species into areas where they were not native. Many, many trees were brought into the parks and botanic gardens of Europe as empires developed (⧫ page 236), with other continents increasingly becoming recipients in this traffic. As a result, we now find East Asian and North American maples (*Acer*) and cherry trees (*Prunus*) coloring the gardens and streets of temperate areas everywhere, Australian bottle-brushes (*Callistemon*) and Mediterranean oleanders (*Nerium*) lining the highways in warmer regions, while Burmese coral-trees (*Erythrina*) and American frangipani (*Plumeria*) adorn cities throughout the tropics and subtropics.

Planting of mixed species

The role of Man in disseminating interesting and attractive tree-resources has been paralleled by the introduction of many economically important species. From the founding of the Malaysian rubber plantations and the extensive cultivation of Monterey pine (*Pinus radiata*) in New Zealand and elsewhere, to the large-scale plantings of spruce (*Picea*) in West Europe and gum-trees (*Eucalyptus*) in all areas

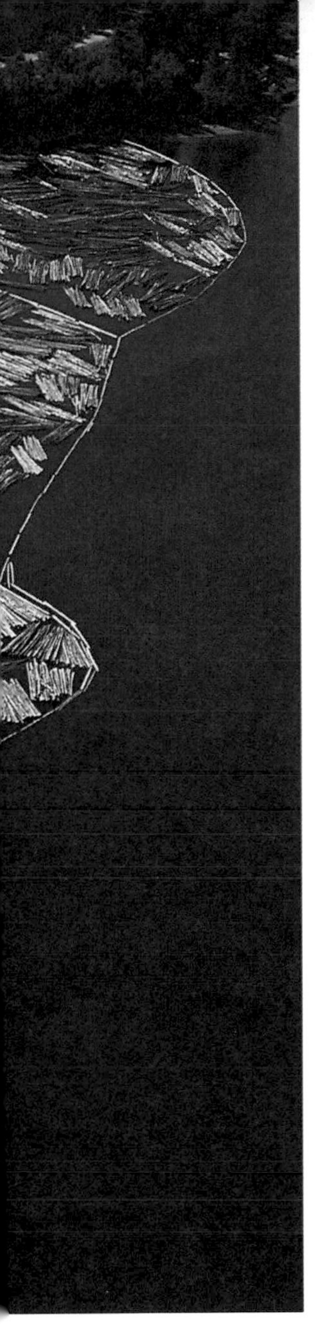

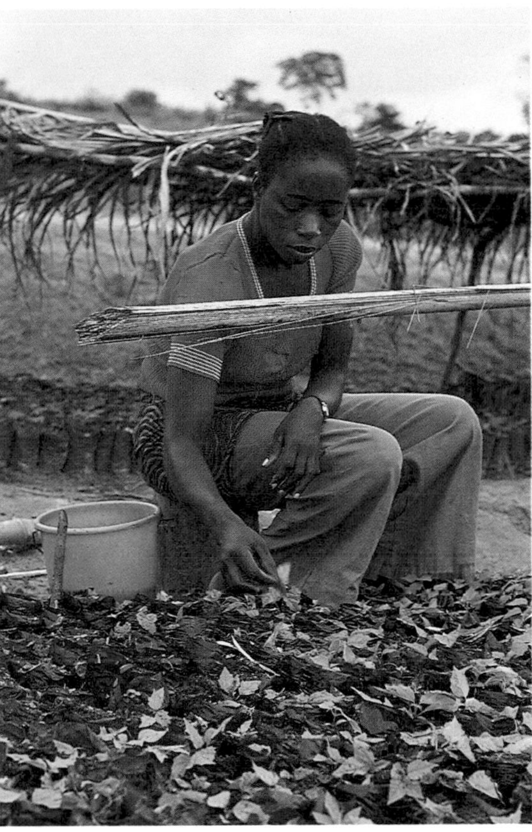

◀ *Most of the wood-pulp used in paper-manufacture and, at least in the Northern Hemisphere, timber for the construction industry is derived from softwoods, especially pines. This lumbermill in British Columbia is flanked by rafts of logs floated here after being felled.*

▲ *Despite advances in technology, the basis for finding new varieties of trees with increased productivity still depends upon selecting the most promising offspring of artificial or natural crosses. This Ghanaian woman is tending seedlings which will be tested in field trials.*

of Mediterranean-type climates, the problems of such monoculture have become apparent. Plantations of single species have proven to be susceptible to diseases of various kinds which, once started, can ravage enormous areas. Increasingly, it has become acknowledged during the past 30 years that plantings of a mixture of species not only reduce the risk of such diseases but also actually increase the overall productivity of the forests. There is also a developing mood that reafforestation with native species is more acceptable to many peoples, and this may also be environmentally and economically a happier situation.

Trees and Man – the prognosis

Whilst, in the current climate of enlightened concern, it is perhaps too pessimistic to envisage a world bereft of trees, the clock cannot be completely turned back. We must try to maintain those wilderness areas remaining and learn to manage the remnants of woodland and forest island elsewhere. Much work is needed to determine the minimal sustainable area of these reserves and the effects of scrub buffer zones between them and agricultural land. In some regions reafforestation will be aided by the use of novel kinds, produced by artificial breeding, and mixtures of trees, with the assistance of new management techniques.

Tree-breeding

*Since the 1740s, when H.L. Duhamel de Monceau began provenance trials on pines (*Pinus sylvestris*) from central and northern Europe for use by the French navy, foresters have been selecting economically suitable varieties of trees from natural populations. The genetic variability upon which such selection depends can be enhanced by artificially crossing plants from different populations or even species and utilizing the best progeny for further hybridization and testing. These techniques, widely used by plant breeders, are inevitably slow and consequently costly for foresters, given the usually long period between germination and flowering of most trees.*

To partially obviate these problems tree-breeders have increasingly become interested in the use of grafting. Scions (cuttings) from a mature tree possessing desirable features are grafted on to root stocks, the provenance of which is immaterial. This effectively results in a clone of genetically identical trees which, if planted in a separate "seed orchard", intercross to give seeds likely to produce a higher proportion of the desirable features amongst the progeny. Problems of graft incompatibility between different populations in Sitka spruce and other conifers are currently the subject of much research.

*Amongst the most recent developments in tree-breeding are those involving tissue culture (**♦ page 151**). By placing pieces of plant tissue in liquid growth media containing vitamins, minerals, sucrose and other substances it has been shown that a "callus" is formed from which roots and, on solid media, shoots are produced. Such techniques are used to produce numerous, genetically similar plants of, for example, orchids on a commercial scale.*

*Trees have proved more intractable but very recently there has been success with the economically important coconut. Undifferentiated floral tissue cultured with the synthetic growth substance 2,4-D (usually a herbicide), with activated charcoal added, produced a callus. By gradually reducing the concentration of the 2,4-D it proved possible to produce small plants in about 9 months. Although some technical difficulties remain to be solved, it seems likely that a way is now in sight to allow rapid propagation of better varieties which could increase yields by some five times. Similar promising results have been obtained from work on the timber tree, Douglas fir (*Pseudotsuga menziesii*).*

Such culture techniques seem likely to open up tree-breeding to the new technology of genetic engineering, whereby genes controlling such economically desirable characteristics as disease resistance can be incorporated into the genetic material – almost a science fiction concept just 20 years ago.

Similar techniques in plant breeding are having an impact on the direction of tree-breeding. Thus the greatest potential for increasing the source of fiber for papermaking lies in nonwoody fibrous plants. Forests which have a 30-year regeneration cycle cannot produce as much fiber per year as suitable annual plants.

See also
Communities of the Land 97-116
An Ape Called Man 129-140
Threats to the Environment 209-224
Saving Habitats 225-234

Agroforestry

Agroforestry concerns the ways in which forestry and agriculture or horticulture, trees and crops or grassland, can be managed within a coherent framework. The aim is to exploit the land in such a way as to obtain a diversity of products within a unified area without exhausting particular environmental components.

Traditionally the forests of temperate regions were largely cleared as Man required increasing land for his crops and grazing animals. Copses or hedgerow trees were left for shelter and fuel, but trees and agriculture at most coexisted. The rich forest soils proved very productive. In tropical forests, on the other hand, shifting agriculture was widely practiced – relatively small areas of forest were cleared sufficiently for crops to be cultivated for a few years and then left to return to forest as other areas were cleared. This safeguarded much of an ecosystem in which most nutrients were locked in the plant cover and little in the soils.

There are various examples of joint management of trees and non-woody crops in temperate regions. Perhaps most spectacularly, the extensive olive groves of southern Spain, for example, bear an early summer harvest of wheat and a mid–late summer crop of melons before the olives are picked in late summer-early autumn. However, it is in the tropics that such multiple use of land has found its widest application.

Dating from at least the 10th century AD, the village-forest-gardens of Java provide an example of the old roots of agroforestry. In these, the crop plants, of many species, are planted in an ecologically agreeable, multistoreyed, close cover vegetation, with usually 3 to 5 strata. Up to 1.5m the dominants are such species as spinach, beans, cucumber and tomatoes, the second layer, up to 5m, is occupied by cassava, banana and papaya, while the upper layers consist of trees whose crowns can reach different levels – a lower stratum of citrus species, guava, coffee and the like, an intermediate layer of mango, sugar palm, mangosteen and bamboos, with the uppermost, such as durian and coconut, reaching 35m. As trees exceed maturity they provide fuel, and open spaces for light-demanding ground layer crops, and so the cycle is maintained.

It is now recognized that such simple descriptions of systems is but a basic step. What agroforesters are trying to discover is how best to achieve maximum production of sustainable resources on a scientific basis. Various scenarios are being actively explored. The tree crop might be underplanted beneath crops such as beans and maize, replaced, as the trees increase in size and shading effect, by cassava, yams and bananas. The trees might then be allowed to develop to sawwood or fuelwood size, clear-felled, and the cycle restarted immediately; or the trees might be maintained as fodder or multipurpose trees – by a combination of pruning, lopping and thinning. The trees can provide fodder from their foliage, fuel and fruit, whilst the managed canopy permits enough light for herbaceous ground crops.

► *A village-forest-garden in Bali.*

The need to preserve

Where former forest ecosystems have been greatly degraded, current attempts at halting the decline will be extended. In this context, much thought and effort is being given to so-called multiple-use trees. For example, the South American palm *Orbignya martiana*, known as the Babaçú (Brazil) or Cusi (Bolivia), germinates well and grows vigorously on eroded tropical soils. It regenerates from the roots when cut down, thus providing a self-sustaining fuel supply. The foliage is useful for grazing, while the fruits, looking like small coconuts, are abundant (up to 0.5 tonnė per tree per year).

Many plant species, including trees, have been brought to extinction by Man in the past 500 years. Very many more will be lost, especially in the tropics, if the current decline continues, and it is estimated that thousands will not even come to our attention before they vanish. Without these plants we cannot determine whether, like *Orbignya*, they hold promise of hope for environments and communities of people. More research is needed on the potential of trees we already know, only a small proportion has been surveyed for drugs, dyes, spices and oils. Trees have long provided Man with innumerable resources. The future will depend upon how well the trees are preserved and our knowledge of them and their resources are expanded.

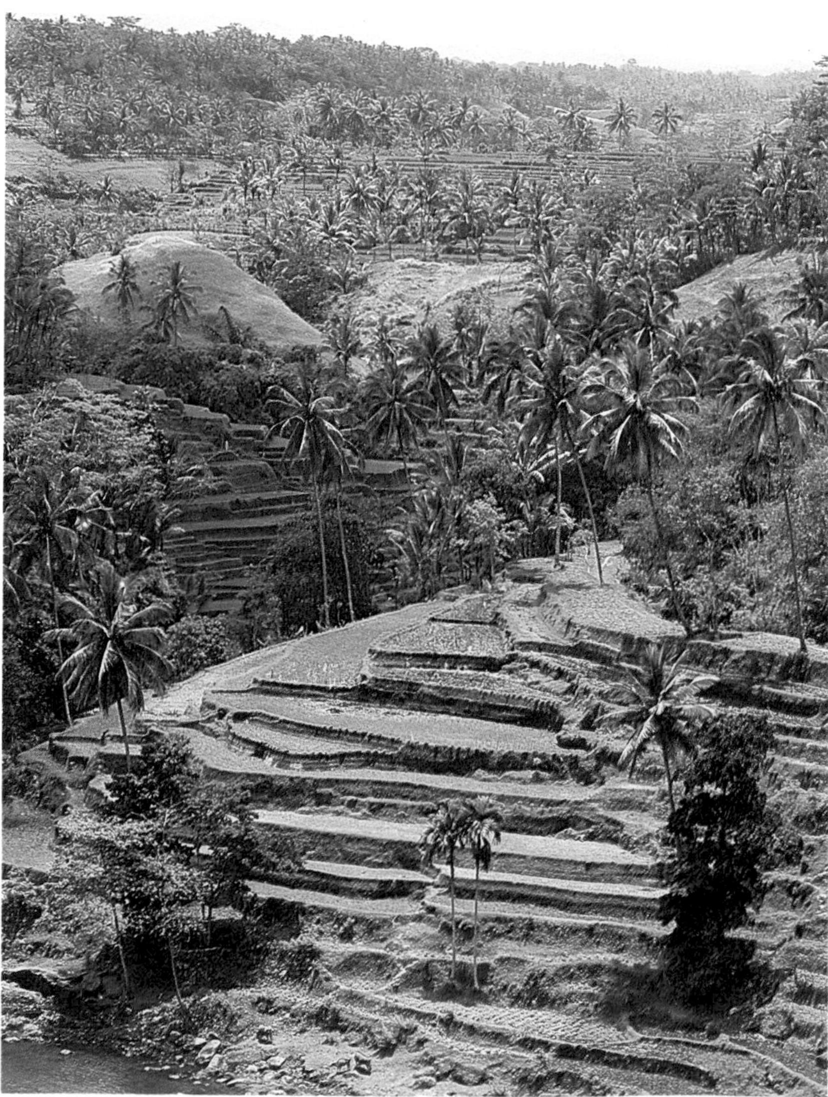

The Last Great Hunt

Fish as a foodstuff...Hunting for fish...Phytoplankton and their predators...Life cycle of sea fish...Fish conservation...PERSPECTIVE...The fish we eat...The world fishing industry...Fishing for different species... The Humboldt current...Fishing technology...Whales and whaling

We have always eaten fish, but for most of history they have normally been freshwater species. There was no way to keep fish caught at sea edible for the time it took to transport it far inland except by salting, drying, or smoking. Preserved fish was eaten, but was of little dietary importance. The building of railways and the development of techniques for producing ice easily and cheaply changed the pattern and by the 19th century, city markets had begun to receive regular supplies of sea fish. Demand for them increased, while demand for freshwater species fell.

Today, in the world as a whole, freshwater and sea fish together provide about 15 percent of all dietary animal protein (◀ page 142) and about one percent of all dietary protein when vegetable foods are included. In some countries the figure is higher: in Japan, for example, fish accounts for about 60 percent of the animal protein eaten.

Fish contains as much protein as other meat. White fish, such as cod or haddock, have less fat and even oily fish, such as herring or mackerel, have much less saturated fat than meat from mammals. Fish meat contains rather less iron than other meats, but is rich in minerals and if the bones are eaten, as they are in the case of some canned fish, they supply useful amounts of calcium and phosphorus.

Each year 65 million tonnes of fish are caught and landed. More than a quarter of the world catch of fish is used as raw material for the production of oils, or for processing into fishmeal, which is then used as feed for farm livestock or as fertilizer. In addition, fish are farmed (▶ page 191), mostly for human consumption.

The fish we eat

Commercial fisheries are based on four main groups of aquatic vertebrates, defined by their habitat as freshwater, diadromous, marine demersal, and marine pelagic. So many species are hunted, many of them only locally, that it is impossible to list more than the most important representatives of each group.

The most important freshwater species belong to the family Cichlidae, which includes the tilapias, and the Cyprinidae, which includes the carps.

"Diadromous" fish are those which live in the sea, but breed in fresh water. The group includes the salmons, trouts, sturgeons, and also the shad, a close relative of the herring. Eels are "catadromous" – they live in fresh water but breed at sea.

Fish which feed on or near the sea bottom are known as "demersal". The young of such species as flounders, plaice and sole are all members of the Pleuronectidae family and are round-bodied when young, and form part of the zooplankton (◀ page 122). They grow flatter in shape as adults and start to swim on their side, eyes uppermost.

*Not all demersal fish are flatfish. The group also includes the Gadidae family, comprising the cods (*Gadus *species), haddock (*Melanogrammus aeglefinus*), and ling (*Molva molva*).*

*The "pelagic" fish are species which inhabit the upper waters. Most are fast, active swimmers with torpedo-shaped bodies. The most important families are the Clupeidae and Scombridae. The Clupeidae includes the herring (*Clupea harengus*), pilchard (*Sardina pilchardus*) whose young are canned and sold as "sardines", sprat (*Sprattus sprattus*) and anchovy (*Engraulis encrasicolus*). All clupeids feed on plankton and occur in temperate and warm waters throughout the world. The Scombridae comprise the mackerel (*Scomber scombrus*), albacore (*Thunnus alalunga*), tunny (*T. thynnus*) and bonito (*Sarda sarda*). They are all fast-swimming predators, found in warm and temperate waters throughout the world.*

◀ Fishing for anchovies off the coast of Peru. Anchovies feed on plankton close to the surface. The Peruvian current brings relatively cool water welling to the surface, supplying nutrients on which the plankton feed and forming the basis of the anchovy fishery. Climatic changes, especially the "El Niño" phenomenon, have increased the depth of warm surface water, preventing cold water welling to the surface. This has depressed the plankton population and drastically reduced the population of anchovies.

Crustaceans, including shrimps, lobsters and crabs, are important because of their high retail value but only account for about 5 percent of the annual world catch from the sea

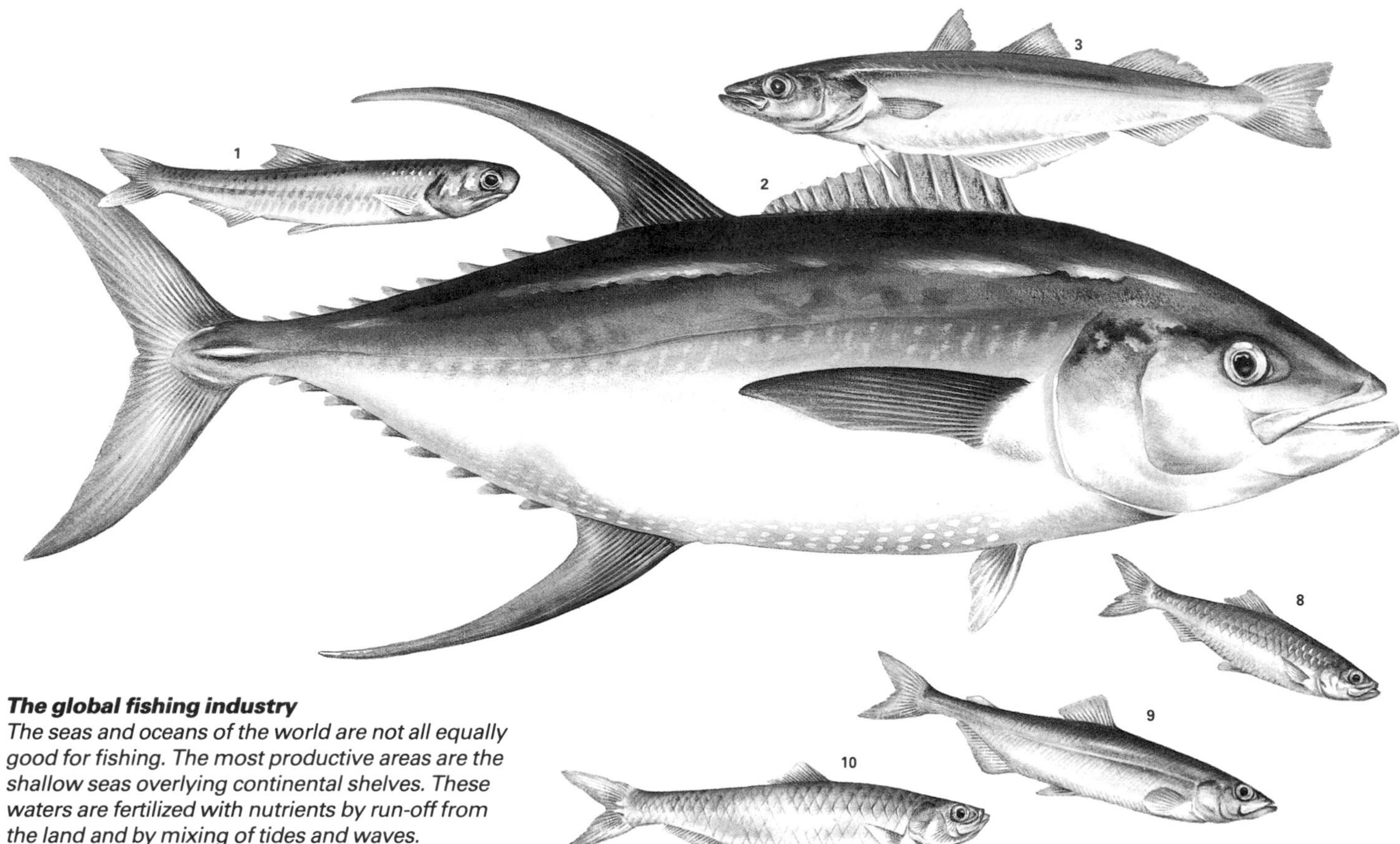

▲ *Important species of commercial marine fish. Peruvian anchovy (*Engraulis ringens*) (1). Yellowfin tuna (*Thunnus albcares*) (2). Poutassou or Blue Whiting (*Micromesistius poutassou*) (3). Atlantic cod (*Gadus morhua*) (4). Skipjack tuna (*Katsuwonus pelanis*) (5). Atlantic herring (*Clupea harengus*) (6). Common (Atlantic) mackerel (*Scomber scombrus*) (7). Sprat (*Clupea sprattus*) (8). Capelin (*Mallotus villosus*) (9). Pilchard or European sardine (*Sardina pilchardus*) (10). These species swim well clear of the sea bed and occupy extensive ranges, limited mainly by water temperature.*

► In lakes and canals in China cormorants have been used for fishing for many centuries. The bird dives for fish from a raft, a cord fastened round its neck to prevent it from swallowing its catch. The cormorant is extremely fast under water and will hunt until its hunger is satisfied.

The global fishing industry

The seas and oceans of the world are not all equally good for fishing. The most productive areas are the shallow seas overlying continental shelves. These waters are fertilized with nutrients by run-off from the land and by mixing of tides and waves. Nutrients support the growth of minute animals and plants (plankton) in the surface waters and this starts the food chain which leads to fish.

The most productive and heavily exploited fishing regions in the world are in the northeast Atlantic and northwest Pacific which together yield nearly half the annual world catch. Both have extensive areas of continental shelf and the adjacent land masses – Europe and East Asia – are densely populated by developed nations with long fishing traditions and well-equipped fishing fleets.

Other important areas include the west coast of South America and southern Africa. In both of these places water from the deep sea, rich in nutrients, upwells at the surface generating strong plankton growth and important fisheries. Some fishing grounds, in particular the Antarctic and Indian Oceans, are underexploited because they are remote from the developed nations. The least productive waters of the world are the centers of the great oceans where there is little mixing and nutrients and plankton are scarce.

Japan and the Soviet Union together take nearly a third of the annual world catch. This is because they have both invested in well-equipped, long distance fleets capable of fishing for extended periods anywhere in the world's oceans, from the Falkland Islands to the Seychelles to the English Channel. Other important fishing nations include those with modern medium-range fleets and good fishing grounds near at hand, such as the United States, Peru, Korea, Norway, Denmark, Thailand and Spain.

*Inshore fishing fleets often provide more employment among coastal communities than the distant-water fleets. Other fishing methods are used, especially in Asia (**➧ page 183**).*

A variety of fishing methods

Fish is the only food of any economic importance that we still obtain by hunting. New technologies have been developed for locating (➧ page 188) and catching it, and there has also been more research into fish husbandry, although with certain exceptions it remains cheaper to hunt fish than to raise it, mainly because the fish themselves are expensive to feed.

In addition to vertebrate fish, we also catch or gather a range of marine invertebrates, including crustaceans and mollusks. All marine vertebrates are carnivores. This places them high on food chains and makes them extremely vulnerable to changes at lower levels.

Rivers supply the lowest level in the food chain. For this reason, most marine life is found in the waters covering the continental shelves. As it drains the surplus water from its catchment area, a river collects fragments of organic debris, soil, soluble plant nutrients washed from the land surface or dissolved from underlying rocks and carried in the groundwater, and discharges all of it into the sea (➧ page 88). There it is available to nourish aquatic plants, but the further it is carried from the river mouth, the more depleted the water becomes, as part of its suspended load settles to the bottom and as its nutrients are used and incorporated into marine organisms.

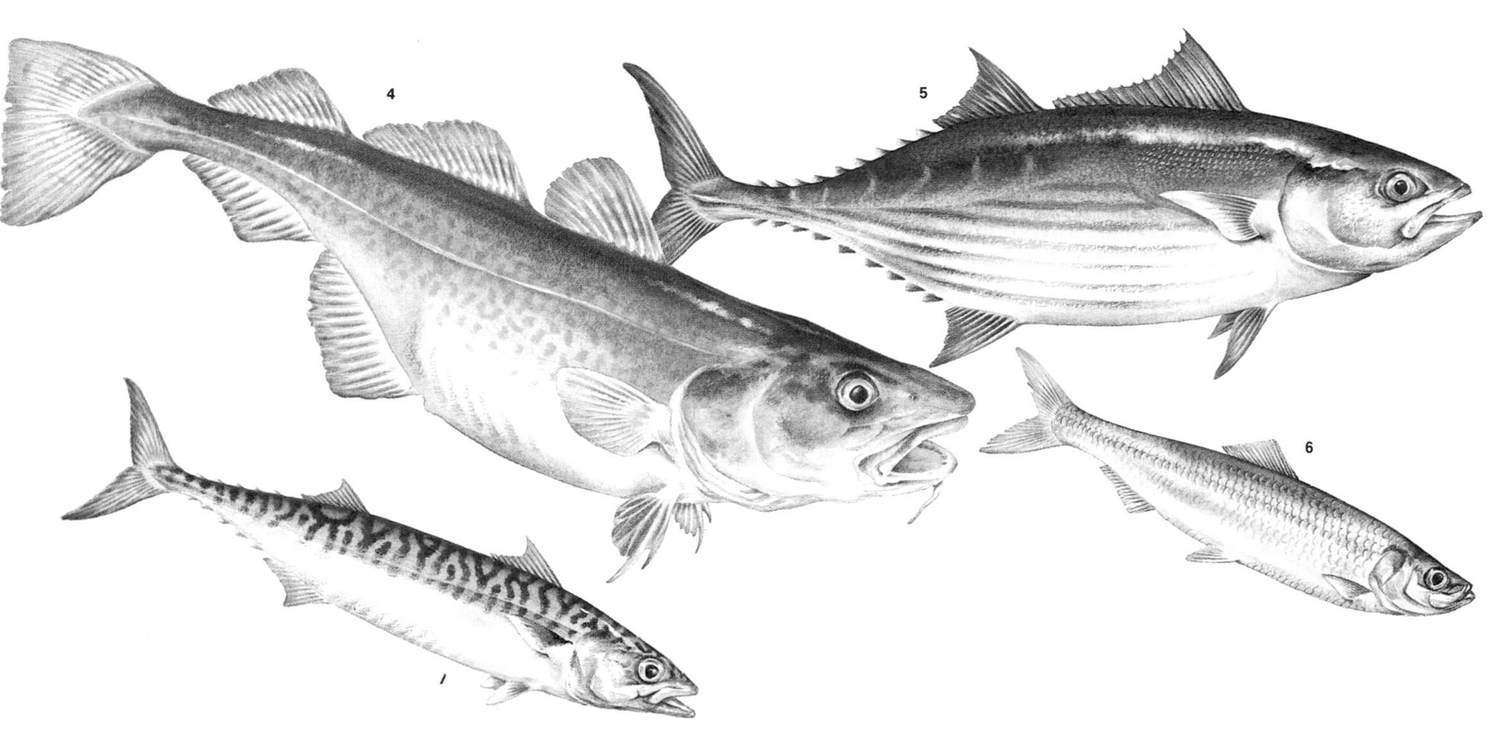
4
5
6

Humans, as well as fish, can eat krill, and the commercial catch of these crustaceans is increasing, to the detriment of Antarctic wildlife

Fishing for different species

*As successful opportunists, we should be prepared to change our choice of prey if our original prey becomes scarce. This might require us to eat different kinds of marine animals, invertebrates as well as vertebrates, and there are several from which to choose. More use might be made, for example, of the blue whiting (*Micromesistius poutassou*), a demersal species that occurs in large numbers in the eastern North Atlantic and Mediterranean, but is caught in only small amounts for consumption in Italy and Spain.*

*It would be more sensible ecologically to move further down the marine food chain and hunt species that were formerly hunted by the top carnivores whose stocks we have depleted. Probably the most common animal in the sea is the squid (*Loligo species*). It, too, is a carnivore, but virtually all predatory fish feed on squid.*

The squid is a mollusk in which the shell is reduced to a "pen" or flattened plate. It is used in Mediterranean dishes but could be exploited more widely because of its remarkable rate of growth. Most vertebrate fish reach a marketable size after not less than four years. Squid hatch as tiny, planktonic, but squid-like larvae, and those that survive the predators reach a length of 25-30cm within 11 months, but live for no more than one to three years. They are caught by trawling and their high rate of productivity should make them commercially more attractive than they are.

Phytoplankton and their predators

Away from the very shallow water where large seaweeds grow are found microscopically small sea plants, known collectively as "phytoplankton". The largest of them is no more than one millimeter across, and the smallest is a thousand times smaller than that (◆ page 117). They make sugars by photosynthesis just as land plants do, but obviously they would gain nothing by growing upward in search of more light. Nor would they gain by growing downward in search of nutrients, for these are present all around them in water that is mixed constantly by currents, tides and the wind. They have no need for stems, branches, leaves or roots. It is these plants which give many seas their green color; blue water is a sign of a marine desert.

The combination of winds (◆ page 97) and the rotation of the Earth move the waters of the Atlantic and Pacific Oceans constantly. Nutrients may be gathered by a current of cold water moving along the ocean bottom and carried for long distances. If the cold current then wells up to the surface, the sea will be enriched locally.

The plants support a community of grazing animals and their predators, the "zooplankton", and most of them are also very small. They include protozoa, but also the larvae of such larger animals as worms and mollusks, and all are equipped with cilia or, in the case of larger animals, with gill slits that have screens through which water is strained and food collected. The zooplankton is hunted by larger carnivores, many of which spend the day far below the surface and drift upward to feed as night falls. They include jellyfishes, comb-jellies, arrowworms, predatory shrimps, small squid, and, largest of

all, fish and mammals. Some of these are very large. The giant manta ray, whale shark, basking shark, and the baleen whales feed exclusively on zooplankton as do many of the fish we eat (◀ page 122).

On the sea bed, life is different. Nutrients drift down constantly from above and accumulate in the sediment. There they sustain a range of microscopic organisms and invertebrate animals and it is these animals that in turn sustain the larger bottom-dwelling fish.

Life cycle of sea fish

Fish reproduce by laying eggs, in most cases for fertilization outside the body of the female, and in coastal waters. The eggs drift with the plankton, the larvae feeding for a short time on the yolk carried with the eggs, and continuing to live among the plankton as juveniles. They may be four or five years old before they mature sexually and it is then that they join shoals. The local stock of a species comprises the shoals living in a particular sea area. Which stock the young fish joins depends on where it was spawned (◀ page 71), for each inhabits its own area and its population fluctuates over the years in response to changes in the number of eggs laid and of the number of its young that survive to maturity.

Overfishing may deplete a stock, although it is inconceivable that it could bring even a local population close to extinction, and it is not the only reason stocks decline. Favorable conditions may lead to heavy spawning in certain years. The population increases, but the young fish do not mature and join the shoals for about five years. It is then that catches start increasing. A few years later, catches fall.

The Humboldt current

The Humboldt (or Peru) current flows from the Antarctic along the Pacific coast of South America, with strong upwellings near to where it meets a warm current flowing south from the Gulf of Panama. This brings its nutrients close to the surface where they support vast populations of fish, including the local species of anchovy (Engraulis ringens). The anchovies sustain a large fishing industry, catching mainly for the fishmeal industry, and also the flocks of seabirds whose droppings on nearby islands provide guano fertilizer. Climatic changes may move the warm current farther south; known as "El Niño", it is associated with a sharp fall in the fish population. In 1971 the Peruvian fishing fleet caught nearly 13 million tonnes of anchovies; by the end of 1973 the catch had dropped to only a little over two million tonnes.

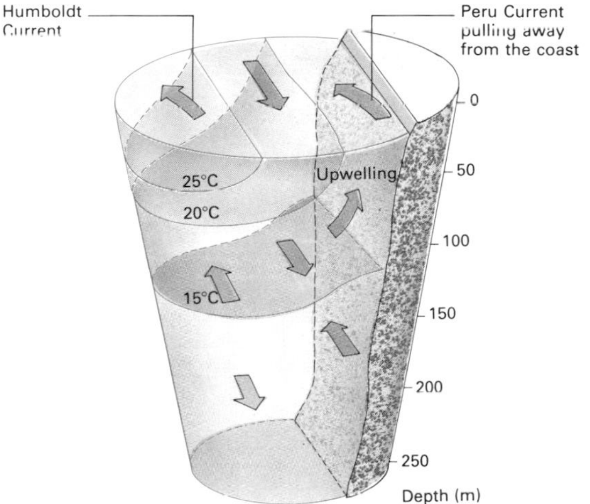

◀ Krill are tiny crustaceans that feed on the rich plankton in the southern oceans. Lacking competition, their population is vast. They support the carnivores.

▼ Phytoplankton determines the distribution of all animal life that feeds on it. Its distribution is determined by temperature, currents and nutrient-rich upwellings.

► The prevailing Trade Wind system off Peru causes an ocean current to pull away from the coast. Cold water from the depths then wells up to take its place.

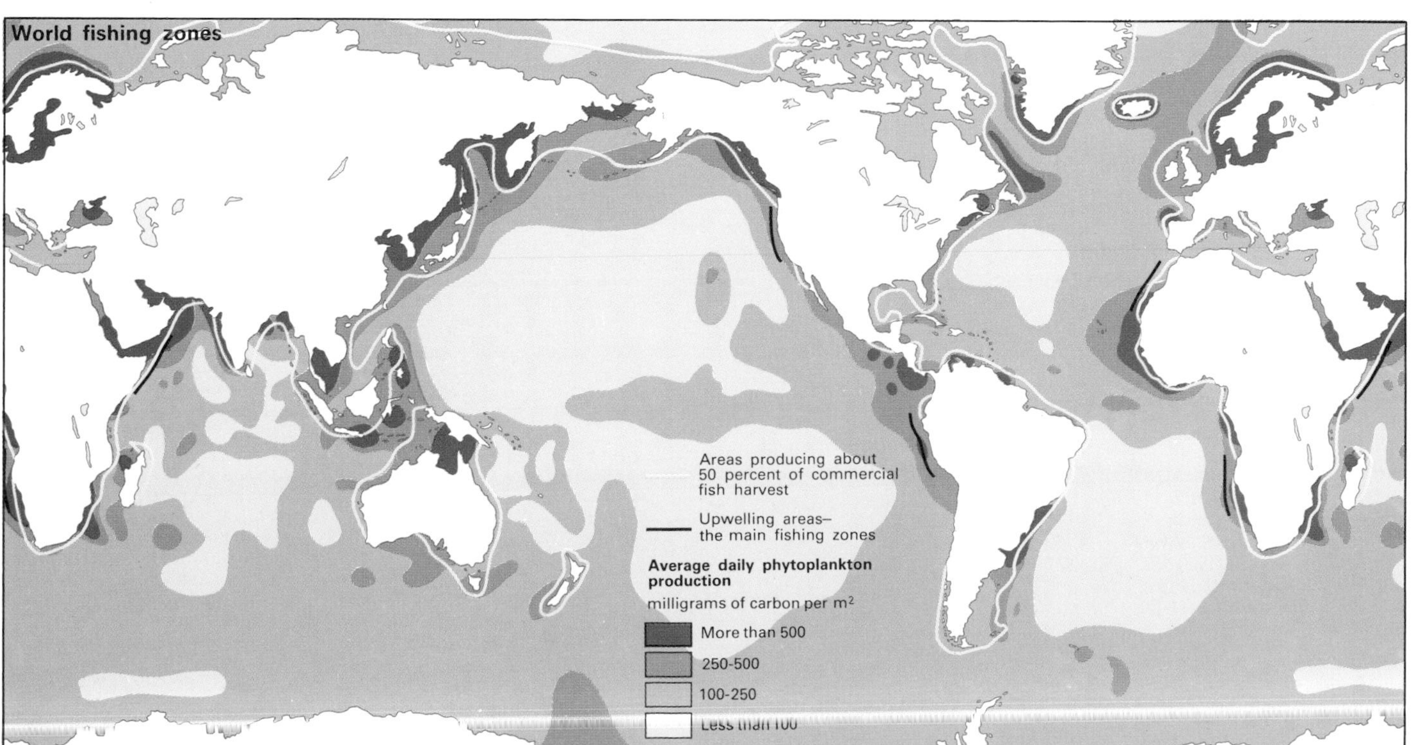

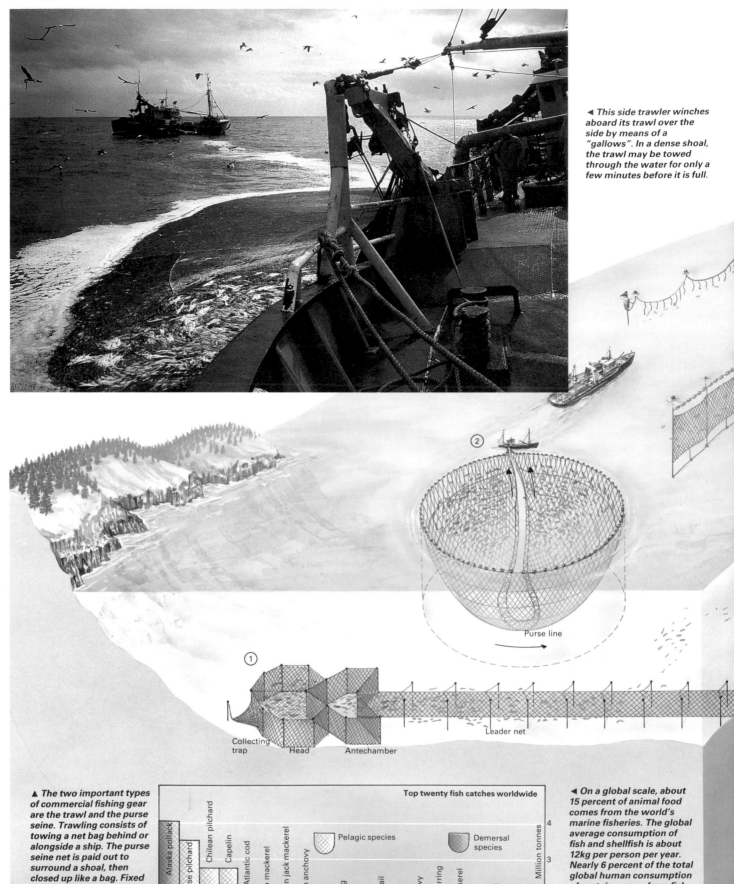

◀ *This side trawler winches aboard its trawl over the side by means of a "gallows". In a dense shoal, the trawl may be towed through the water for only a few minutes before it is full.*

(2)

Purse line

(1)

Collecting trap

Head

Antechamber

Leader net

▲ *The two important types of commercial fishing gear are the trawl and the purse seine. Trawling consists of towing a net bag behind or alongside a ship. The purse seine net is paid out to surround a shoal, then closed up like a bag. Fixed nets, which catch fish by entanglement, and long lines, which aim to catch large fish on a series of hundreds of baited hooks strung out at regular intervals, both use many kilometers of nets.*

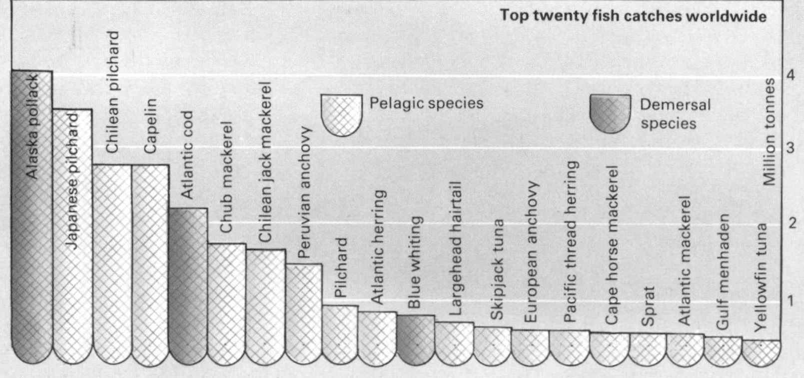

◀ *On a global scale, about 15 percent of animal food comes from the world's marine fisheries. The global average consumption of fish and shellfish is about 12kg per person per year. Nearly 6 percent of the total global human consumption of protein comes from fish and shellfish. Many countries are heavily dependent on fish for protein, and in developing countries fish may form a vital component of a barely adequate diet.*

Line and Net Fishing

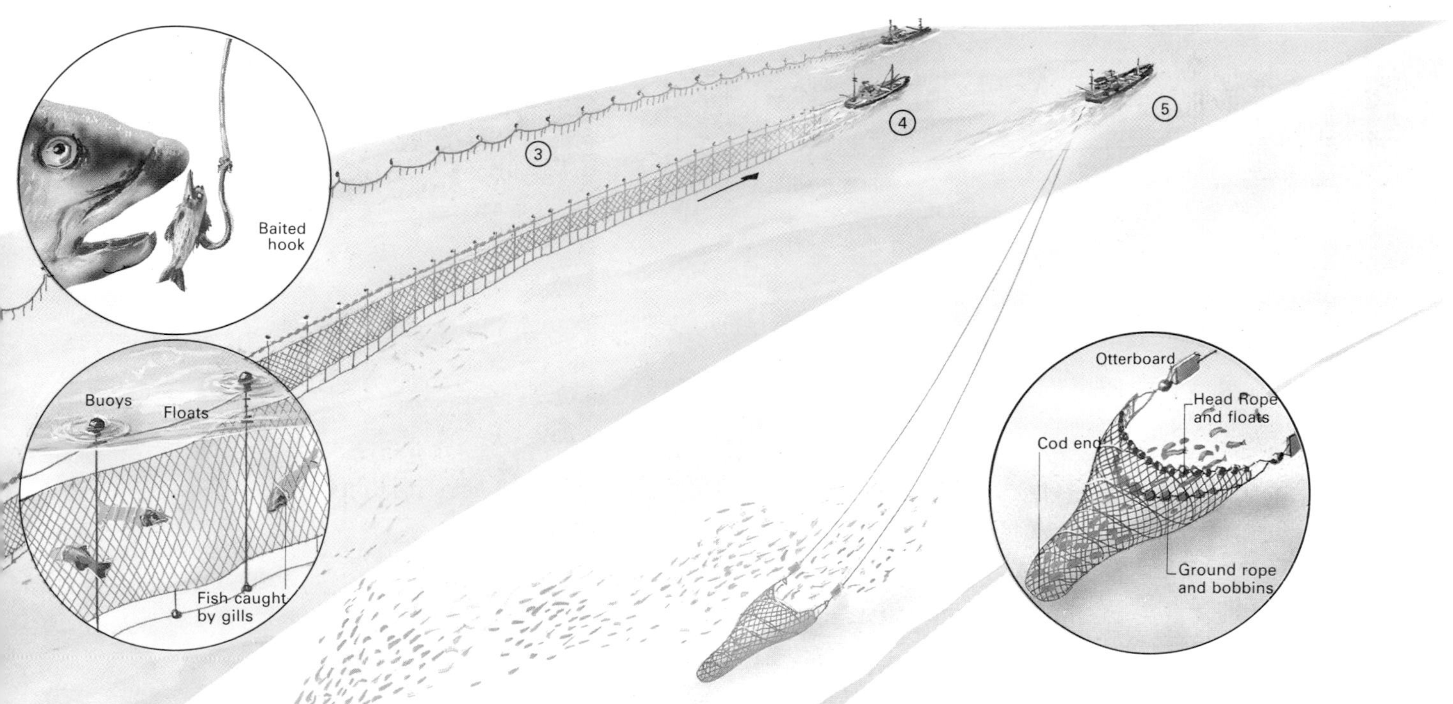

Fishing gear
1 Fixed net
2 Purse seiner
3 Long-lining
4 Drifter
5 Factory stern trawler

Fishing technology

There are two types of fishing fleet: "inshore" fleets use small vessels, relatively primitive techniques, and work in their own local waters; "distant water" fleets comprise much larger vessels, sometimes including "mother" ships that receive the catches and process the fish at sea.

All modern fishing vessels carry sonar equipment to locate shoals, measure their size, and estimate the quantity of fish they contain (▸ page 188). Large vessels have catching equipment that can take an entire shoal from the sea in one haul.

It is now possible to catch more fish each year than are spawned and so to deplete a stock, and because distant water vessels can travel far from their home ports, those fleets move from one fishing ground to another, taking fish that would otherwise be caught by local inshore fleets.

Demersal fish are caught mainly by nets, but spears are still used in many parts of the world, especially in estuaries with a large tidal movement. The incoming tide brings sea water flowing beneath the outgoing fresh water and flatfish can be found close to the bottom in shallow water. Fishermen impale fish with a three-pronged spear or pike (like that carried by Poseidon).

Line fishing is more efficient. It is used by small vessels working directly above shoals. Hand lining involves lowering a line with up to about 30 unbaited hooks, then winching the line in by hand. It does not disturb the shoal, can be used in small shoals or those close inshore, and can be stopped easily if the fish it catches are too small. It is hard work but productive.

The trawl net is used for pelagic and midwater fishing. Very large trawls may be towed by two vessels ("pair trawling"). The trawl is a funnel-shaped net towed through the water and held open at the neck by "otter boards", attached in such a way that they tend to move outwards. The closed "cod" end of the net tapers to a narrow, sealed tube. Small trawls may be lowered from the side of a vessel ("side trawlers"), but larger ones are lowered from the stern ("stern trawlers"). The trawl is winched aboard over a "gallows". The mesh size can allow small fish to escape, provided they are not trapped by the larger fish around them.

The beam trawl, used for demersal fishing, is a trawl the lower edge of whose open end is fixed to a heavy beam, which is dragged along the sea bottom.

The drift net is used for pelagic fishing. It consists of a curtain of net, commonly 15m deep and in sections joined to make a length of up to 2.5km. Weighted at the bottom and floated at the top, it drifts in the water and catches fish by their gills. The size of the mesh is fixed by law to allow smaller fish to escape.

The ring-net, or seine net, is a variant of the drift net, used from a beach or by vessels working in pairs in calm water. The net, usually 300m long and 36m deep, is paid out in a circle to surround a shoal, then hauled in by lines attached to the bottom as well as the top of the net, so the net is closed like a sack or bag.

The purse seine net, also used for pelagic fishing, takes this principle a stage further. The net is paid out by one vessel, in a circle to surround a shoal, then closed, or "pursed", by hauling in a cable running through rings on the edge of the net so it forms a basin in which the fish are trapped. A modern large purse seine net may be 1.2km in circumference and 150m deep. It is wholly indiscriminate and the fishermen cannot see what they have caught until the net is alongside. If the catch is unsuitable the net is opened and the catch "slipped", although many of the fish will have been crushed or fatally injured.

Despite most nations ceasing to hunt whales, several populations of these giant mammals are endangered by the activities of those that interpret "conservation research" to include culling on a large scale

15
30
45
56
60

Asdic
4°
12°
Distance from ship
Echo sounding (depth)
Shoal

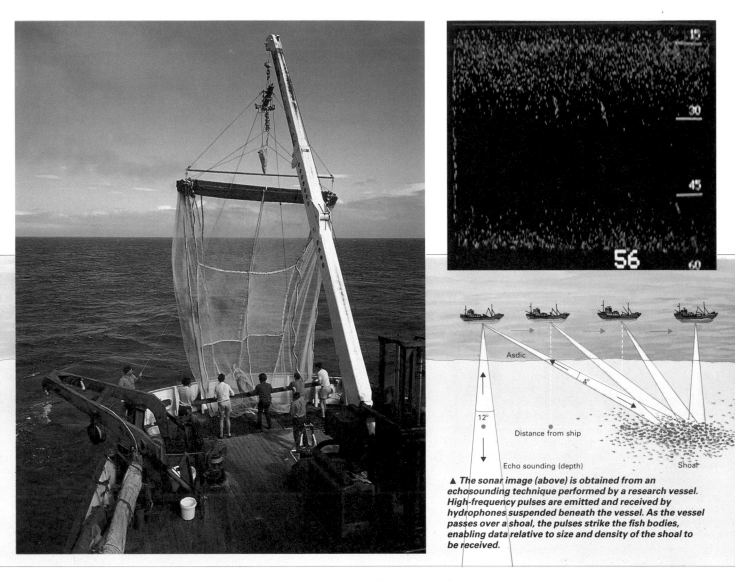

▲ The sonar image (above) is obtained from an echosounding technique performed by a research vessel. High-frequency pulses are emitted and received by hydrophones suspended beneath the vessel. As the vessel passes over a shoal, the pulses strike the fish bodies, enabling data relative to size and density of the shoal to be received.

▲ The oceanographic research vessel, R.S.S. **Discovery. *Successful conservation measures employ sophisticated scientific techniques, including the sampling of individuals and computerized imaging of shoals.***

▼ This smolt, a young salmon of about two years old, is marked with a numbered tag. Tagging fish enables scientists to estimate the size of a "cohort" population (a group of fish all of the same age) and the number in each which die every year as well as their migration route (♦ page 74).

Monitoring fish populations

Conservation of stocks aims to ensure that the catch does not exceed the rate at which a stock is replenishing itself, and to share the catch fairly among different fishing fleets. Before any conservation measure can be proposed, the stock must be assessed. This is a complex undertaking which involves estimating the total number of fish in the population being assessed, the age distribution of the fish, and the size and age distribution of the catch.

Scientists make acoustic measurements, using sonar and sophisticated imaging equipment, to locate the shoals, estimate their sizes, and track their movements. They sample commercial catches and arrange the number of fish in the sample in "cohorts", a cohort being a group of fish all of the same age. Fish are also caught, tagged, and released again. When tagged fish are caught commercially, the tags are returned to the scientists. Because the ratio of tagged fish caught to the total number of fish caught should be the same as the ratio of tagged fish to the number of fish in the population, the size of the population and the number in each cohort which die every year can be estimated. The plankton is sampled in the areas where the fish breed, and the number of eggs and young fish indicates the rate at which the stock is replenishing itself.

Whales

Whales are mammals (order Cetacea*), which long ago abandoned their life on land and returned to the sea. The earliest fossil whales are known from Eocene rocks, some 40 million years old. They are entirely aquatic but, being mammals, they give birth to live young which they suckle, and they breathe air with lungs, not gills. The forelimbs have become paddles with no visible digits, the hind limbs have disappeared, and the tail has become a horizontal fin used for propulsion (all fish have vertical tail fins).*

All whales are carnivores, but the two groups feed differently. The toothed whales (suborder Odontoceti*) have teeth, but the baleen whales (suborder* Mysticeti*) have in their mouths "baleen" plates made from sheets of keratin through which they strain food from the water.*

The whaling industry

*Whales have long been hunted for their flesh, blubber, skin and bones. In modern times whale products have found industrial uses. The baleen plates have been used as "whalebone" in corsets. Ambergris, from the intestines of the sperm whale (*Physeter catodon*), the largest of the toothed whales, has been used as a fixative for perfumes in toilet soaps, and oil made from the fat of the sperm whale has been used as a high quality lubricant and in many cosmetic preparations such as hair oil and lipsticks. Also used in cosmetics is spermaceti, the vascular tissue filled with oils which solidify at about 32°C up to one tonne of which is contained in the enlarged snout of a sperm whale to provide the animal with neutral buoyancy. Tendons from whales have been used to string tennis rackets. Substitutes now exist for all these products.*

Most species of medium-sized and large whales are hunted – Right, Gray, Humpback, Blue, Fin, Sei and Sperm – but only the Right, Gray and Blue whales seem to be threatened.

Conserving the whale

Whaling became an economically important industry early in this century and by the 1930s it was a pelagic enterprise, based on seagoing factory ships which processed animals brought to them from their fleets of whale catchers. Fears of over-hunting led to the International Whaling Convention, signed in 1937 and intended to allow quotas to be set. Whaling ceased during World War II, but the postwar world shortage of oils and fats compelled it to resume, but to conserve the stock by a system of quotas set by the International Whaling Commission, a body established under UN auspices in 1946.

Popular opposition to the hunting of these attractive and highly intelligent animals grew throughout the 1960s and 70s. Anti-whaling campaigns, combined with scientific fears that whales were being overexploited, led to a demand for a ten-year moratorium on all whaling at the 1972 UN Conference on the Human Environment. This proposal received wide support from nations whose own whaling industries had ceased, but could not be implemented because of opposition from countries still engaged in whaling. In 1985 Japan, the principal whaling nation, agreed to a five-year moratorium starting in 1986. It was expected that the Soviet Union, most of whose whale meat was sold to Japan, would follow and that Norway would be isolated as the only nation with a fleet still hunting whales. If Norway accepted the moratorium effectively whaling would end because it would be unlikely to resume after a gap of several years during which the ships would find other uses and the employees other jobs.

Stocks of some species had been depleted, and the great difficulty of assessing the size and reproductive behavior of whales caused much disagreement among rival experts. Nevertheless, no species has actually become extinct as a result of commercial whaling.

▲ **Whales have been hunted throughout history and for some people, such as certain Inuit (Eskimo) communities in Alaska, they are an important source of food and raw materials. Their flesh has provided food, the thick subcutaneous layer of fat ("blubber") with which they are thermally insulated has provided oils, their bones have been used as building material, and their skin has been used as leather.**

▼ **In 1925, the factory ship "Lancing" was fitted with a stern slipway up which whales were hauled for processing on deck. This heralded the era of open-sea whaling with expeditions extending far round the Antarctic ocean.**

See also
Sex and Species 35-46
Communities of the Water 117-128
Farming the Waters 191-194
Threats to the Environment 209-224

Non-fish products

Seawater is rich in chemical substances. The salinity of the sea varies from place to place, but averages 35 parts per thousand of salt. "Salt" includes all mineral salts, although all our common salt comes from the sea either directly, by evaporating seawater or, more usually, from mining deposits left by the evaporation of ancient arms of the sea that were trapped behind high ground when the sea level fell. In years to come, we may obtain some of our uranium from seawater, which contains about three parts per million of it, and fusion reactors will use deuterium, an isotope of hydrogen also obtainable from sea water.

In many areas, the sea bed is littered with roughly spherical nodules made of mixtures of metals. No one knows how they form, but they occur on the bed, not beneath its surface, so their metals are probably collected from the water rather than the sediment. They can be collected by dredging, although the technology is advanced and expensive.

*People living near sheltered coasts have always made use of the large seaweeds that grow there and, in modern times, new uses have been found for some traditional commodities. Carragheen is a complex sugar, obtained from some red seaweeds and especially from Irish moss (*Chondrus crispus*), used in various Irish dishes. Today it is also used commercially as a stabilizer in sauces and creams and also in pharmaceutical products and paints.*

The number of fish that may be caught in a year without depleting the size of the stock is known as the biological optimum yield or the "maximum sustainable yield" (MSY). The economic optimum yield is usually smaller, because it is less profitable to fish the smaller shoals than the larger ones, so fleets turn to another species before the biological limit is reached.

The biological optimum yield provides the basis on which a "total allowable catch" (TAC) is calculated. Conservation measures are agreed by the governments whose fleets fish particular species or waters. It may be forbidden to fish a stock between certain dates, or at any time in certain places; to use certain types of fishing gear; a minimum mesh size of nets may be fixed to allow younger, smaller fish to escape; quotas may be set beyond which no fish from that stock may be landed. Governments can enforce these measures only within sea areas under their control, and until recently these were very restricted. Concern over the impossibility of regulating fishing was one of the main reasons for convening the first United Nations Conference on the Law of the Sea in Caracas, Venezuela, in 1974. By 1977, most countries had accepted the concept of the "Exclusive Economic Zone" (EEZ), developed during Conference negotiations. In theory, this allows coastal states to regulate all economic activity over the continental shelf extending from their shores as far as the 200m contour. In practice, it allows them control within a 320km band, or up to a median line agreed with their neighbors in narrow seas or straits, and it has made the task of conserving fish stocks much simpler and considerably more effective.

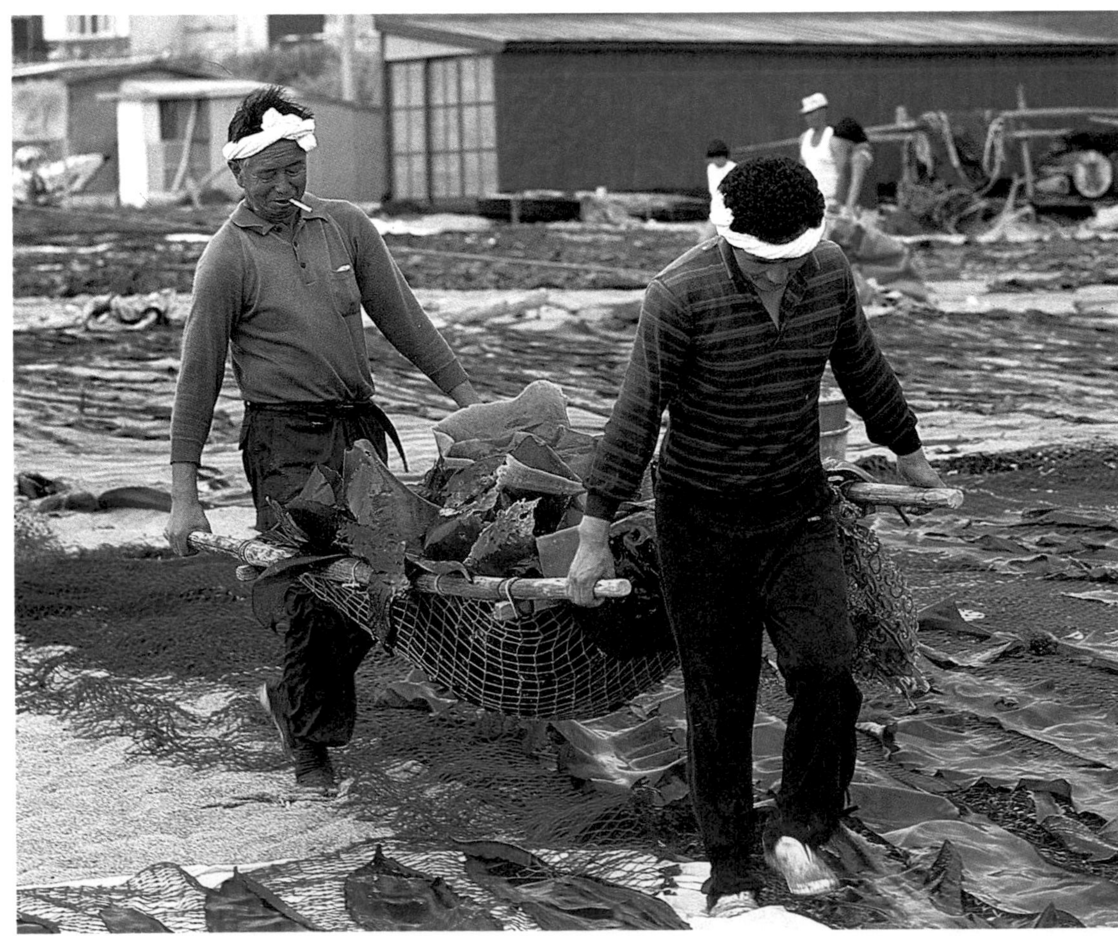

► The seas produce plants as well as fish, and coastal communities have long harvested seaweeds. Kelp, in particular, forms extensive, dense "forests". Most seaweed is used as a mineral-rich fertilizer, but it is also eaten, and used as a raw material for the extraction of chemicals. In Japan, where it is part of the traditional diet, kelp is now being farmed. The picture shows kelp being harvested in Japan from a bed where it is cultivated. The sheets are laid out in the open where they will be washed by the rain and hosed down with fresh water, to dissolve and remove excess salt. They will then be dried. Kelp is the basis of the dish kombu.

Farming the Waters

Fish farming compared to meat production...The economics of fish farming...Flexibility in feeding... Flexibility in management systems...PERSPECTIVE... Farming marine vertebrates and invertebrates...Salmon and trout...Carps, catfish, milkfish and yellowtails...The economic side effects

Tilapia were being raised in ponds in Egypt by 2000 BC, oysters have been grown in managed beds in Europe since Roman times, and the Chinese have raised carp in ponds for at least 2,000 years. The farming of fish is not a new idea. Where the species are herbivorous they can be fed easily and cheaply to provide a reliable supply of animal protein. Invertebrates may require no feeding at all. Today fish are farmed mainly in Asia, probably because fish farming is ingrained more deeply in Asian than in European or African cultures.

The annual yield from farmed fish amounts to only a tiny proportion of the amount of fish eaten each year, but fish farming could be expanded greatly. Economically, the farming of fish is compared unfavorably with the hunting of them; however, it compares very favorably with the production of meat from other farmed livestock. Fish are poikilotherms ("cold-blooded"). They live close to the temperature of their surroundings and do not expend food energy maintaining a constant body temperature. They live in water, which supports their weight, so they expend much less energy than a land animal in opposing gravity and they have no need of a strong, rigid skeleton, whose construction and maintenance is also paid for in food. Because they can live above and below one another as well as alongside, fish can be stocked much more densely than land animals. Intensively farmed cattle may be stocked at about 500-700kg/ha. Fish can be stocked at 2,000-3,000kg/ha.

Farming marine vertebrates

*Many marine fish can be bred and raised wholly in captivity. Plaice (*Pleuronectes platessa*), common sole (*Solea solea*) and turbot (*Scophthalmus maximus*) have been cultivated in captivity experimentally in Britain. In France scientists have also raised bass or sea perch (*Dicentrarchus labrax*).*

*Like livestock breeders throughout history, the scientists select for breeding those individuals which grow fastest and produce the best meat. In time this may result in much larger fish, but it may also lead to new "breeds", produced by hybridization. Fish hybridize more readily than birds or mammals because their eggs are fertilized externally. Plaice hybridize naturally with flounders (*Platichthys flesus*), which share their breeding grounds. Such hybrids may grow more slowly than their parents, but a plaice-halibut cross might grow as fast as a halibut (*Hippoglossus hippoglossus*) and to a large size, but be as easy to raise as a plaice. Plaice and halibut are closely related but breed in different sea areas and so never meet.*

Farming marine invertebrates

Oysters are grown in brackish water and the water must flow, or at least be changed regularly. They require a salinity of 23-28 parts per thousand, the water must be well oxygenated, and if they are to spawn temperatures should be between 20-25°C although the animals can live and grow in much cooler water. They feed on phytoplankton (◆ page 122*). Commercially they may spawn in captivity and be raised in tanks until they are "spats", large enough to be moved to the beds where they will feed until they are harvested. Traditionally, and more simply, spats are collected from the natural spawning grounds and installed in the beds.*

Mussels grow in sea water, so their beds must be located far enough from the shore for them not to be adversely affected by inflows of river water, or even of heavy rain. Usually they are grown on ropes or chains suspended beneath rafts moored in sheltered, well oxygenated, tidal waters. They are not spawned in captivity. "Seed mussels" are collected from natural mussel beds and moved to the rafts. They feed on phytoplankton, which they obtain for themselves. Shrimps can be raised in much the same way, but as scavengers prone to cannibalism they must be fed meat.

*Experimentally several other species have been "ranched" by collecting juveniles from the wild, raising them in captivity until they are past the stage at which they are most vulnerable to predators, then releasing them. Lobsters, crabs, shrimps and scallops (*Pacton species*) have been raised in this way. Eventually the adults are caught by fishermen, but fishing is forbidden in the areas where the animals are released until they have had time to grow to their full size.*

◄ Oysters are the most widely cultivated marine invertebrate, yielding more than 700,000 tonnes a year compared with about 250,000 tonnes of clams and mussels. This farm, in Ago Bay, Japan, grows oysters on ropes suspended beneath rafts. These are not edible oysters, but are being cultured for pearls. Oyster farming is highly labor-intensive.

Dense stocking of fish means that there is a constant risk of disease

Salmon and trout

*Salmon and trout are carnivores, but have a high market value and so can be raised economically. Those trout which migrate to sea probably remain close to the shore throughout their lives. Some never migrate and most trout can spend all their lives in fresh water (**♦ page 75**). The most popular species for farming is the American rainbow trout (*Salmo gairdneri*), marketed either at about 230g or grown to a larger size. Fish weighing up to 9kg have been produced.*

Rainbow trout are usually grown in fresh water ponds, but in Japan and Denmark they are raised in brackish water. Eggs and "milt" (sperm) are removed manually from adult fish and the eggs fertilized and held in trays through their larval stages, then transferred as young fish to the ponds in which they grow to market weight. They will grow slowly provided they are fed and the water is clean and well oxygenated, but if they are to attain their maximum rate of growth they should be kept in running water at a constant temperature of about 15°C (59°F). Their diet should contain 60 percent protein, 10 percent carbohydrates, 25 percent fats, 5 percent minerals, and a range of vitamins. Dense stocking means there is a constant risk of disease, so hygiene is important. Yields can be impressive. In the United States rainbow trout have been produced at a rate of 2,000 tonnes per hectare per year, which is equivalent to 170kg/l/s.

Salmon cultivation is more difficult and two sites are needed. At the freshwater site eggs are fertilized and the young raised until they are ready to migrate. Then they are transported to the marine site, usually in sheltered water in a tidal inlet, where they are either released or kept in cages suspended from rafts. If released, they may be allowed to escape to sea, in the knowledge that they will return when it is time for them to spawn. This is a variety of ranching. Alternatively the site may be closed by nets to prevent their escape.

Carps, catfish, milkfish and yellowtails

Carps are usually raised in ponds fed by rain. The fish feed on plankton so fertilizer is often added to promote plant growth, and the stock is augmented with young fish caught wild in nearby rivers. When raised intensively, however, in running water and extra feeding Japanese fish farmers have produced annual yields of 1,000-4,000 t/ha.

*Catfish are more difficult to raise because they are carnivores and must be fed. They are kept in freshwater ponds and in the United States channel catfish (*Ictalurus punctatus*) raised in ponds stocked with young fish with added fertilizer and extra feeding have yielded 3t/ha a year. In Asia walking catfish (family Clariidae) are raised in ponds supplied by rain water and are fed waste fish and rice.*

*Milkfish (*Chanos chanos*) provide the most important fish-farming enterprise in the Philippines, where they are a popular food. Young fish are caught wild, raised in ponds where they are fed, and sold when they weigh up to 1kg. Even without extra feeding they have yielded 1t/ha a year in Taiwan.*

*Yellowtail (*Seriola lalandi*), popular with anglers off Pacific coasts, is the fish most widely cultured in Japan, but it is difficult to raise. It requires brackish water, a water temperature of 18-29°C, and a diet that includes fishmeal and whose cost amounts to about half the total production cost. The Japanese raise yellowtails in floating cages 35-100m square and 3-6m deep, made from net.*

► The more widely cultivated vertebrate fish include: Atlantic salmon (1); Catla, an Indian carp (2); Tilapia (3); Ayu (4); Yellowtail (5); Milkfish (6), of which about 170,000 tonnes are produced each year; Thick-lipped gray mullet (7); European eel (8); Common carp (9); Channel catfish (10). The catfishes are bottom-feeding scavengers, and carps and tilapias are herbivorous. Herbivorous species can be fed more cheaply than carnivores and tend to be more highly productive. They need a minimum of management. Methods for harvesting freshwater and diadromous species grade from catching wild, naturally-occurring stocks to full-scale breeding.

6

7

8

9

10

▼ *Unlike trout, which spend all their lives in fresh water, salmon must be hatched in fresh water and moved later to salt water, where they can be kept in moored pens, as here in the United States.*

See also
Animal Migration 71-80
Communities of the Water 117-128
The Last Great Hunt 181-190

► Here, in Java, villagers protect their fish by keeping them in cages. This prevents their escape and keeps out predators, but it also make it difficult for the fish to find their own food. Penned fish must be supplied with food and so require more management, and there is a danger the fish may injure themselves against the sides of their cages as they move about, especially if they struggle to be free.

Side effects

Fish farming is generally beneficial, but there are dangers. Where it has been practiced for centuries attempts to "modernize" methods may favor the richer landowners. This is a problem that occurs in all agricultural modernization programs in developing countries (♦ page 144), but in the case of fish farming it is more acute because the cost of modernization is higher when farmers are expected to move from working natural ponds with the minimum of attention or feeding to intensive systems. Costs include siting and preparation of the tanks and a diet suitable for each stage of the development of the fish. Costs become more critical with the increased output from the more productive farms tending to push down prices.

Ecologically, fish farming requires clean, usually well oxygenated water, so fish farmers are likely to complain if water is polluted by other users.

*If exotic fish species are introduced for farming the local ecology may be disturbed. This has happened in Lake Victoria, East Africa, for example, where overfishing had depleted the native stock of cichlids, including tilapias, and the Nile perch (*Lates niloticus*) was stocked in nearby ponds to become an alternative source of food. A fish-eating species, individuals escaped into the Lake and are now widespread. It feeds on the cichlids but can never replace them since the total mass of a predator species can never equal the total mass of its prey, and it is less popular with local consumers.*

Fish farming is flexible. While some species must be fed a diet rich in protein, and often made from fish wastes or cereals, others have more modest requirements. The addition of a little phosphate fertilizer to encourage aquatic plant growth may be all that is needed for herbivorous species, and most invertebrates require no additional feeding at all. Provided their beds are sited carefully they will filter all they need from the natural constituents of the water around them.

It is no less flexible in the range of management systems it offers, which is at least as wide as that available to other livestock farmers. Depending on the species and its market value fish can be ranched, fattened in ponds or lagoons with a feeding regime designed to bring them to marketable weight by a particular time, or raised intensively, sometimes in cages, throughout their lives, from egg to market size.

In practice, farmed fish are sold alongside hunted fish rather than other meats. A candidate for domestication must therefore be a familiar and accepted hunted species that commands a price high enough to make farming it profitable. Tilapias, for example, are natives of Africa, where natural stocks are fished from lakes and rivers, but they have been introduced, and are cultivated, in many other parts of the world. Freshwater vertebrate fish are farmed more widely than diadromous (occurring in both fresh and salt water at different stages in their life cycle) or marine species partly for this reason, but also partly because of the difficulty of choosing sites. Freshwater ponds either exist already or can be made easily. Sea water is very difficult to synthesize from fresh water, and it is only on certain coasts that suitable sites are found.

The Shadow of Extinction

18

The mechanisms of extinction...Extinct, vulnerable and endangered species...The importance of preserving wildlife...Genetics of conserving endangered animals ...Effective population size...PERSPECTIVE...The plight of the wolf and the giant tortoise...Time-chart of extinction...Ratites in danger...Gene pool and inbreeding depression...Mutations and natural selection...Introduction to breeding in captivity

We share this planet with perhaps 10 million species of living creatures. By the end of this century at least 1 million of these will be extinct. Within another five centuries probably only a million or so species will persist. That is the best that can be hoped. Unless conservation is pursued vigorously, far fewer than that will survive.

Of course, all species eventually become extinct. Those that now exist probably represent less than one percent of those that have lived since life began, some 3.5 billion years ago. The fossil record shows that species of mammal persist for an average of only 600,000 years before they give rise to something else, or simply die out. The fossil record shows, too, that there have at times been periods of mass extinction. All the dinosaurs disappeared within a few million years of each other, at the end of the Cretaceous period, around 65 million years ago.

Mass extinctions in the past, however, have never occurred on anything like the present scale, or at the present rate (➧ page 198). The current wave of extinction is not "natural". It is caused primarily by human beings. Human beings are more intelligent than other animals, and better able to communicate (➧ page 137). They organize, and they pursue long-term plans. If humans choose to regard another species as a competitor, they do not allow it simply to retreat. Instead they conduct campaigns against that species, and may pursue it long after it has ceased to be a threat, until it is eliminated. For example, humans have generally regarded the wolf as a competitor and have already wiped out several subspecies (➧ page 197).

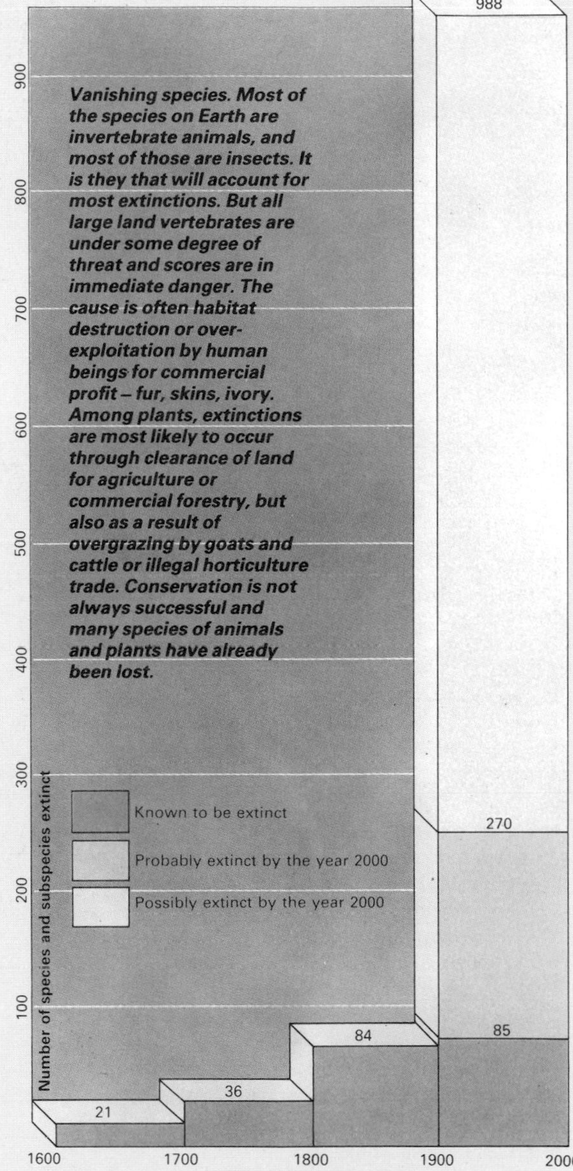

Vanishing species. *Most of the species on Earth are invertebrate animals, and most of those are insects. It is they that will account for most extinctions. But all large land vertebrates are under some degree of threat and scores are in immediate danger. The cause is often habitat destruction or over-exploitation by human beings for commercial profit – fur, skins, ivory. Among plants, extinctions are most likely to occur through clearance of land for agriculture or commercial forestry, but also as a result of overgrazing by goats and cattle or illegal horticulture trade. Conservation is not always successful and many species of animals and plants have already been lost.*

Why are so many species threatened?

Animals and plants are driven to extinction because of various kinds of conflict with other living things – that is, because of biotic factors – or because their environment is changed by physical factors. Biotic factors include competition – when, for example, two different species compete for the same food, or the same nesting hole (➧ page 111) and predation – when one animal feeds upon another. Physical factors include changes in climate, or the rise and fall in sea level. Human beings are now by far the most significant of the biotic factors, and are also responsible for more destructive physical changes than are ever likely to be caused by natural means.

Thus, except on islands, where animals and plants are particularly vulnerable (➧ page 196), direct competition between animals does not generally lead immediately to extinction. On continents there is plenty of room, and when an established animal is faced with competition it simply retreats.

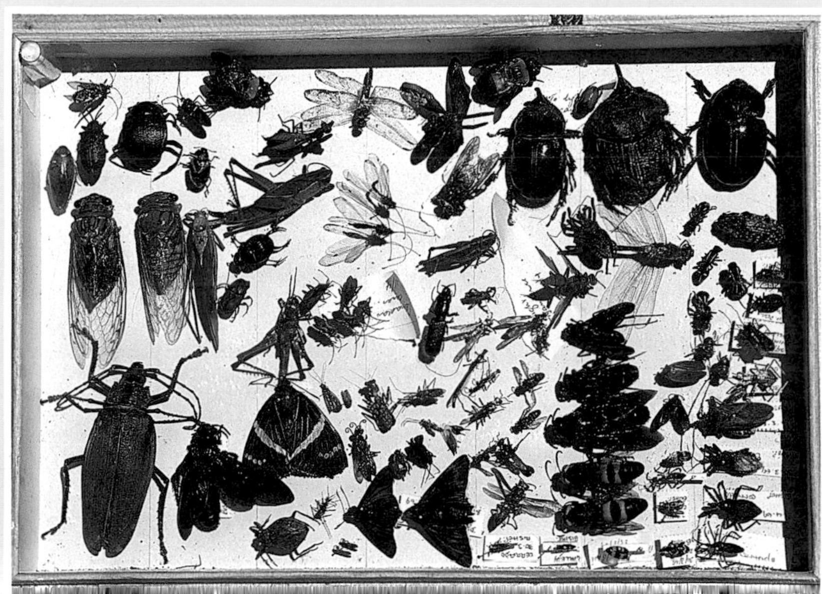

◀ Insects are the most numerous and diverse of creatures.

Of the 30 subspecies of wolf that once lived in North America, 20 have, in the last 80 or so years, been driven to extinction through shooting and poisoning by professional bounty hunters

Human beings are extremely accomplished predators

Whatever the animal, humans can devise a way to catch and kill it. But it is their memory and persistence, rather than their technical skill, that makes human beings so very dangerous. Other predators rarely if ever drive their prey to extinction. If they did, they would become extinct themselves. As a particular prey animal becomes rare, the predator generally switches to hunting another animal until the original kind begins to recover. But humans continue to hunt a favored prey even after it has become rare; indeed, the rarer the prey becomes, the more highly it is valued. In historical times, humans have accounted for many an island species (➧ page 198) simply by catching them for food.

However, the main reason why we are so dangerous to animals and plants is because we change their physical environment and so rob them of their habitats (➧ page 131). Natural changes in the physical environment tend to occur fairly slowly (➧ page 210), so that animals and plants have time to adapt and evolve – as many did, for example, when the Ice Ages began – or they are local, as when a volcano erupts. We now change the environment far too rapidly to allow other species to adapt, and on a vast scale.

But why should we care about the loss of wildlife?

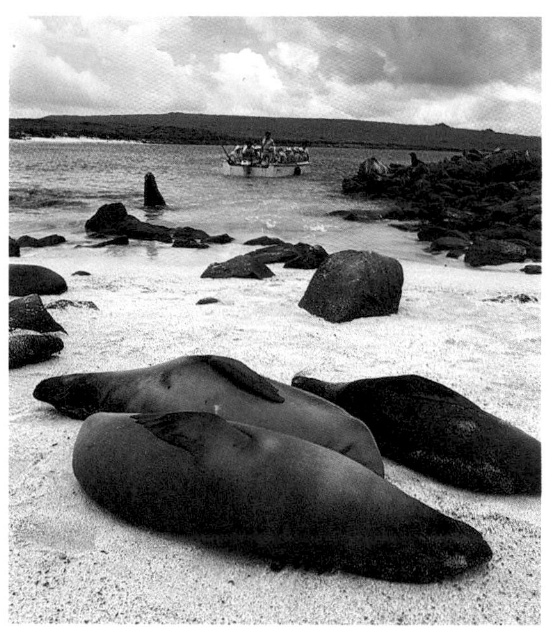

▲ **Tourists are the latest threat to Galapagos wildlife.**

Vulnerable islanders

By definition, islands are isolated and small. If they are large, they qualify as continents. Because they are small and isolated, the animals and plants that live on them are particularly prone to extinction.

The small wildlife populations on islands are in constant danger of extinction even when they are under no obvious threat. Islands offer few "hiding places"; when continents are invaded by new kinds of animal (such as starlings in North America) the native animals simply retreat. But when islands are invaded there may be nowhere to retreat to.

Because islands are isolated, the animals that inhabit them frequently evolve ways of life or body forms that they could not have evolved on continents, when they would be harassed by competitors and predators. Thus island animals are often far smaller or larger than their mainland relatives. Five million years ago Cyprus had tiny elephants (adapted to the lack of food), but also had giant dormice (which would have been rapidly gobbled up if the island had contained any suitable predators). Birds on islands are frequently flightless. After all, birds evolved flight to escape predators, and to feed; if there are no predators, and if they happen to feed on the ground (as many birds do) then flight can be a disadvantage. It requires enormous energy and, on small islands, can be dangerous, as flying animals may be blown out to sea.

Not surprisingly, then, a large proportion of extinctions in the past few hundred years have occurred among island animals (➧ page 198). Overall, scores of island species have become extinct or are endangered. But the best that conservationists can now do to conserve animals in the wild is to put them into reserves. And reserves, separated from one another by farms and cities, hostile to wildlife, are in effect, islands. The animals within them are in some ways as vulnerable as the dodo, the extinct giant pigeon of Mauritius.

◄ ***A rock crab on the Galapagos. By studying species on these islands, Darwin explained how species evolve and the process of natural selection. Similarly, Alfred Wallace (1823-1913), studying islands of the Malay archipelago, shed light on wildlife distribution.***

◄ ***Giant tortoises on the Galapagos Islands – an 18th century print. These animals illustrate the vulnerability of island species. Giant tortoises could not have evolved on continents. As their size has increased so have their proportions changed. Change of proportion with size is a common phenomenon in nature, known as allometry. The consequence of allometry for tortoises is that the large ones cannot retreat their heads into their shells. A hyena could crush their heads with no trouble. But there were no hyenas on the Galapagos islands in the Atlantic, where 14 species and subspecies of giant tortoise evolved. Yet 4 of 14 Galapagos giants are now extinct (harassed by imported dogs, cats, and goats, but mainly eaten by sailors) and most of the Indian Ocean types have also gone, including two of the marvellous saddleback types, whose recurved shells allowed them to lift their heads high to browse. This is the inevitable fate of small populations.***

Vulnerable predators

Large carnivorous animals are in particular danger of extinction in the wild for three reasons. First, human beings destroy their habitats and eliminate their prey – for example, by replacing forest with farm (➧ page 246) or hunting their food for sport. Second, carnivores compete with humans – occasionally killing people, but far more often killing livestock. Humans, accordingly, persecute them. Thirdly, large carnivores require large territories, and therefore they can never be particularly numerous. But populations of animals need to be surprisingly large if they are to be "safe" from extinction (➧ page 204).

Wild canids – dogs and wolves – have fared even worse than the big cats. African hunting dogs, which hunt in large packs over huge areas, now exist in the wild only in a small area of Botswana, and must be among the most endangered of all species. Two subspecies of "red" wolf, Canis rufus, *have become extinct in this century: the Texan red and the darker Florida black. Charles Darwin noted the wolf-like Warrah, or "Antarctic wolf", in the Falkland Islands in 1833. By 1876 the islanders had destroyed the last one.*

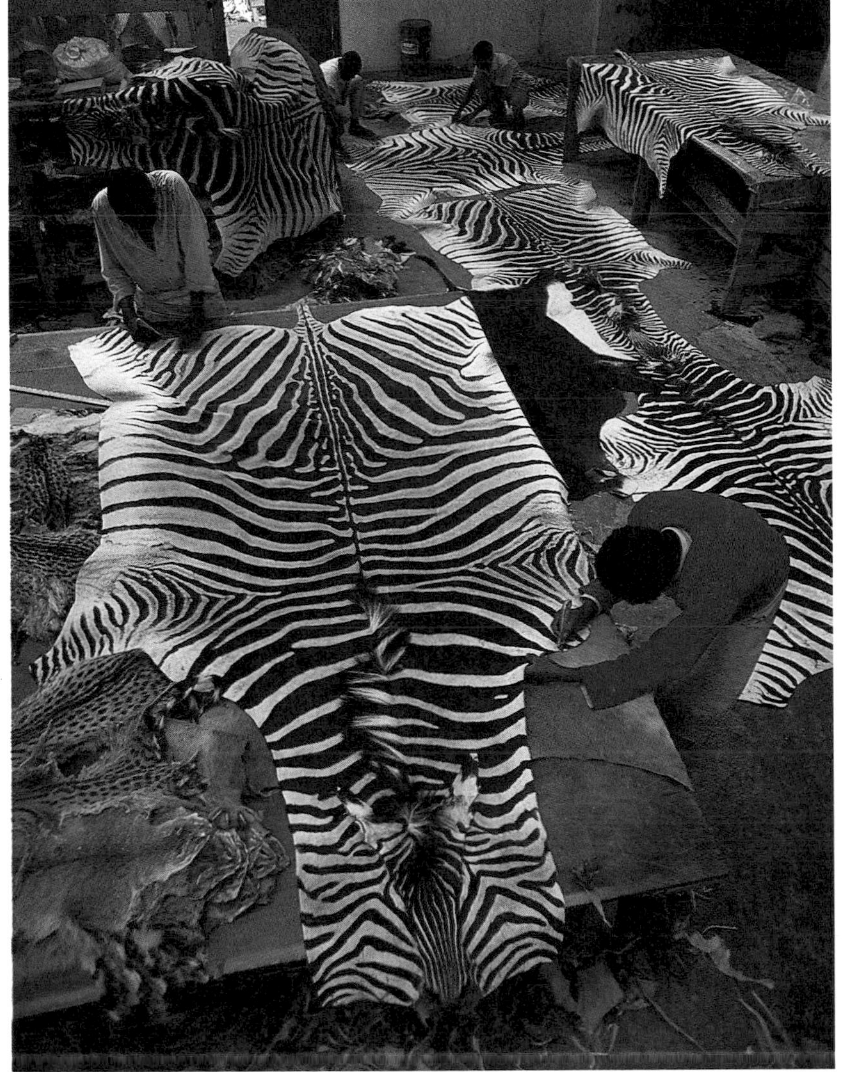

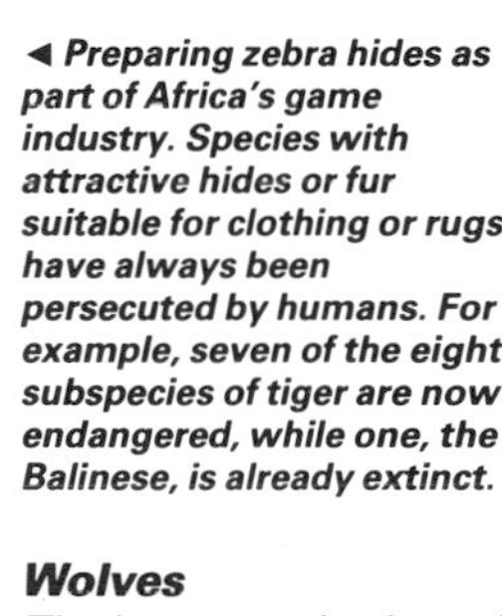

◄ Preparing zebra hides as part of Africa's game industry. Species with attractive hides or fur suitable for clothing or rugs have always been persecuted by humans. For example, seven of the eight subspecies of tiger are now endangered, while one, the Balinese, is already extinct.

▲ The most widespread of the canids is the grey wolf,* Canis lupus. *Once there were 30 distinct subspecies, ranging from the Arctic tundra to Arabia. Now there are less than 10. The huge Newfoundland white wolf was the first North American species to be made extinct, in 1911.

Wolves

The human attitude to the wolf is most unjust. Biologists are now showing that wolves are highly intelligent animals with an intricate and well-mannered social life. Their hunting technique is complex, and very discriminating. When they attack large animals such as caribou, they select old and lame individuals whenever possible (➧ page 57). But in some areas, through much of the year, they subsist in large part upon voles and mice.

Wolves may kill livestock but rarely if ever people. Wolves that are forced into the company of human beings generally learn not to attack them. Thus about 100 wolves are still living in the Abruzzi National Park in Italy, and they never trouble people. Feral dogs, on the other hand – tame dogs that have gone wild – are often extremely dangerous because they have learned no fear or respect for human beings. Where feral dogs are common (as in much of Italy), they often help to give the wolves a bad name.

Wolves indeed are exterminated not for rational reasons, but because of mythology. A medieval writer asked, "For what can we mean by the wolf except the devil?", and this ancient prejudice persists. Thus, in Germany, when wolves escaped from their enclosure into the Bavarian National Park, it was planned, at first, to leave them in the wild. But one of them snapped at a small boy (it had not had time to learn not to behave in this way) and the entire population was immediately eliminated.

Extinct and Endangered Wildlife

1 Mammoth
2 Woolly rhinoceros
3 Cave lion
4 Cave bear
5 Irish elk
6 American mastodon
7 Imperial mammoth
8 Giant ground sloth
9 Saber-tooth "tiger"
10 Dire wolf
11 Reunion solitaire
12 Dodo
13 Guadaloupe amazon
14 Elephant bird
15 Auroch
16 Steller's sea cow
17 Blue buck
18 Hispaniolan hutia
19 Green and yellow macaw
20 Moa
21 Dwarf emu
22 Rodriguez little owl
23 Sandwich rail
24 Great auk
25 Spectacled cormorant
26 Atlas bear
27 Tarpan
28 Sea mink
29 Portuguese ibex
30 Quagga
31 Warrah
32 Palestine painted frog
33 Abingdon Island tortoise
34 Round Island boa
35 Passenger pigeon
36 Carolina parakeet
37 Pink-headed duck
38 Lord Howe Island white eye
39 Hawaiian O-O
40 Madagascar serpent eagle
41 Kauai Nukupuu
42 Greater rabbit bandicoot
43 Arizona jaguar
44 Schomburgk's deer
45 Caribbean monk seal
46 Thylacine "wolf"
47 Jamaican long-tongued bat
48 Barbary lion
49 Newfoundland white wolf
50 Bali tiger
51 Italian spade-footed toad
52 Chinese alligator
53 Central Asian cobra
54 Geometric tortoise
55 Hawksbill turtle
56 Japanese crested ibis
57 California condor
58 Red-billed curassow
59 Black robin
60 Reunion petrel
61 Abbott's booby
62 Hawaiian gallinule
63 Mauritius pink pigeon
64 Western ground parrot
65 Hawaiian crow
66 Leadbeater's oppossum
67 Ghost bat
68 Woolly spider monkey
69 Mountain gorilla
70 Blue whale
71 Humpback whale
72 Indus dolphin
73 Northern kit fox
74 Baluchistan bear
75 Giant otter
76 Siberian tiger
77 Asiatic lion
78 Mediterranean monk seal
79 Grevy's zebra
80 Przewalski's horse
81 Mountain tapir
82 Great Indian rhinoceros
83 Swamp deer
84 Giant sable antelope
85 Indri
86 Orangutan
87 Sumatran rhinoceros
88 Mountain anoa

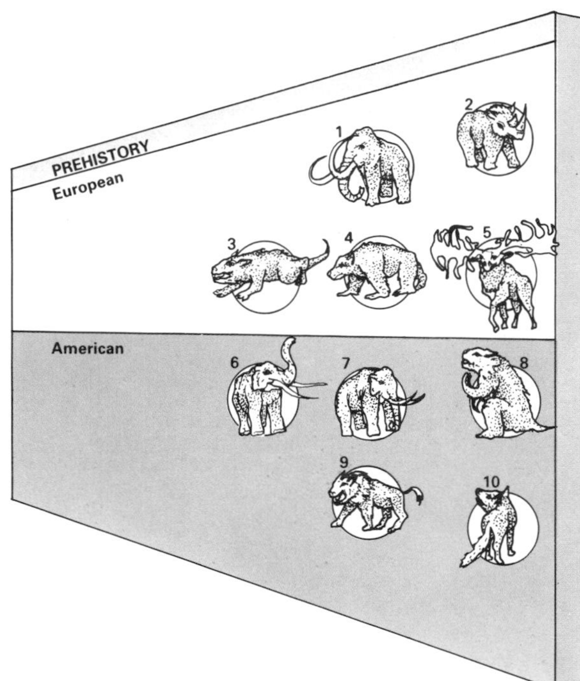

The crescendo of extinction

Biologists have so far described about 1.4 million species of invertebrate, most of which are insects. An estimated five million species, at least, have yet to be discovered. But most of these live in the tropical rain forests, which are rapidly being destroyed (▸ page 214). So most of them will become extinct before the scientists have had a chance even to give them names.

The fate of the world's 50,000 or so species of vertebrate is known with greater certainty. In AD 1600 naturalists recognized 4,226 species of mammal. Since then 34 are known to have become extinct, and the International Union for the Conservation of Nature (IUCN) estimates that at least 120 more are in danger of extinction. In addition, many mammalian species are divided into distinct races, or subspecies, and many subspecies are threatened even when the species as a whole is not. For example, the world's total of leopards is high, but five of the individual subspecies are endangered. Birds numbered around 8,000 species in AD 1600. 94 full species and around 164 subspecies have become extinct since then, while at least 160 species and almost 300 subspecies are now endangered.

These two pages highlight the better-known examples of the total destruction. There is a high number of island species compared with mainland ones, both extinct and currently threatened (▸ page 197), and today species with fewer than 100 individuals are in extreme danger. At least as many species have become extinct in the last two centuries as in prehistory or in the previous 2,000 years. The same number again are in danger of dying out in only the next 50 years!

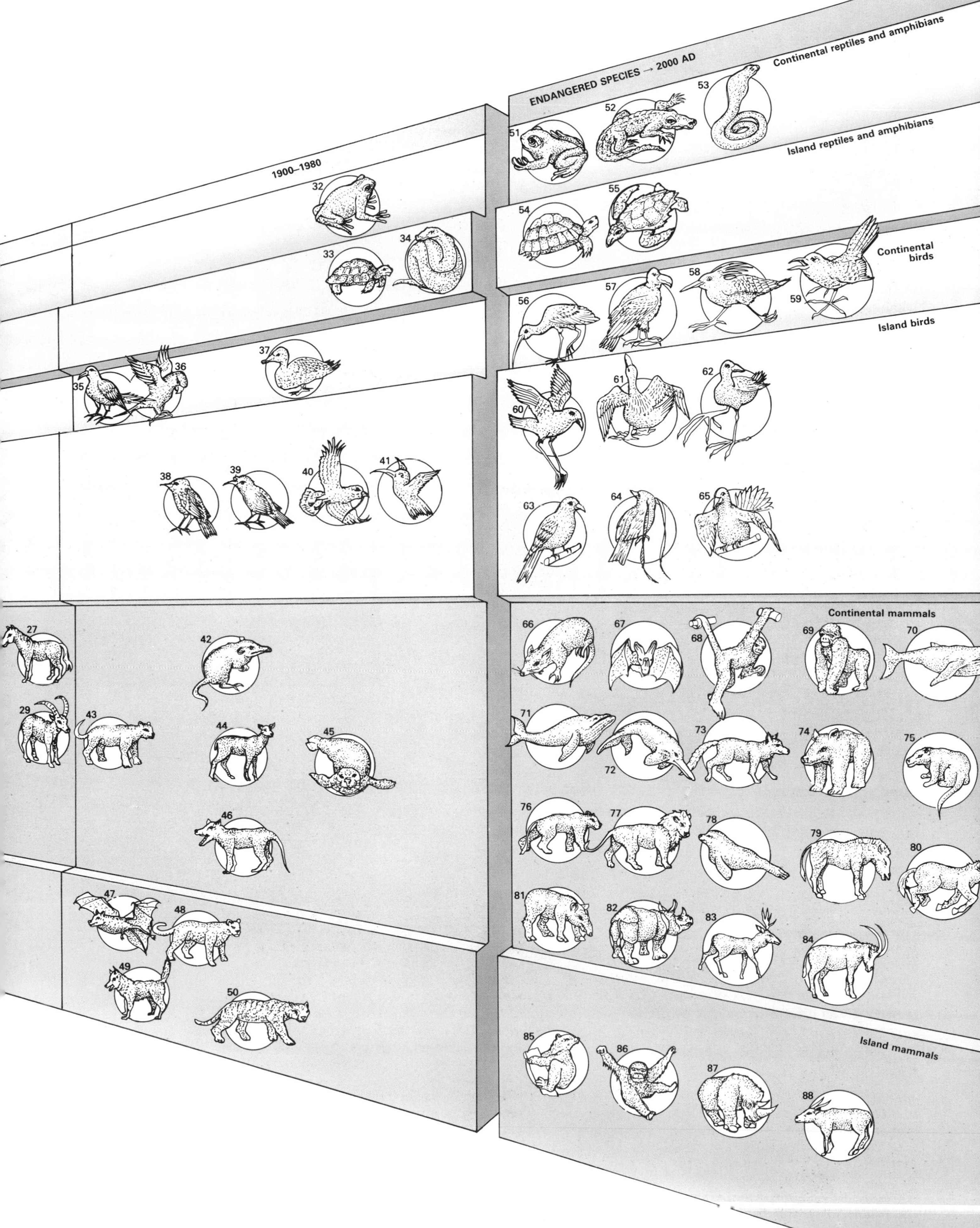

1900–1980
ENDANGERED SPECIES → 2000 AD
Continental reptiles and amphibians
Island reptiles and amphibians
Continental birds
Island birds
Continental mammals
Island mammals

The conservation of many animals, such as the orangutan, is complicated because the species may be divided into distinct races, or subspecies, each of which must be conserved separately

Why conserve?

Conservationists recognize four main reasons for wanting to save other animals and plants. Three of them are selfish reasons, and one is unselfish.

First, many ecosystems benefit human beings simply by existing. The forests on the sides of hills in Asia act as a sponge. They absorb the huge amounts of rain that fall during the brief rainy season (the monsoon) and allow it to run only slowly to the valleys below. When the trees are removed, the water rushes off the hills and takes the topsoil with it. The valleys are flooded, the rivers are polluted with silt (which destroys the fisheries) and the hillsides are reduced to bare rock (▶ page 214). The minute plants that float in the plankton of the sea, the diatoms, may be even more valuable (◀ page 122). They provide most of the oxygen in the atmosphere. Severe pollution could wipe them out. Then all animals would suffocate.

Second, many individual species of wild plant and animal supply people with food, drugs, dyes, or other industrial chemicals. Others have valuable qualities, such as resistance to pests, that could be bred into existing livestock and crops (◀ page 146).

Third, people are prepared to pay for the pleasure of being in wild places, and looking at wildlife. Tourism generated by nature reserves plays an increasing part in the economy of developing countries.

The unselfish reason is that we should act as guardians of this planet, and of the plants and animals it contains.

The scale of the problem

In theory, animals and plants may be saved from extinction by protecting their habitats in the wild, and breeding them in captivity (▶ page 235). But the wild habitats are under severe threat, and breeding in captivity is often far from easy. Yet, to ensure the long-term survival of any one species of animal, we must maintain scores or even hundreds of individuals. So it is not enough simply to maintain small patches of natural habitat, or maintain small captive colonies.

How many is enough?

In driving an animal to extinction it is not generally necessary to wipe out every individual. Usually, if a population falls below a certain critical level, the population ceases to be viable. The reduced population cannot produce enough healthy offspring to replace the ones that die, and so is doomed. A crucial question for conservationists is: how many animals must be conserved if the species is to be safe?

In practice the answer varies from species to species, and with circumstance. However, in recent years mathematicians have combined with biologists to provide two general principles that do apply to all species: first, the number of individuals that needs to be conserved to guarantee safety is remarkably high, and, second, it is very important to conserve the right individuals.

Species, alleles and gene pools

Two animals are said to belong to the same species if they can breed together to produce fully viable offspring, and, when one animal breeds with another, the offspring contains a mixture of genes from both parents (▶ page 202). Of course males cannot breed with other males, or females with females, but a male can breed with another male's daughter, or a female with another female's son or father, so the principle holds.

▼ A traditional medicine stall in China. Plants are the main source of medicines. Species that have not yet been studied may yield many more drugs – a good reason for conserving them.

► In Africa, large numbers of pangolin are killed for their meat and scales by the native people. The future of one species, the Cape pangolin, is seriously endangered.

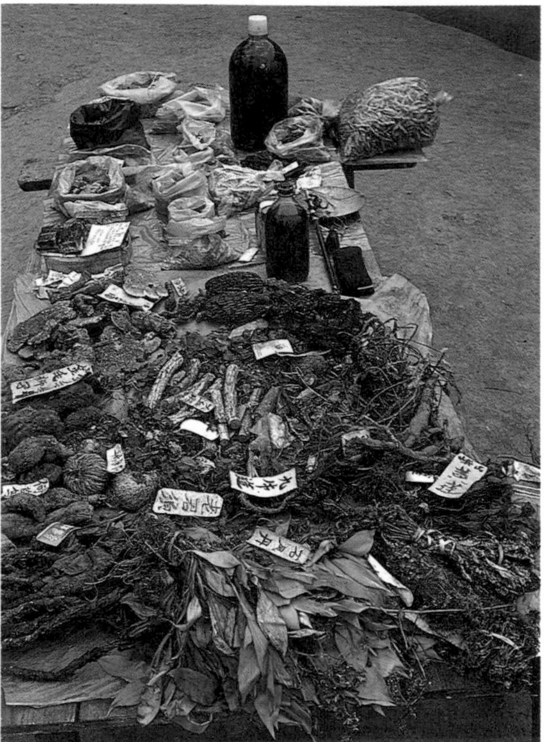

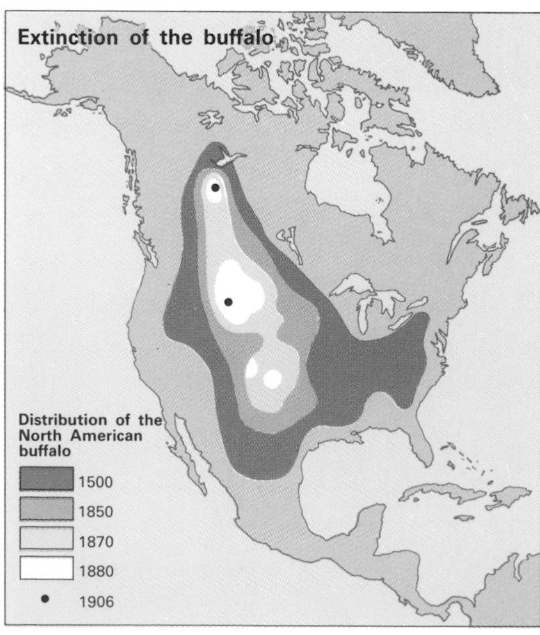

▲ North American buffalo were driven almost to extinction as a matter of policy: to eliminate the Indians' main food supply and thus open the plains to farming and to the railways.

► Emus are reasonably safe on the Australian mainland, but the Tasmanian emu was driven to extinction. Animals that live on islands have nowhere to run when threatened.

Extinction of the elephant birds

The ratites are the world's biggest birds, now represented by the ostrich (Africa), rhea (South America) and emu and cassowary (Australasia). These, however, are pale shadows of the elephant birds that once inhabited Madagascar, and the moas of New Zealand. The last of the various species of elephant bird was wiped out in 1700. It was also the biggest bird of all times; 3m tall and weighing 0.5 tonne. Human beings stole its magnificent eggs, each with a capacity of 7 liters: the largest single cells in all nature.

In truth, most species of elephant bird died out before human beings arrived. But when the first humans came to New Zealand in the 10th century AD – the Maoris, from Polynesia – the islands still contained about 25 species of moa. Some were fairly small but one, 4m high and weighing 0.3 tonne, was the tallest bird ever. The Maoris established effectively a "moa" culture; they ate the birds, hunted them, involved them in magic. The last species of moa – the tallest – was wiped out in 1850. Incidentally, it is argued that "primitive", meaning "pre-technological", people live in harmony with nature. The Maoris and the moas demonstrate that life is not so simple.

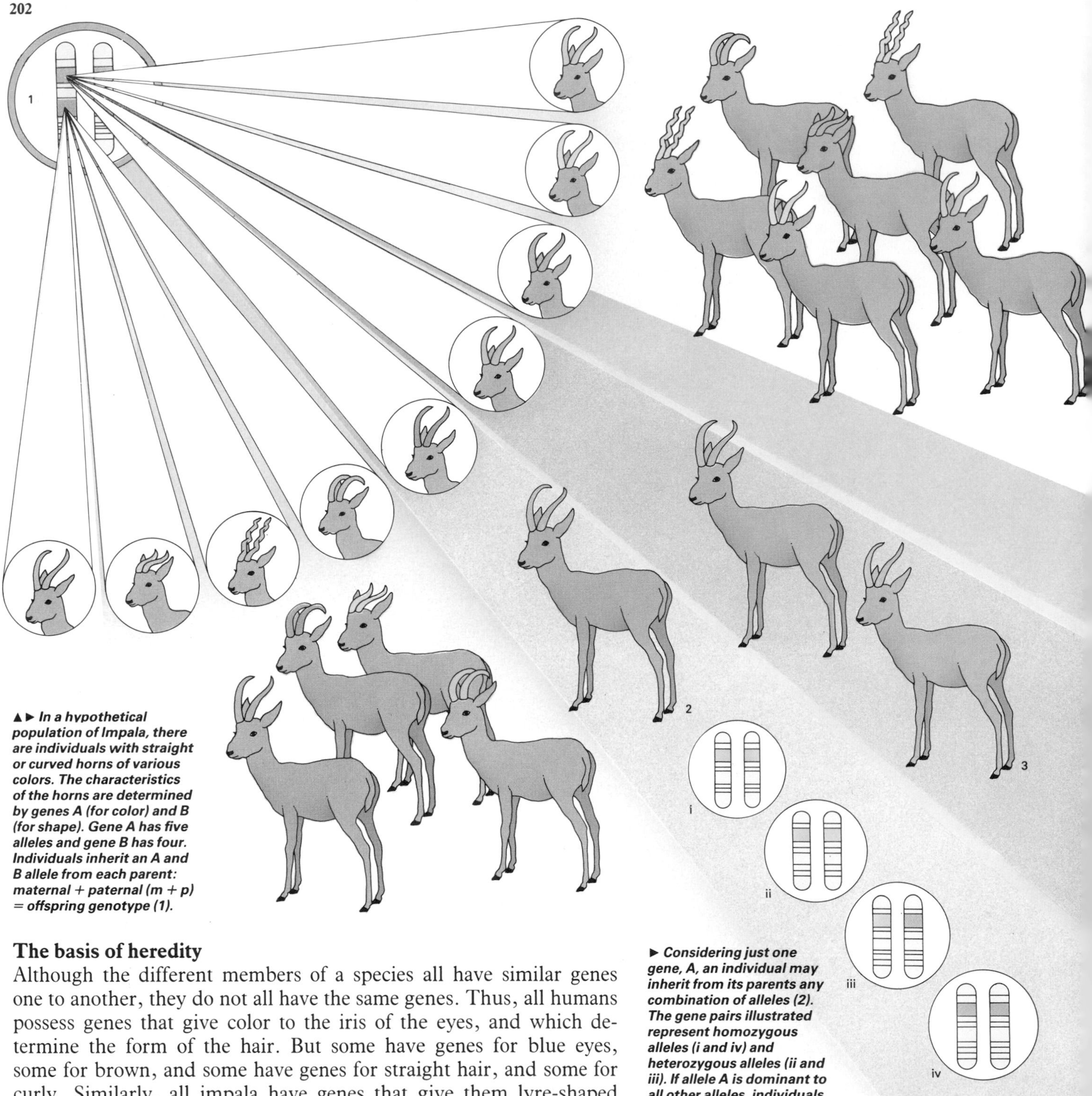

▲► In a hypothetical population of Impala, there are individuals with straight or curved horns of various colors. The characteristics of the horns are determined by genes A (for color) and B (for shape). Gene A has five alleles and gene B has four. Individuals inherit an A and B allele from each parent: maternal + paternal (m + p) = offspring genotype (1).

► Considering just one gene, A, an individual may inherit from its parents any combination of alleles (2). The gene pairs illustrated represent homozygous alleles (i and iv) and heterozygous alleles (ii and iii). If allele A is dominant to all other alleles, individuals with pairs i, ii and iii will have identical color horns to one another.

The basis of heredity

Although the different members of a species all have similar genes one to another, they do not all have the same genes. Thus, all humans possess genes that give color to the iris of the eyes, and which determine the form of the hair. But some have genes for blue eyes, some for brown, and some have genes for straight hair, and some for curly. Similarly, all impala have genes that give them lyre-shaped horns. However, some have genes that produce extremely curved horns, and others have genes that produce only slightly curved horns. Different varieties of genes that do the same job – different variations of eye color genes for example – are termed alleles.

In practice, some genes come in only one form in a given species while other genes are highly polymorphic: there may be five, ten, or in a few cases scores of variations – alleles – of any one gene. Each individual animal contains two sets of genes – one inherited from the mother, and one from the father. At most, then, it can contain only two variations – alleles – for each kind of gene. When it has inherited the same allele of a particular gene from each parent it is said to be homozygous for that gene. When it has inherited a different allele of a particular gene from each parent, it is said to be heterozygous for that gene. Within an individual's genotype, there is a random mixture of homozygous and heterozygous genes.

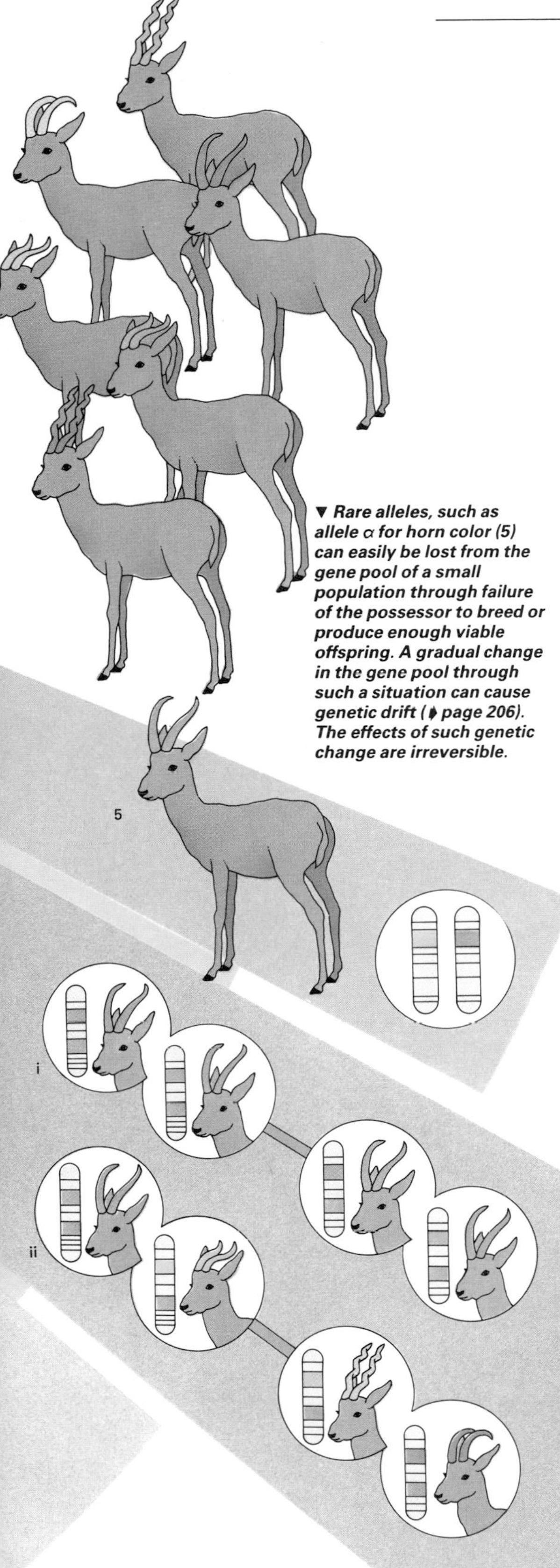

▼ *Rare alleles, such as allele α for horn color (5) can easily be lost from the gene pool of a small population through failure of the possessor to breed or produce enough viable offspring. A gradual change in the gene pool through such a situation can cause genetic drift (➧ page 206). The effects of such genetic change are irreversible.*

▲ *The impala pairs indicated (the linked circles) are four possible breeding pairs (3). The two top pairs (i) contain identical alleles for the horn genes: a and B. When these breed a genetic bottleneck may arise. The lower two pairs (ii) each contain a good, and different, mix of alleles. These pairs would each make excellent founders.*

In the wild, a species may be broken up into a number of different breeding populations – although, of course, when a species becomes rare, there may be only one population left, and that very small. Each individual in the breeding population is potentially able to combine its genes with any other (or with any other's daughter or son) so all the individuals together can be thought of as partaking of a share of one big collection of genes that all the individuals contain between them. This collection of genes, including all the variations – the different alleles – within each individual, is termed the gene pool.

The aim of species conservation is to conserve as large a proportion of the gene pool as possible (➧ page 240). The trouble is that conservationists have room for only a limited number of animals. The art, then, is to retain as many different alleles within the least number of individuals. This presents various problems.

Bottlenecks

First, an animal that is rare by definition contains only a few individuals, and when a conservationist begins breeding an endangered species, he inevitably begins with only a limited selection of individuals. Suppose he begins with only two individuals: a breeding pair. Those two animals, between them, are most unlikely to contain all the alleles available to the species. If one particular gene existed in four variations, then it is just possible that two heterozygous animals beween them would contain all four – but it most unlikely. If there were five variations of a given gene then two animals could not possibly contain all of them. For many genes, one or both animals might be homozygous, and any other alleles that exist for that gene would simply be left out. In addition, all animals subsequently bred from the original pair will at best contain only those alleles present in the founders. When a big population is reduced to a small one, or a conservationist takes a few animals to begin a new breeding population, the species is said to go through a genetic bottleneck.

On the whole, animals survive genetic bottlenecks very well. A dozen animals or so – provided they are not all of the same family – could contain a fair proportion of the total number of alleles in the species' total gene pool. Rare alleles are unlikely to be represented in such a small population, and this might affect the population's fitness in the long term. But in the short term even a small founder group can be an adequate representation of the whole. Small populations, however, suffer from two quite different handicaps.

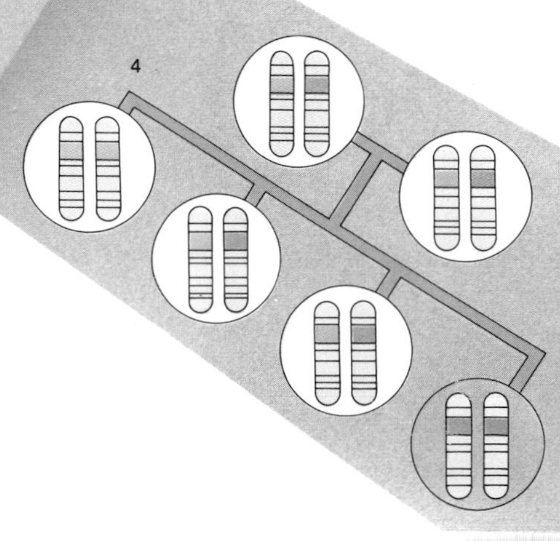

◀ *A reduction in the number of alleles in the gene pool means an increased chance of any individual allele being passed on (4). If two parents are genetically similar, as shown here, not only will the gene pool be reduced to very few alleles, but any deleterious alleles, such as a, will be passed on to all offspring.*

Deleterious alleles in humans include those that cause cystic fibrosis and the blood disorders hemophilia and thalassemia

▲ **By chance, mutant genes can get into many offspring and so increase in frequency in the gene pool (◆ page 202). Chance genetic irregularities can also determine why, for example, the white rhino has evolved two horns, the Indian rhino only one, the number of horns appearing to make no specific adaptive difference.**

The trouble with being rare

Animals get killed, they starve, and they suffer from diseases. In a population of, say, twelve animals, only six, on average, will be female, and perhaps two of these will be too old to breed. If two more come to grief, the breeding capacity of the population is halved. If one of the eggs that is produced is addled and another falls out of the nest (disasters that befall many birds of prey), then the population is finished. It can be shown statistically that unless a population of animals in the wild contains at least 50 animals it is almost bound to become extinct through time and chance alone. Many populations of wild vertebrates are at or below this number (◆ page 244).

Slow breeding and inbreeding depression

The second handicap is "genetic drift". When animals breed, they pass on only half their genes to their offspring. If they have only a few offspring, there is a fair chance that through all their breeding lives a fair proportion of the genes they contain will not be passed on at all. In addition, in any one population a proportion of individuals die before they have bred at all – and they take their genes with them. In a very large population, there are so many individuals that between them they do pass on nearly all the alleles within the pool to the next generation. In small populations the inefficiency does not matter provided the animals breed very quickly indeed. But most of the animals that conservationists are most concerned with are slow-breeding, and genetic drift is a very big factor.

Reduction in the number of different alleles in the gene pool raises another problem. In general it is advantageous to be heterozygous rather than homozygous (◆ page 202). To be heterozygous you have to inherit different alleles from each parent. Some alleles are deleterious – that is, weakening. Most of them do not have serious ill-effects unless they are inherited from both parents.

Genetic drift

Some invertebrate animals, including many insects, go through genetic bottlenecks as a matter of course, being all but wiped out every few years – by weather, fire, lack of food resources or other similar natural disasters. But each surviving individual then breeds while it is still very young (and so is unlikely to die before it has bred) and produces hundreds or thousands of offspring. Between them these offspring will inherit a very large proportion of their parents' genes. But in small populations of animals that breed slowly, the chances of dying or being killed before breeding are high, and the proportion of genes that are not passed on with each generation is appreciable. (When mating occurs a great many gametes are wasted.) Such populations rapidly lose alleles from the gene pool. Conversely, those alleles that are not lost, being the only ones left, are bound to be passed on. The gene pool of the small, slow-breeding population rapidly becomes very different, and in particular much less varied, than that of the original population. This change in the gene pool with time, caused just by time and chance, is called genetic drift (◆ page 202).

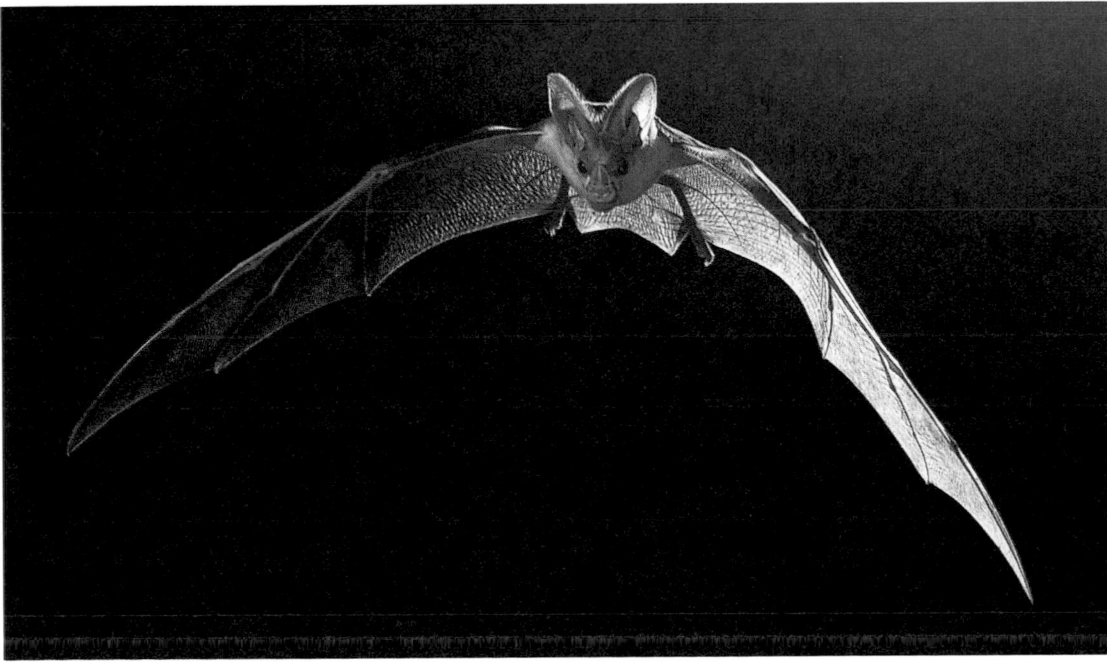

▲ *The vicuña of South America was, in 1969, classified as being rare and endangered. As the population fell from several millions in the 1500s (as reported by the early Spanish and Portuguese explorers), to 400,000 in the early 1950s, then to less than 15,000 in the late 1960s, conservation measures became imperative. World vicuña population now stands at around 80,000 and the animal's status was, in 1981, changed to "vulnerable".*

◄ *The Australian ghost bat has two notable statistics to its name. First, it is the largest microchiropteran bat and, second, it is listed as vulnerable to extinction in the Red Data Book due to loss of its forest habitat – the fate of many bats with restricted distribution. It is carnivorous.*

In animal populations of less than 500 individuals, genetic drift outstrips natural variation and evolution ceases

When two parents are closely related to each other, and so genetically very similar, there is a fair chance that any deleterious alleles present in one will also be present in the other. There is then a high probability that their offspring will inherit a double dose and be badly affected. In practice, animals, and many plants, go to some lengths to avoid matings between brothers and sisters. If they do occur, the resulting offspring are often weak. This is called inbreeding depression.

In a small population, in which a great many alleles have been lost through genetic drift, the remaining animals will inevitably become more and more genetically similar as the generations pass; the animals are effectively inbreeding. The population may eventually die out. One particular problem is that "weakness" often takes the form of sexual sterility, which of course puts an end to all reproductive efforts. Therefore, breeding programs that at present seem successful may yet fail in the future.

The end of evolution

Animals in the wild evolve – in general they adapt to changing circumstances. They do this, as Charles Darwin described, by means of natural selection. The animals within a population are varied, and some are inevitably "fitter" than others (◀ page 14). Only the fittest survive.

But natural selection can operate only if there is a great deal of variation to begin with. Genetic drift tends to reduce variation. Counter to that is that new variation should constantly be arising by mutation. If the population is so small that genetic drift erodes the gene pool quicker than mutation can replenish it, then evolution effectively comes to a halt. The population may change (through genetic drift) but it will not adapt.

There are complications, however, some of which work in the conservationists' favor and some that do not. When it comes to establishing breeding populations, some individuals are far more valuable than others and this point is one that has received considerable and most attention (▶ page 240).

Effective population size

An animal is of no use to the breeder if it is too old to breed or if it is the same sex as others in the captive population. In the future, the breeder may, for example, be able to obtain some viable sex cells (▶ page 242) even from ancient animals in order to increase the gene pool, but at present this generalization holds. In addition, as his task is to maintain the greatest possible genetic variation, and as he also wants each individual animal to be as heterozygous as possible, he clearly requires that his breeding animals should be genetically different to a very high degree.

Even from these two observations it is evident that the actual number of animals in a breeding population is not necessarily the same as the number that is in practice useful for breeding purposes. Indeed, breeders have produced the concept of effective population size – the number of individuals that makes a useful contribution to maintaining the gene pool. Generally, this number is much less than the actual population size. The conservationist may well find that the 100 animals he has at his disposal are unsatisfactory, perhaps because they do not include any females or young males, or they are all siblings. However, breeders of captive animals can take steps to ensure that the few animals they are able to keep do have the largest possible effective population size.

▶ One type of mutation involves duplication of the entire chromosome set, as in the marine grass Spartina townsendii. *This fertile hybrid arose from a cross between* S. maritima *(60 chromosomes) and* S. alterniflora *(62 chromosomes). It is more successful in colonizing tidal mudflats than either of its parents.*

▲ In some species of plant such as bee orchids elaborate mechanisms have evolved to ensure that self-pollination, and therefore inbreeding depression, will not occur. Once the bee has removed the pollinia from the flower, the stigma becomes unreceptive. In other plants, stigma and anthers ripen at different times.

▶ Although a few giant pandas have been kept in captivity for many years, breeding has been poor. The first cub actually reared was born in 1963 in Peking Zoo and since then there have been fewer than 15 successes. Most nations have only a pair or a single animal, and will not send them to places where they might breed better.

The critical number

On mathematical grounds, it can be shown that the minimum number of animals needed in a population that would allow some measure of evolution by means of natural selection is around 500. This is a very large number, far more than are involved in the breeding pools of most zoos or among many wild populations of large vertebrates. For many large vertebrates, then, we must conclude that evolution by natural selection has already come to a halt.

Furthermore, in the wild, for a species to avoid extinction just by chance, there must be at least 50 animals. A breeder, taking an adult male and female from such a population is likely to find that the two animals are very closely related. Their offspring are likely to be largely homozygous and suffer immediately from inbreeding depression (◀ page 204).

See also
Sex and Species 35-46
The World's Livestock 163-172
Saving Habitats 225-234
Saving Wildlife 235-248

A compromise solution
Arranging marriages between individual animals, and trimming family sizes, often runs counter to what is desirable on behavioral or ecological grounds. Curators of modern zoos like to keep "natural" breeding groups – which may well mean one male and many females – but the geneticists demand only one female per male. Managers of wildlife parks may sometimes have to cull some animals – notably elephants – and when they do they prefer to remove whole families, rather than take out a random selection of individuals, which is what the geneticist feels is ideal.

▲ An Akita Inu, a rare dog breed. Breeders of domestic animals, and in particular pedigree dogs, select for the particular genes that produce the characters they require.

▼ Conservers of species such as Przewalski's horse seek to preserve as much as possible of their genetic diversity. Despite the number of domestic horses, wild horses are at risk.

Increasing the effective population

Within a captive population, the breeder should ensure that the sex ratio is, as far as possible, one to one. If one stud male serves all the females then the offspring are clearly much less varied genetically than they would be if each female were served by a different male (◀ page 202). By the same token, the breeder should try to keep as many different families as possible. He does not want thousands of siblings of the same parents but siblings of many different parents. If, as is always the case, he is able to keep only a limited number of animals then he should ensure that each family contains exactly the same number of siblings so that all families are equally represented in the next generation.

Similarly, and in general, the breeder is able to arrange only a limited number of matings in each season (◀ page 38), and any one individual is able to breed for only a limited number of seasons. He must therefore arrange the matings so that as great a variety as possible of the animals at his disposal have a chance to pass on their genes, preferably through marriages with individuals that are not closely related to themselves.

If breeders of captive animals do all these things, they can improve on nature. After all, in nature many animals, particularly large mammals, are polygynous: one male, many females (◀ page 42). In the wild this may be beneficial – it ensures that the biggest and toughest survive – but it reduces variation. When animals are rare, variation is what is important. Similarly, in the wild, family sizes may vary enormously and nothing prevents this: the seven pups of the dominant wolf may all survive, while the subordinate animal's three all die. Thus, the effective population size of a well-managed captive population may be far larger than that of the same-sized wild population. By arranging marriages and planning families very carefully, the breeder can make far more use of the small number of animals available to him than at first seems possible. This has been the case with the great apes in particular (▶ page 244).

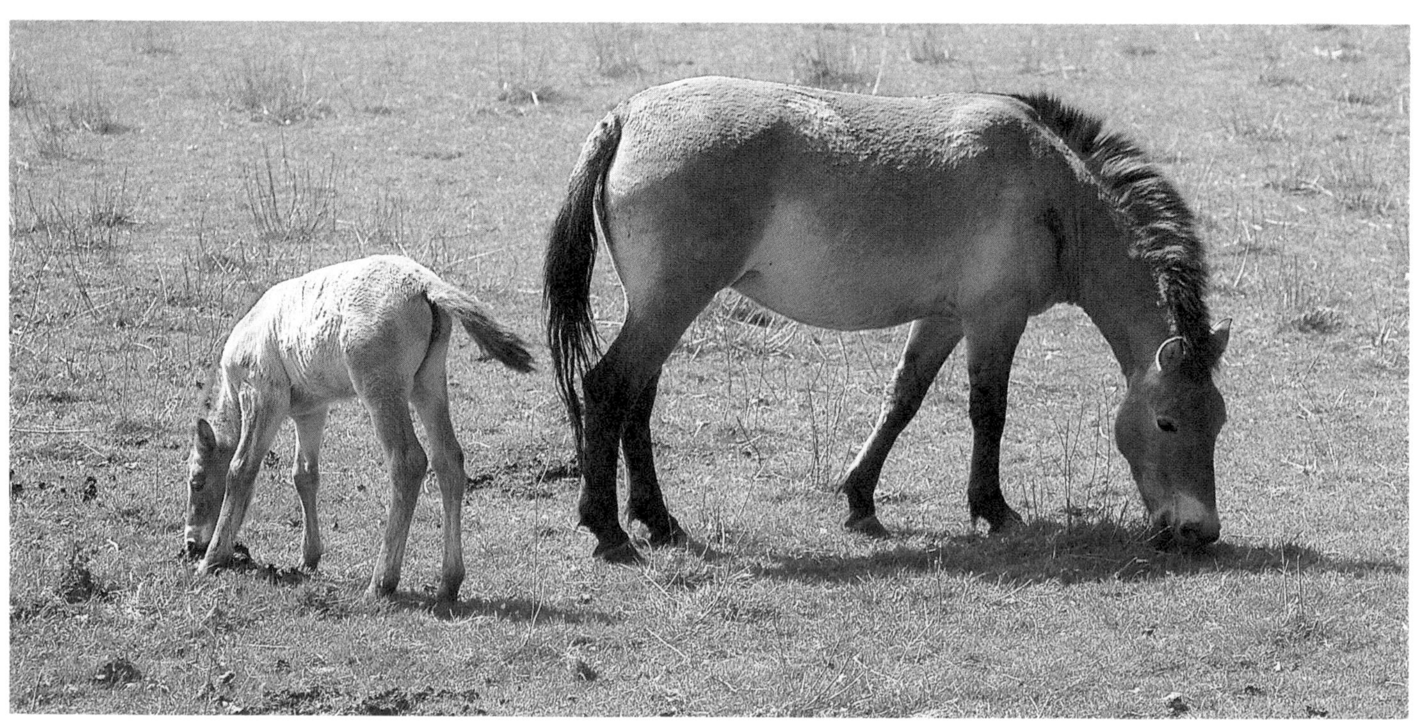

Threats to the Environment

Natural change...Manmade change...Unintentional threats...Farming malpractice...Pollution... Desertification...Assessing the damage...PERSPECTIVE ...Destruction by farming...Glaciation...Loss of habitat ...Loss of land...Eutrophication...Deforestation... Environmental pollution...Acid rain...Why are the forests dying?...Dams and reservoirs...The greenhouse effect...The ozone layer

All living organisms depend on the resources of food, water, space, and shelter that together comprise their habitat. If the habitat should change, the supply of resources may be reduced and species may not survive. Loss of habitat is by far the gravest threat facing plants and animals today, and habitat is being lost in many parts of the world.

Habitats may change because the climate itself changes, eliminating those species which cannot tolerate the change, and encouraging those which can. In high latitudes, repeated glaciations have scoured away the soil and buried vast areas beneath ice, allowing tundra communities to thrive at the edges of the ice sheets (◆ page 112).

Sometimes a population of organisms may find itself isolated and trapped. In 1986, the Hubbard Glacier, which flows into Disenchantment Bay, on the southern coast of Alaska, began advancing rapidly. The bay is bounded on its eastern side by a narrow peninsula extending northwards; to the east of the peninsula lies Russell Lake, a long, narrow body of water inhabited by seals and porpoises. The glacier blocked the seaward end of the lake, so trapping the lake animals, and the lake began to fill with fresh water, altering their habitat drastically. A rescue attempt was mounted, but no such intervention could have aided animals trapped in earlier glacial episodes.

Although glaciations eliminate many species locally, they do, however, cause very few extinctions on a global scale. Species that once inhabited glaciated areas continue to live in lower latitudes, returning when the ice sheets retreat.

Reductions in rainfall may produce desert conditions in places that once supported grassland. In classical times, much of the northern part of what is now the Sahara grew farm crops. Today, the Sahara is expanding southward due partly to a change in the climate, whose effects are aggravated by excessive grazing pressure on poor rangelands (◆ page 160). Of 19 countries along the southern edge of the Sahara, 14 reported in the early 1980s that the desert was extending into their territories and 5 of them said that the desert was advancing rapidly and significantly. By the late 1980s, 88 percent of the productive land along the edge of the Sahara had become desert, as well as 90 percent of the rangelands.

Certain habitats are especially susceptible to rapid change. Beaches and coastal sand dunes are often altered dramatically in the course of a few winter storms. Lengths of unstable cliff may collapse; as much as two meters of sand may be scoured from a beach overnight; elsewhere a similar amount of sand may be thrown ashore. Few organisms can survive such violent changes, but as conditions stabilize, recolonization begins, and since material scoured from one stretch of coast is often deposited on a shore elsewhere, habitat is not lost permanently but merely moved from one place to another. Such change kills individuals, but does not cause extinctions.

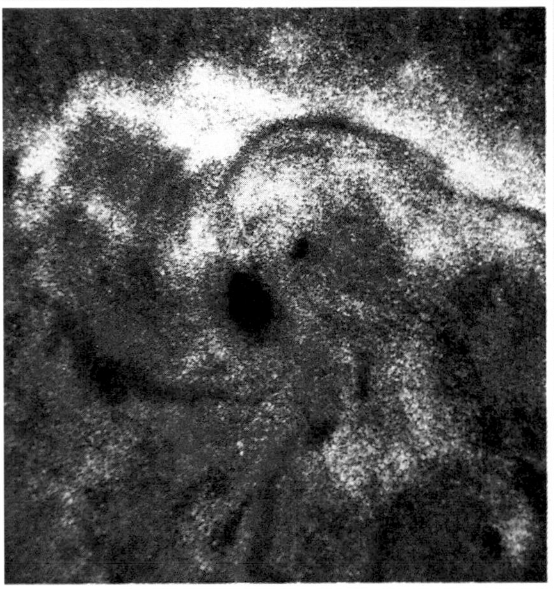

▲ *A false-color photograph of a region of sunspot activity. The amount of activity indicates the intensity of radiation being emitted, and this is believed to affect climates on Earth.*

▼ *The Colorado River flooding low-lying ground, and a road, in Arizona. Such floods can improve the fertility of land by leaving behind a deposit of silt rich in plant nutrients.*

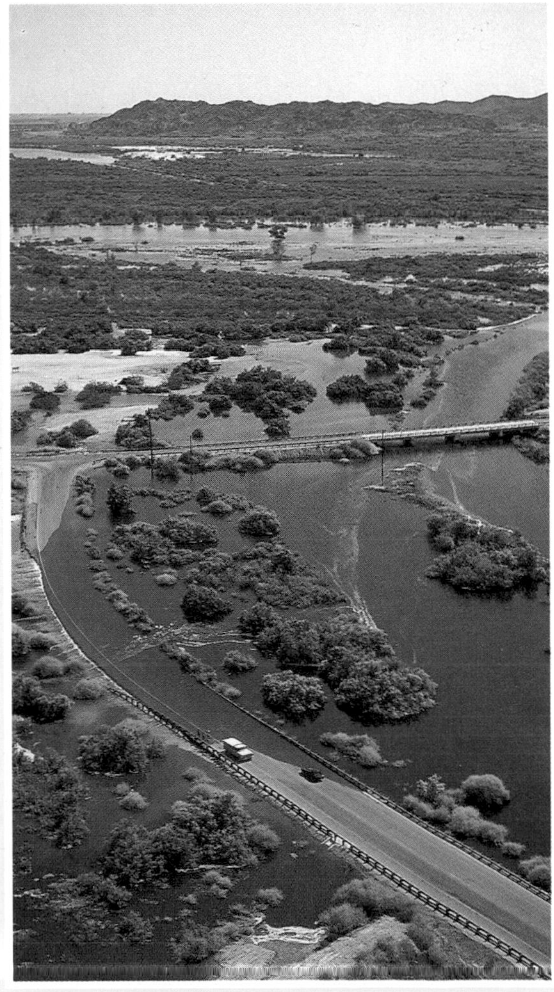

Lowland habitats are those most threatened by manmade change

Glaciation

About two million years ago, ice caps and glaciers advanced over much of the northern hemisphere and parts of the southern hemisphere. This was the first glaciation, or "ice age", of the Pleistocene Epoch. During this, and further episodes of extensive glaciation, most of the land in latitudes higher than about 50° was covered by ice to depths of up to 3,000m. At times, more than 25 percent of the Earth's surface lay beneath ice, compared with around 10 percent today. Each glaciation lasted tens of thousands of years, but during interglacial periods, of similar duration, the ice retreated, conditions improved, and the land was recolonized. The last glacial retreat occurred about 10,000 years ago.

A reduction of 2 percent in the amount of solar heat reaching the surface of one hemisphere of the Earth is sufficient to start an ice age. Part of the area that is covered by ice and snow in winter fails to thaw in summer, because it is white and reflects more solar radiation back into space. This reduces the temperature still further, and the ice advances. As more of the Earth's water is held as ice, sea levels fall. At one time, America and Asia were linked across what is now the Bering Strait, and Britain was joined both to continental Europe and to Ireland.

During glaciations, species retreated to lower latitudes and land animals were able to move freely across bridges that disappeared as the ice melted and sea levels rose. In the Pleistocene, many species of mammals disappeared, but new species also evolved, notably including humans over the last two million years. Tundra conditions close to the ice sheets favored large mammals, such as mammoths, mastodons and saber-toothed tigers. Some may have lived during interglacials, but in the end many became extinct. Horses and rhinoceroses made their first appearance, however.

▼ The Columbia Ice Field glacier in Jasper National Park, Canada. All that remains of the latest ice age are the polar ice caps, and the mountain glaciers of Canada, northwestern Europe, Chile and New Zealand. Much of the Earth's landscape has been transformed by the action of compacted ice flows and their meltwaters.

World land use

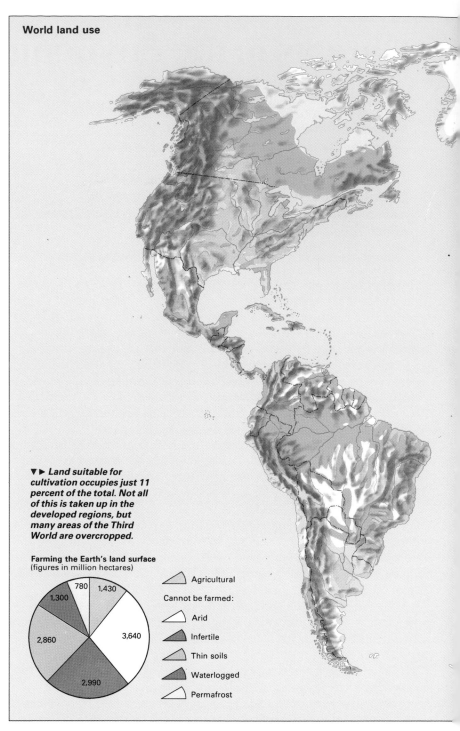

▼► Land suitable for cultivation occupies just 11 percent of the total. Not all of this is taken up in the developed regions, but many areas of the Third World are overcropped.

Farming the Earth's land surface
(figures in million hectares)

780, 1,430, 3,640, 2,990, 2,860, 1,300

Agricultural

Cannot be farmed:

Arid

Infertile

Thin soils

Waterlogged

Permafrost

Manmade change

Changes to the natural habitat may also be wrought by humans, deliberately or inadvertently. Change is deliberate whenever such an area is claimed for human use and altered radically. Forest may be cleared or waterlogged land drained to make way for farming, mining, or urban development. Buildings, highways, airports, and harbors provide few opportunities for wild species to thrive, but their total area is small. In the United States, only three percent of the land area is covered by buildings and highways. In Britain, which is highly urbanized and densely populated, buildings and highways occupy about eight percent of the total land area. Elsewhere in the developed world, urban expansion is now slow or has ceased altogether, with new development taking place largely on land reclaimed from former urban use. Conversely, in many developing countries, the urban area is expanding and in time may occupy a proportion of the total area equivalent to that of the developed countries.

Despite the small area devoted to urban development, competition for desirable land is often keen because land that is most suitable for one use may be equally suitable for others. Farms, forest plantations, cities, roads, airports, and many tourist facilities are all sited in the lowlands, on plains, and in river valleys. Natural habitats in such preferred areas may be under greater threat than those in the less hospitable mountains or deserts and some lowland species have a very local distribution.

The el segundo blue butterfly (*Philotes battoides allyni*), for example, occurs in only two small areas of Los Angeles County, California, one amid the buildings of a Standard Oil Refinery, the other beside Los Angeles Airport.

Threatened natural habitats may also include popular tourist areas, far from urban or agricultural development. Plants may be destroyed by trampling, and wildlife may be driven away by the pressure of human numbers.

Farming
When land is brought into cultivation for the first time, the original vegetation is cleared thoroughly and prevented from returning. The habitat is altered drastically and species that might feed on crop plants or livestock are persecuted. Where small areas of original habitat remain, they are often isolated from one another; divided into small groups and doomed to inbreeding, the native plants and animals may not survive.

Throughout history, the ceaseless expansion of the area of farmland has destroyed natural habitats on a vast scale. Most of lowland Europe was once forest, but today almost all of it is farmed. The cereal-growing areas of North America were originally prairie, supporting communities based on a wide variety of grasses and herbs. The threat remains, as farmers in developing countries plow land that was once forest.

Unintentional threats

Many threats to habitats can be foreseen when changes in land use are proposed, but not all. Throughout the world, wetlands in coastal areas (◆ page 124) are under threat from development. They are not difficult to drain, and provide land for farming or for tourist facilities. Species peculiar to them may disappear if the habitat is destroyed, and altogether 297 sites, occupying a total of nearly 19 million hectares, have been listed as especially important under the Convention on Wetlands of International Importance. It is not just the obvious species, such as birds, which depend on wetlands, however. So do many invertebrates and the microflora and microfauna that live in the mud trapped by wetland plants. These organisms, in turn, play an important part in the cycling of nutrients (◆ page 84) on which the young of many commercially important fish species rely. Thus, damage to the wetlands may also injure fisheries.

Habitats may also be threatened by construction projects. Coral reefs in tropical waters may suffer from the building of harbors and marinas, and the construction of large dams involves the flooding of valleys to form lakes and alters the flow of rivers. The combined effects are not large, but they tend to be concentrated in particular areas. There are 175 large dams in the world, with a sizable proportion of them in California and the Soviet Union, especially along the Volga River. A further 38 are planned, with several more to follow, and many of them will be located in developing countries: the Tucurui Dam, for example, is the first in a program of 30 dams to be built in the Brazilian Amazon Basin.

▼ In semi-arid regions, watering places are used by livestock coming long distances. The sparse vegetation is liable to be overgrazed, as it has been in the scene below, in Australia. Plants are cropped so severely that they die and trampling prevents the establishment of seedlings. The desert moves closer to the oasis.

▲ The bacterial decomposition of organic matter in water consumes dissolved oxygen, as carbon is oxidized to carbon dioxide. A sudden influx of organic matter, or of plant nutrients, can cause mass deaths of fish, as in the German river seen above.

► Land in Upper Egypt poisoned by salination.

Eutrophication
Freshwater plants are fed mainly by nutrients draining from surrounding land, but agricultural fertilizers, some of which are highly soluble in water and therefore washed from soil easily, may provide too much nourishment. Plants grow luxuriantly, but when they die their decomposition depletes the water of the oxygen dissolved in it on which aquatic organisms depend. This is "eutrophication", the excessive enrichment of water. In extreme cases, affected water can support only simple algae, slimes, bacteria, and the larvae of insects such as mosquitoes, which breathe at the surface rather than taking oxygen from the water itself. The discharge of sewage, also rich in plant nutrients, has a similar effect and Lake Erie, receiving wastes from Buffalo, Cleveland, Toledo and Detroit as well as from many smaller towns, suffered badly in the 1960s.

Loss of habitat
During World War II an airstrip was built in Bermuda. The construction caused silting offshore and altered the pattern of water movement. These minor changes caused damage to patches of coral reef in Castle Harbor from which the reefs have still not recovered. Coral reefs grow very slowly and are highly susceptible to changes in the temperature of the water surrounding them. Rich in species of invertebrate animals and fish, all over the world they have suffered from the unforeseen consequences of commercial or industrial development.

Today, the 1,500 kilometer-long Australian Great Barrier Reef, the largest and ecologically richest reef in the world, may be threatened by proposed oil exploration off the Queensland coast, endangering many of the unique species it supports (▶ page 227).

Loss of land
At present about 1.4 billion hectares, or 11 percent, of the Earth's land surface is farmed. Probably this area could be increased to more than 3 billion hectares, but every year we lose about 11 million hectares of land through soil erosion, waterlogging, the poisoning of soil by the accumulation of salts, the conversion of farm land to nonagricultural uses, and the spread of deserts, which alone destroys 7 million hectares of grassland each year. To some extent the loss is due to unavoidable climatic change, but this is exacerbated by poor farming. Overgrazing of land bordering deserts, especially the southern edge of the Sahara, destroys plant cover, and allows the soil to dry out and blow away. Windborne dust and sand bury and kill other vegetation nearby.

Without proper drainage to remove surplus water, irrigation may allow water tables to rise, eventually turning fields into marshes. Where the climate is hot, dry and windy, irrigation water evaporates quickly from the ground surface, drawing more water from below by capillary pressure. As it passes through the soil, water dissolves out mineral salts which are deposited near the surface until the soil becomes so saline it poisons crops

Clearance of tropical forests may have serious climatic consequences extending beyond their borders

Bad or merely inappropriate farming may lead to soil erosion, which alters the habitat of species living in farmed areas. At present about 11 million hectares are lost in this way each year. The clearance of forests on steep slopes also encourages soil erosion as rains wash into the valleys soil that has lost the plant cover by which it was held in place. The 1978 floods in the Ganges Plain of North India were the results of such a process.

Changes in farming practice can have subtle consequences for a species' habitat. In September, 1979, the large blue butterfly *Maculinea arion* was declared extinct in Britain. It had lived on the chalk downs, where its larvae began their lives feeding on wild thyme plants. Later the larvae are carried into ant nests where they complete their larval stages and pupate. The wild thyme grew on grassland and was intolerant of shading, but grazing by sheep and rabbits kept the grass short. When the number of sheep was reduced, and myxomatosis decimated the rabbit population, grazing declined, the grasses grew taller, and the wild thyme was shaded and disappeared, so breaking an essential link in the complex life cycle of the butterfly.

If an area of forest is cleared of standing trees it will usually recover, given time, but cleared land is land that readily finds other uses. Thus a legitimate and traditional form of exploitation can lead to the permanent loss of habitats. At present about 4.5 million hectares of natural tropical forest is cleared each year, most of it in Asia.

Pollution

Habitats may be poisoned through environmental pollution, although the effects on wildlife are usually local, and often much less severe than they may seem. It is unlikely that any species has been brought to complete extinction as a consequence of pollution. At present the only form of pollution that might produce severe global effects arises from the release of carbon dioxide, raising the possibility of a "greenhouse effect" climatic warming (▶ page 222).

Spoil heaps in ore-mining areas often contain metallic compounds in concentrations too low to make it worthwhile recovering metals from them, but high enough to exclude many species. Land close to factories, or once occupied by them, may also be poisoned, often by metallic compounds. Such pollution is usually persistent, but over a number of decades species that can tolerate the conditions will colonize industrial wastelands.

Agricultural chemicals, such as fertilizers and pesticides, may find their way into groundwater or rivers. Pesticides may poison susceptible species directly, and fertilizers may encourage the exuberant growth of aquatic plants, leading to the deoxygenation of the water.

About six million tonnes of oil are discharged into the world's seas each year. It has little lasting effect on marine organisms, but when seabirds alight on the surface of the water, oil clogs their feathers and some is ingested as they try to preen themselves.

Sewage, including liquid industrial effluent, is also discharged into coastal waters. It can injure human bathers, but it has little effect on marine organisms. Wastes discharged into the deep oceans, beyond the limits of the continental shelves, are dispersed in such a large volume of water it is doubtful whether they can be harmful. Radioactive wastes, some discharged into coastal waters and some formerly dumped in deep oceans in sealed containers, are harmless to marine organisms. No food chain has been discovered leading from the deep oceans to surface species that might be consumed by humans.

Tropical forests

The most diverse communities on land are found in the tropical rain forests that once covered the lowlands of Central and South America, West and Central Africa, Southern Asia, many Pacific islands, and part of northeastern Australia (▶ page 100). Today, their combined area is about 830 million hectares, approximately 6 percent of the land surface of the Earth, but they contain nearly half of all the growing trees on Earth and more than one-third of the Earth's species of plants and animals. Many tropical species are found nowhere else. In the Malaysian forests, for example, nearly half the 8,000 species of native plants occur naturally only in those forests.

The forests have been exploited for centuries as sources of timber and other products, but the intensity of that exploitation increased rapidly from about 1950. Between 1975 and 1980 they were being cleared at a rate of about 10 million hectares a year, partly to supply timber but also to provide land for farming and fuel for local people. As the total forest area decreases, pressure increases on what remains, not least from the demands of forest tribes practicing primitive "slash-and-burn"

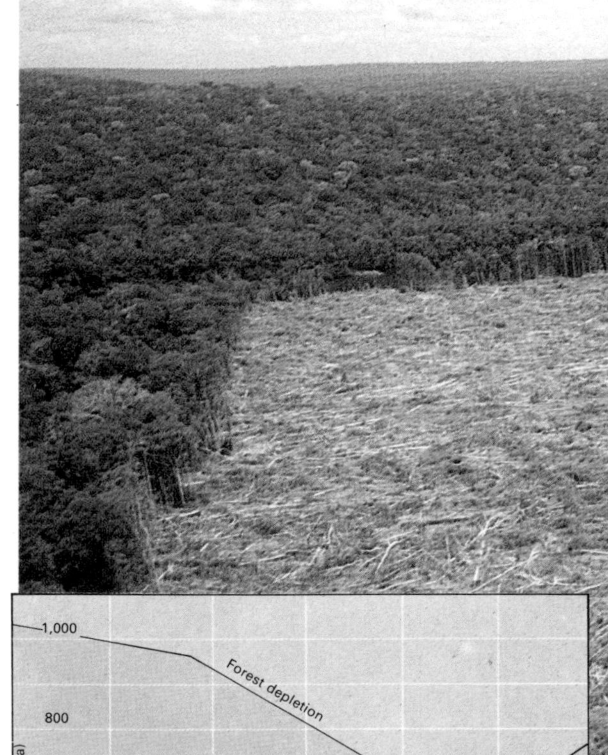

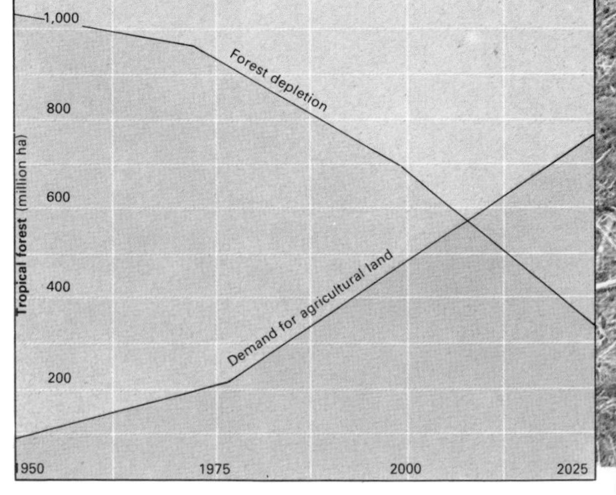

agriculture. They fell the trees in an area, burn the foliage, surface vegetation and some of the wood, then grow crops in the ashes. As the fertility of the soil declines and yields fall they move on to a new site. The system can be sustained indefinitely as a rotation in which each site is allowed about 12 years to recover, but the rotation has been accelerated, sites are revisited more often, and today the forest in affected areas has insufficient time to regenerate.

Apart from the loss of species, the clearance of tropical forests may have serious climatic consequences extending far beyond their borders. Tropical forest soils contain pockets of air enriched with carbon dioxide produced by the respiration of soil organisms and the decay of roots. A column of this soil can contain 30 times more carbon dioxide than the column of air above it. Forest clearance causes much of that carbon dioxide to be released into the atmosphere, where the concentration of atmospheric carbon dioxide is known to be increasing, leading to fears of a "greenhouse effect" climatic warming.

Up to half of that increase may be attributed to tropical forest clearance.

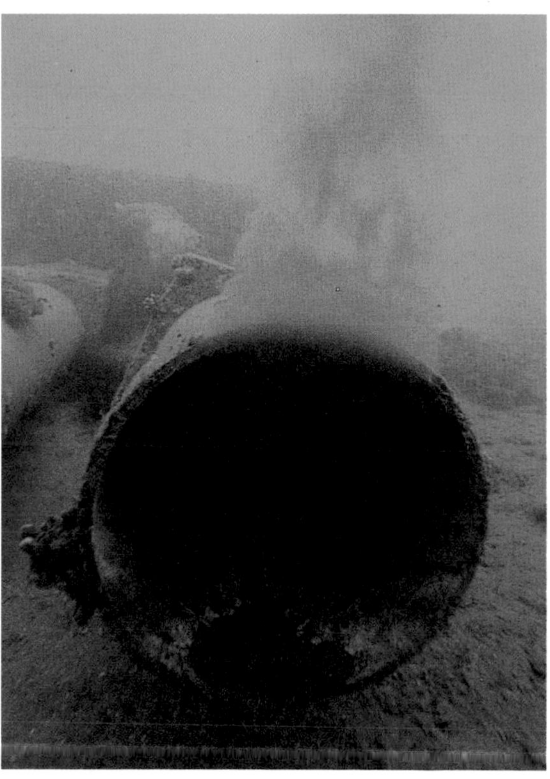

▲ Garbage can be disposed of efficiently, for example by burning it to generate power in specially-designed incinerators, but much of it is simply dumped.

◄ The clearance of tropical forests is a matter of global concern. By far the richest natural habitats on Earth, their clearance threatens the survival of innumerable species. The clearances may also produce climatic effects. This clearance is in the Amazon region.

► Sewage adds plant nutrients to water into which it is discharged, sometimes leading to excessive growth of aquatic plants. The bacterial decomposition of the organic matter in the sewage depletes the water of oxygen, and fecal microorganisms can present a health hazard to humans. When discharged into the sea sewage tends to float upwards because it consists mainly of fresh water, which is less dense than sea water.

Air pollution	
Pollutant	**Effect**
Factory smoke	Nitrogen oxide and sulfur emissions damage all forms of life and thereby disrupt ecosystems; also damage to property
Car exhaust	– as for factory smoke
Power-station emissions	Carbon dioxide and waste heat contribute to the greenhouse effect, sulfur emissions to acid rain
Nuclear power stations	Potential contamination with radioactive waste or liquid coolant
Oil refineries	Waste gases damage all living things
Pesticides	Wind carries chemicals to healthy plants
Domestic fuels	Smoke and chemical contamination
Jet aircraft	Noise; kerosene combustion products and soot damage animal and plant life
Space launchings	Exhaust and propellants jettisoned in the ozone layer, debris left orbiting in space
Volcanoes	Sulfur, metals, heat and particulate matter spread over a wide area

Land/water pollution	
Pollutant	**Effect**
Coal mining	Spoil heaps look untidy and are not easily colonized by wildlife
Ore mining & quarrying	Unsightly spoil heaps and scars; exposure of toxic materials
Motorway construction	Destroys landscape and restricts movement of wildlife
Garbage disposal	Unsightly waste tips and contamination of land
Chemical disposal	Contamination of land; run-off from land can pollute rivers and lakes
Industrial effluent	Kills all river plants and animals
Cooling water from power stations	Heating of rivers kills natural flora and fauna
Fertilizers	Toxic if used in excess; as rivers discharge into the sea, kills seawater life
Oil tankers	Jettisoned or accidental release of oil kills seabirds, fish and invertebrates
Sewage	If untreated or released in excess, kills wildlife or causes blooms of toxic marine plankton

Oil and heavy metals

On New Year's Day 1981, 5 tonnes of dead seabirds were washed ashore at Stockholm, the first of a total of about 100,000 birds whose bodies were collected from Swedish coasts by the end of the month. In the same month a further 100,000 birds died in the Kattegat, 32,000 in the Skagerrak, 10,000 in coastal waters from southern Norway to northwestern France, and in February more than 11,000 more came ashore on the east coast of England and on the coasts of West Germany, Holland, Belgium and northern France. All these deaths occurred because the birds had been covered in oil spilled at sea from tankers.

Pollution occurs when a substance is released into the air, water or soil and is harmful to living organisms either because it is not present naturally and is poisonous, or because it is present naturally but in much smaller amounts. For example, each year about 30,000 tonnes of mercury enters the environment from volcanic eruptions or by evaporating from the ground surface and returning as a common constituent of rainwater (◆page 88). This mercury is distributed thinly over the Earth,

Pollution

however. The 20,000 tonnes a year that enters the environment as a result of human activities is concentrated in certain places, and sometimes poisons humans.

Many species can tolerate high concentrations of mercury and probably of other heavy metals. Tuna fish caught in the Mediterranean and in parts of the Pacific have been found with up to 300 parts per million of mercury in their tissues, and humans have eaten them without coming to harm. Apparently healthy fulmar petrels have been found with 46 parts per million of cadmium in their livers. Mine spoilheaps often contain high concentrations of metallic wastes but some grasses colonize them (▶ page 232), and the presence of tolerant plants is used in prospecting for metallic ores.

Many metals are essential to organisms in trace amounts, being involved in the induction or synthesis of proteins. Where metals occur in large amounts they may interfere with these processes, especially with the large protein reserves accumulated by female birds prior to egg-laying. They may also disrupt cell membranes and the transport of substances across them.

◀▲ The Amoco Cadiz oil tanker disaster released 223,000 tonnes of crude oil just off the coast of Brittany. Seabirds were killed in their thousands.

▲ Sulfur emissions from factories pollute the atmosphere, producing acid rain (▶ page 218). This is a problem not only in industrialized nations but also in the Third World, where pollution control is only patchy. These ferns in Java have been choked by sulfurous air.

◀ Almost any manmade product is a potential pollutant, and natural products can harm the environment if deposited in large quantities in unlikely places. Pollutants of the air are well-defined but those of the land may have originated in the air and, in turn, may pollute the water.

Organochloride insecticides

The organochloride group includes such formulations as DDT, aldrin, dielrin, and lindane, but also polychlorinated biphenys (PCBs), which are industrial chemicals used at one time mainly for insulation in transformers. Organochlorides are insoluble in water but soluble in fats and oils, and are chemically stable, so they break down only slowly. Most are harmless at the concentrations sprayed by farmers, but because they are soluble in fats and eliminated from the body only slowly they tend to accumulate in the tissues of animals that consume them. The bird that eats ten sprayed insects may store the pesticide from all of them. The hawk which eats smaller birds may store the pesticide from all of their bodies. In this way pesticides accumulate along food chains (▶ page 86) until they reach concentrations large enough to be harmful. In Europe and North America, many species of hawk, eagle and falcon decreased drastically in number during the 1960s and 70s, mainly because the pesticides in their bodies affected their eggs; either the shells broke because they were too thin, or the eggs failed to hatch

Sulfur dioxide has been blamed for killing trees in the 1970s, but industrial emissions have been falling for years

Acid rain

All rain is slightly acid, but "acid rain" is a term used today to describe three ways in which pollutants may be deposited. Rain and snow may carry acids; dissolved acids may be carried in mists, moistening surfaces with which the mist comes into contact; (possibly the most harmful of all) solid particles carried in dry air may adhere to surfaces.

In the 1960s, Swedish scientists became worried about the high concentration of sulfur in the air in Stockholm and the increasing acidification of lakes and groundwater; in the early 1970s damage to forests in West Germany was attributed to the deposition of acids. Over the following years, acid damage was reported from the northeastern United States, southeastern Canada, Czechoslovakia and other parts of Central Europe, the Soviet Union and China.

Apart from the corrosion of mainly limestone buildings, common in many industrial cities, acid damage affects trees and lakes. Lake water is often rather acid, especially on peaty upland soils or where seawater sometimes intrudes. Elsewhere, certain substances, most commonly bicarbonate, held in solution in the water tend to neutralize acids, a process known as "buffering". Many Scandinavian lakes are poorly buffered and so prone to acidification.

Salmon and trout cannot tolerate water that is even slightly acid and their disappearance may be the first sign that water is polluted. If the acidity continues to increase, one by one, the other groups of fish disappear, first carps, and eventually the most tolerant of all, pike and eels. Invertebrates, too, disappear, one group at a time, until only a few plants and microorganisms remain.

Aluminum compounds formed under acid conditions may enter the water from surrounding land. They irritate the gills of fish and may kill them directly by asphyxiation; they replace some of the phosphorus nutrient taken up by plants and so kill them; they cause humates, compounds produced by the decomposition of organic material, to precipitate to the lake bed depriving the aquatic ecosystem of essential nutrients (➧ page 118).

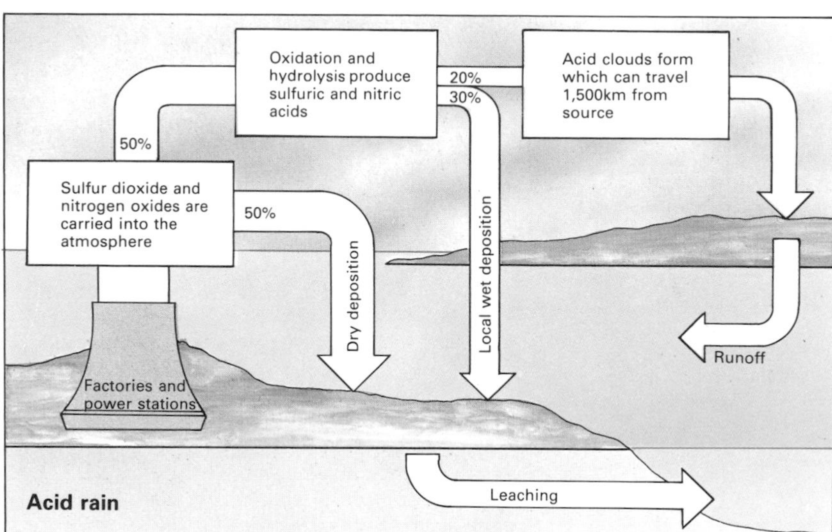
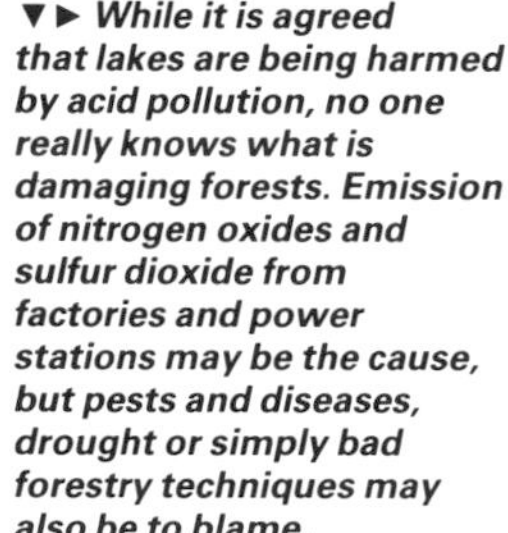

▲ Sulfur dioxide and nitrogen oxides from factory and power station smokestacks may fall to land as solid particles ("dry deposition") or as dilute acids, in rain and snow, which may then leach important nutrients from the soil. Acids may also be dissolved in clouds, which can travel long distances from source.

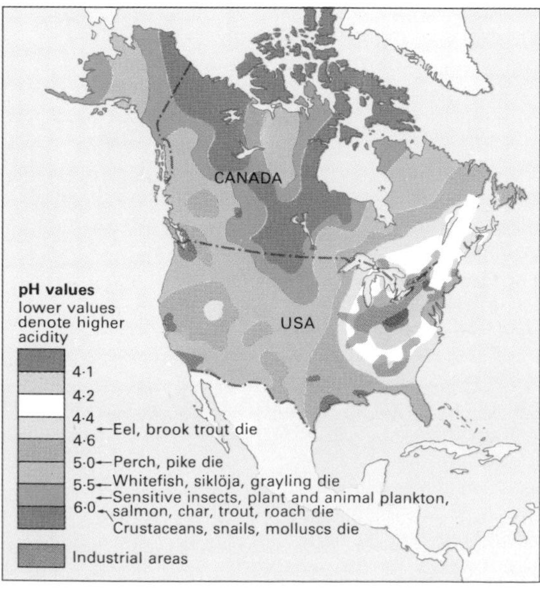

◄ The acidity of precipitation in North America. Contours represent lines of equal pH value, derived from data monitored at sample sites. Lake waters which lack natural buffers, or neutralizers, are particularly susceptible to acidification, which may exceed the tolerance limits of certain aquatic species.

▼► While it is agreed that lakes are being harmed by acid pollution, no one really knows what is damaging forests. Emission of nitrogen oxides and sulfur dioxide from factories and power stations may be the cause, but pests and diseases, drought or simply bad forestry techniques may also be to blame.

Why are the forests dying?

Sulfur dioxide and nitrogen dioxide damage cells on the surface of leaves. Depending on the concentration and the amount of water present, they cause stomata either to close or to open, in either case disrupting the flow of water through the plant. Taken up by plant roots, they encourage the growth of shoots rather than roots, and sulfur dioxide makes plants more susceptible to frost damage.

The first tree to be affected in Europe was the silver fir (Abies alba), which began to die back, the needles becoming discolored and falling first from the lower and inner branches but moving upward and outward until only the crowns were left standing on bare stems. By that stage, it is only a matter of time before the trees die. By the late 1970s Scots pine (Pinus sylvestris), Norway spruce (Picea abies) and finally beech (Fagus sylvatica) were affected, together accounting for more than 75 percent of all the forest trees in Germany.

At first, sulfur dioxide, emitted mainly from coal-fired power plants, was blamed. Many countries agreed to reduce their emissions by one-third over a number of years, but closer investigation showed the problem was not so simple. German forests were rich in lichens that are known to be intolerant of sulfur, and air in the forests contained little sulfur dioxide. Nor was it easy to explain why industrial pollution should cause harm when emissions had been falling for years. In the United States sulfur dioxide emissions have fallen by 7.6 percent since their peak of 32 million tonnes in 1970, and nitrogen oxide emissions have fallen by 10.6 percent from their 1973 peak of about 25 million tonnes.

Air in German forests did, however, contain ozone, which can be produced by the action of strong sunlight on unburned hydrocarbons from vehicle exhausts. Ozone damages trees directly, but it is also very reactive. It will oxidize sulfur and nitrogen dioxides to sulfates and nitrates which may then take part in further reactions that convert them to sulfuric and nitric acids.

If sulfur dioxide emissions are to be reduced, coal-burning power plants, their major source (♦ page 88), will need to be modified or replaced, one option being nuclear plants. Modification is possible, but expensive, environmentally as well as economically. Removal of the sulfur before coal is burned would mean the loss of up to 10 tonnes of coal for each tonne of sulfur recovered, and removal of sulfur from flue gases requires passing the gases through water containing pulverized limestone. Because of the way pollutants are carried in the air, Britain, for example, would have to reduce its sulfur dioxide emissions by 53 tonnes to reduce deposition in Sweden by one tonne, and by 46 tonnes to reduce deposition by one tonne in Norway.

A reduction in emissions of nitrogen oxides and hydrocarbons would be achieved most effectively by reducing driving speeds or modifying automobile engines, especially in West Germany which is the largest West European contributor of nitrogen oxides. A catalytic converter, for example, fitted to the engine exhaust removes most of the nitrogen oxides.

Desertification

If the evaporation of water from the ground is high enough to remove more water than falls as precipitation, the soil will be too dry to support any but the hardiest of plants adapted to arid conditions, and the land will be desert (◀ page 106). An annual rainfall of less than 25cm is likely to produce a desert anywhere in the world; settled agriculture may lead to soil erosion where the annual rainfall is less than 40cm.

At the edges of deserts there are semiarid regions where evaporation exceeds precipitation over the year as a whole, but where seasonal rains are sufficient to support plants temporarily. Since the 1960s, the world climate has been changing, the climatic tropics becoming compressed into a narrower belt than they occupied formerly (◀ page 97). This has restricted the seasonal northward movement of moist tropical air that brought rain to the desert margins. The rains have become unreliable and regions that were once semiarid are now desert. The effect is most marked along the southern edge of the Sahara, in the so-called Sahel Zone.

Up to about 700 million people live along the desert margins. Most are nomadic pastoralists (◀ page 144), following their herds and flocks in the perpetual search for pasture. In the past, deteriorating conditions forced them to migrate further, but today there are national frontiers to halt them and the better land that might once have fed them is being farmed, much of it to grow cash crops for export. Instead of migrating, the nomads have been encouraged to settle permanently on the best land available, but this has led to overgrazing of the pasture. At the same time, markets have been found for their livestock produce and better veterinary care has improved the health of their stock. This has encouraged the number of stock to increase, leading to further overgrazing. Between 1955 and 1970, the number of livestock in Mauritania increased by 125 percent.

When semiarid land is overgrazed, plants that could survive being nibbled are uprooted and destroyed altogether. Plant roots once bound the soil together, but when they are gone the soil is blown by the wind as a dry dust, some of which falls on neighboring land, burying and killing its plants. Deserts are advancing in this way at a rate of about 5,000km^2 each year.

Dams

Dams are built to provide supplies of fresh water and a renewable source of electric power. However, new large dams (defined as more than 150m high, or with a reservoir whose volume is more than 25 million cubic meters) may also have harmful effects on the environment – and not just by drowning natural habitats contained in river valleys.

Where the rocks beneath a dammed reservoir are faulted water may seep through fissures. In seismically active regions, seepage may lubricate masses of rock whose stresses push them in opposite directions, and the weight of water may cause one side of a rock fault to move in relation to the other, resulting in an earthquake. The first such catastrophe occurred in the 1930s while the Hoover Dam in the United States was filling.

Dams hold back river water and release it at a controlled rate. This alters the flow downstream. Where rivers such as the Nile used to flood seasonally, the floods no longer occur, and neither do the relatively arid conditions between floods. This allows certain aquatic species to proliferate, because there is no longer a dry season to check their numbers. Among those aquatic species are certain snails that act as intermediate hosts of trematode flatworms for which humans, or in some species humans or monkeys, are the final host. They cause schistosomiasis, a debilitating disease that has increased since the large dams were built.

Reservoirs hold water before releasing it, and silt carried by rivers has time to settle. The water leaving the reservoir is "cleaner" than the water entering it, but for thousands of years the farmers of the lower Nile depended on the silt deposited on their land by the annual floods. It supplied nutrients to their crops and texture to their soil. Deprived of the silt they must buy fertilizer which they can ill afford. Meanwhile, the silt accumulates in the reservoir until eventually the dam is made useless.

Dams are not new, but modern large dams are, and it may be years before their full effects are known.

► Increasing aridity, most notably in the Sahel region along the southern border of the Sahara, is due mainly to a change in the climate. Seasonal rains are no longer dependable and drought may persist for years. Where groundwater is close to the surface, plants can grow, forming an oasis. This is the oasis of Amtoudi, in Morocco.

◄ Dams to supply hydroelectric power are fed from artificial lakes and their construction involves flooding large areas. The Tucurui dam, in Brazil, is one of the largest in the world and its lake has destroyed the rain forest in the valley of the Rio Tocantins, along with unique species of plants and animals.

The greenhouse effect

Rather more than 40 percent of the radiation we receive from the Sun has a wavelength between 400 and 700 nanometers. It is shortwave radiation, and the Earth's atmosphere is transparent to it. The radiation warms the ground (♦ page 97), which then radiates heat, but at longer wavelengths, of 800 to 4,000 nanometers. Most gases are transparent to this longwave radiation, but carbon dioxide is not. It absorbs any radiation with a wavelength of more than 1,000 nanometers, and is especially absorptive of radiation between about 1,200 and 1,800 nanometers.

When carbon dioxide absorbs radiation, its molecules move faster, which increases the probability and violence of collisions between molecules, with the result that much of the absorbed energy is expended in warming the air itself. Carbon dioxide acts as a heat trap, allowing radiation to pass inward but not outward, rather like the glass in a greenhouse. This warming is the "greenhouse effect" and if the amount of carbon dioxide in the atmosphere increases, we may expect the atmosphere to grow warmer.

If the atmosphere is warmer, more water will evaporate into it. Water vapor is also a "greenhouse gas", absorbing most strongly at wavelengths slightly longer than those at which carbon dioxide absorbs, and so thickening the thermal "blanket".

Carbon dioxide is released into the air whenever a carbon-containing substance, such as wood, peat, coal, natural gas or oil, is burned (♦ page 88). It is also released from soils when forests are cleared. The amount of atmospheric carbon dioxide is increasing. Analysis of air trapped in the polar ice sheets at levels that can be dated has shown that in the fifteenth century the air contained about 270 parts of carbon dioxide to a million parts of other gases. In 1984 it contained 345 parts per million.

Other gases, including nitrous oxide, methane, and chlorofluorocarbons are released industrially or as a consequence of farming or forest clearance. The amount of them is small, but they, too absorb longwave radiation (nitrous oxide and methane at wavelengths of 700 to 1,300 nanometers) where carbon dioxide and water vapor are not strongly absorptive.

Climatologists believe they have detected a very slight increase, of less than one degree Celsius, in the average atmospheric temperature throughout the Northern Hemisphere. They cannot say whether this is due to the greenhouse effect, but it may be. In the United States, the Environmental Protection Agency and the National Research Council have both predicted a gradual climatic warming, with temperatures by the year 2100 some 5°C higher than they are now.

A slight warming might alter climates, making continental interiors drier and coastal areas wetter, but an increase of as much as 4°C might trigger the melting of the polar icecaps. If they should melt, sea levels would rise by about 50m, enough to inundate the low-lying areas that contain many of the world's major cities.

An opposite effect, of cooling the Earth's surface, is predicted should there be a nuclear war.

▲ *Usually the air temperature decreases with altitude. Air warmed close to ground level rises because it is less dense than the cooler surrounding air, cooling as it does so. In sheltered places, such as a valley or bowl surrounded by higher ground, air temperature sometimes increases with height. This is a temperature inversion. Warm air becomes trapped at low levels, along with pollutants, which then accumulate. Particles of smoke or dust reduce visibility and if the air is saturated, water may condense onto them to produce the mixture of smoke and fog known as smog. "Photochemical smog", seen here over Los Angeles, is formed by the action of strong sunlight on unburned hydrocarbons from vehicle exhausts and oxides of nitrogen.*

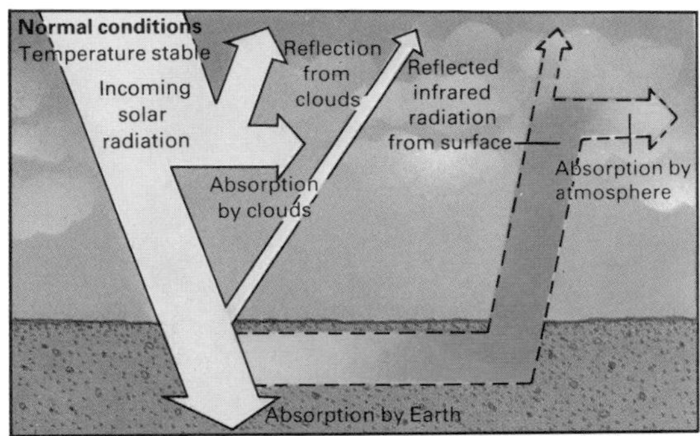

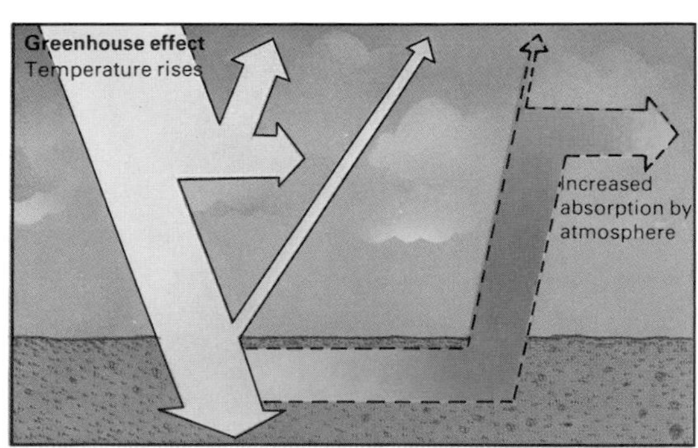

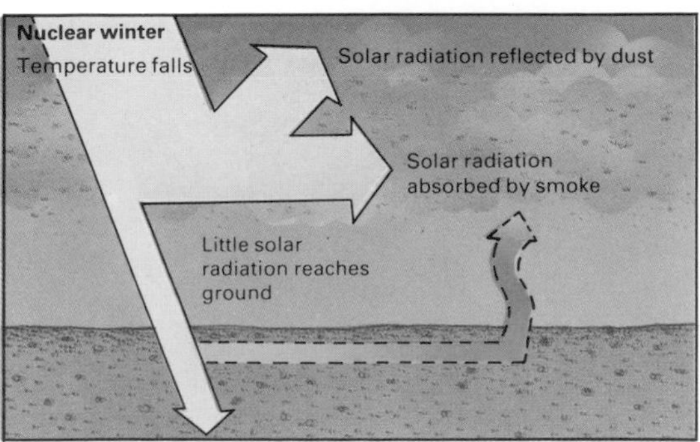

◄ *Carbon dioxide, water vapor, methane and certain other atmospheric gases absorb and reradiate long-wave radiation, but are transparent to short-wave radiation. Solar radiation passes through them, warms the ground surface, and this cools by emitting long-wave radiation. The "greenhouse effect" occurs when this outgoing radiation is trapped, and the air is warmed.*

► *One source of excess carbon dioxide in the atmosphere: burning off oil in the Mexican oilfields.*

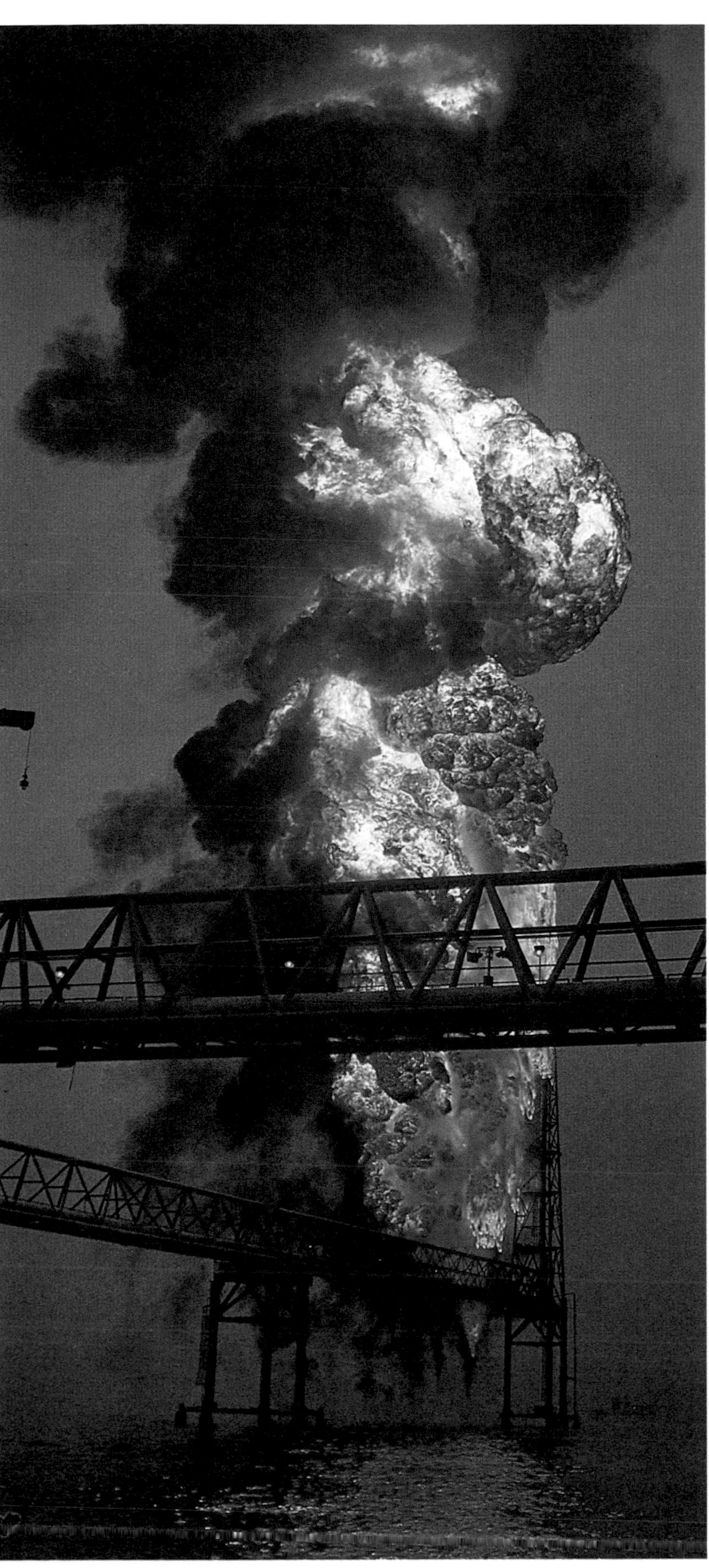

The ozone layer

At a height of about 10 to 50km, solar radiation with a wavelength between 4 and 400 nanometers, the "ultraviolet" (UV) waveband, provides energy to dissociate oxygen molecules (O_2), some of which re-form as ozone (O_3). The ozone is unstable, breaks down, re-forms as oxygen, and dissociates again. The effect is to absorb some of the UV radiation, and the ozone itself absorbs more.

The UV radiation that penetrates this layer of ozone reaches the surface. It causes suntans in pale-skinned humans and provides energy for the synthesis of vitamin D in human skin. Too much exposure to UV is believed to cause a mild form of skin cancer, and it can damage some plants. Because our exposure to UV is limited by the absorptive effect of the ozone layer, we are protected.

In the 1970s, it was feared that large fleets of supersonic passenger aircraft, flying at high altitude through the ozone layer, would deplete the ozone. Jet engines emit oxides of nitrogen, and these can react with the ozone to form stable compounds, so removing ozone faster than it could form. The large fleets failed to materialize, and it transpired that the amount of nitrogen oxides released by Concorde enhanced the ozone layer rather than depleting it.

Another fear was that nitrogen oxides might be released in large amounts as a result of the agricultural use of nitrogen-based fertilizers. Again, the threat was examined and dismissed. It has been suggested that the atmospheric explosion of nuclear weapons would release large amounts of nitrogen oxides and deplete the ozone layer, but atmospheric nuclear testing actually increased and thickened the ozone layer.

In the mid 1970s a new threat was identified, this time from "freons", chlorofluorocarbon (CFC) compounds used as propellants in aerosol cans, as refrigerants in freezers, refrigerators and air conditioners, and in the manufacture of plastic foams. CFCs are extremely stable chemically, but they are destroyed by UV radiation, to yield atoms of chlorine which form compounds with ozone. Estimates of the predicted extent of the ozone depletion from this cause within the next century have varied from between 2 and 16.5 percent. Several uses of CFCs have been banned in some countries and restricted in others; their use in aerosol cans was banned in 1978 in the United States. The dangers may, however be illusory. It is estimated that we release about 26,000 tonnes of CFCs a year, but about five million tonnes of a rather similar substance, chloromethane, is released each year by wood-rotting fungi, yet the ozone layer has survived.

The concentration of ozone in the ozone layer varies from day to day and season to season. Ozone production depends on sunlight, and ceases at night and during the polar winter. The ozone layer is thickest over the poles during their summers, and thinnest over the equator. Most of the predicted depletion would take place over the poles during their winters, so any effect on organisms at the surface would, in any case, be either insignificant or small.

See also
Communities of the Land 97-116
Communities of the Water 117-128
The Resource of Trees 173-180
The Shadow of Extinction 195-208
Saving Habitats 225-234

How serious is the threat?

If deserts continue to advance at their present rate (◀ page 220), and the erosion and poisoning of soil caused by bad farming continues, farm output will be reduced. It is possible that, of the 1.4 billion hectares of cultivated land available to the world's farmers in 1975, by the end of this century 25 million may be desert, 50 million may be poisoned, 50 million may have eroded away, and 150 million may have been converted to non-agricultural uses, a total loss of 275 million hectares or nearly 20 percent. If no remedies are applied even then, by 2025 the loss may be repeated, so that within a mere half century nearly half the Earth's land available for agriculture will have been made useless (◀ page 154).

This will be extremely harmful to humans, but the effect on wildlife is more complex. Desert species, for example, are likely to thrive, while those of the semiarid grasslands will suffer. This group includes many of the large mammals, such as the zebras, rhinoceroses, giraffes, elands, impalas, wildebeests, gazelles and elephants of Africa, and the bisons and pronghorns of North America, as well as the carnivores, the cheetahs, lions and coyotes, and the innumerable insects, reptiles and birds that also inhabit the savanna and prairie grasslands.

In less extreme climates, the loss in soil fertility that reduces farm output often encourages wildlife. Where soil is fertile, the habitat tends to be dominated by a small number of plant species which grow rapidly. If the nutrient level falls, their growth is inhibited and the plant cover is more varied, supplying the needs of a greater variety of animals. The best nature reserves are found on land that would need improvement before it was of use to farmers.

Human loss is not always nonhuman gain, however. If the damage to temperate forests continues, many species may be reduced in number as their habitat deteriorates. Tropical forests, too, support species that live nowhere else, and forest clearance leads directly and inevitably to the extinction of many.

▲ In Britain, many kilometers of hedgerow have been cleared, making fields large enough for machine harvesting. This stretch has been destroyed by fire caused by badly controlled burning of cereal stubble.

► Close-cropping of grassland in southern England by sheep and rabbits has prevented aggressive plant species from dominating the sward, allowing many small herbs to flourish. A decline in sheep farming and reduction of the rabbit population through myxomatosis has threatened these plants and some areas have begun to revert to shrub.

Saving Habitats

The objectives of conservation...Approaches to saving ecosystems and wildlife...From National Parks to Nature Reserves...People and wildlife living in harmony ...Reclaiming land...PERSPECTIVE...Interactions between animals in wildlife parks...Yellowstone National Park ...Culling elephants...South African wildlife areas... Maintaining manmade wildlife refuges

There are two ways to save species of animal and plant. The first, and best, is to conserve their habitats so they may continue to live in the wild. But there are many species that cannot be conserved just by being left alone. These include virtually all large land vertebrates. In addition to habitat protection, then, we must strive to protect and breed a wide range of the most vulnerable species.

Wild places can be saved simply by leaving them untouched. Far more often, however, they need actively to be managed or else they rapidly become degraded. Furthermore, though conservationists can sometimes exclude humans altogether, they generally have to compromise with human enterprises and ambitions (♦ page 130).

Why habitats must be managed – and how

Many ecosystems (♦ page 87) change qualitatively if left alone. Also, wild areas devoted to conservation are generally small relative to the requirements of at least a proportion of their denizens. Small habitats tend to be physically vulnerable and the populations of animals and plants within them tend to be too small for safety (♦ page 206).

The main reason that ecosystems tend to change with time is that the plants create conditions that are more suitable for other kinds of plants than for themselves. For example, heather binds the peat of moors and then is encroached by shrubs and later by trees.

Succession

The phenomenon whereby one kind of flora gives way to another and then to another is known as succession. The flora that persists at the end of the succession is called the climax vegetation. The climax vegetation in temperate lands is forest (♦ page 98) (dominated by conifers in the highest latitudes, and generally by oak in more temperate climates). All temperate lands will eventually become afforested unless the natural succession is halted. Physical forces such as rainfall may sometimes halt the succession. More usually, the succession can be interrupted only by biological forces; downland stays as such only so long as it is grazed. However, on the downlands in England fewer and fewer sheep are allowed to graze, not least because they are worried by domestic dogs. Conservationists, therefore, must apply the mower if they are to preserve the ancient flora.

▲ On bare rock, blue-green algae pave the way for lichens and mosses and eventually for flowering plants. Rocks on the sea shore, however, may never experience succession if they are constantly pounded by the waves – physical forces are often stronger than biological ones.

◄ Wicken Fen Nature Reserve in England. Many wetland habitats, such as this fen and reed swamp, support a wide assortment of plant species which in turn are the food and home for an equivalent diversity of insects, birds and other animals. Such areas must not be drained.

Overpopulation of elephants, even for a few years, can destroy the habitat irrevocably such that a population crash occurs – numbers can easily be reduced to a point where extinction is inevitable

Animals, particularly individuals of large kinds, also tend to change and sometimes destroy the reserves that are designed to protect them. If they are too well protected, they quickly become too numerous. Thus the elephants in the National Parks of South Africa have to be culled to maintain food supplies within the reserve (▸ page 239). The question of which animals to cull raises serious theoretical conflicts between ecologists and geneticists (◂ page 208).

Wildlife management raises interesting philosophical questions too – esthetic, logistic, and moral. Clearly a managed environment cannot be truly wild because there is human influence. But we must either settle for managed wilderness – in reality a series of artificial parks – or we must reconcile ourselves to having no large animals at all, and, indeed, none of the specialist plants and insects of downland and fenland. But what is the conservationist trying to achieve? Why is fenland better than the woodland it would change into if it were left to itself?

Choosing between alternatives

There are no simple answers to such questions. In general, conservationists feel that it is desirable to save as many different species of animals and plants as possible, and to maintain habitats suitable to contain that variety. They also seek to conserve as many different kinds of ecosystem. Nonetheless, some species seem more worthy of conservation than others. Everyone likes the large vertebrates – gorillas, rhinoceroses, elephants, giraffes – and wants to save them. Scientists in addition tend to emphasize animals or plants of particular biological interest.

It is of course high-handed of humans to decide that some are more worthy of salvation than others, but it is also inevitable. The saving clause, though, is that successful attempts to save a habitat for some "star" species such as the Sumatran rhinoceros (▸ page 245) will inevitably help a great many others that share that habitat.

Why the conservationist must be a good ecologist

In one National Park in Natal in South Africa the waterbuck seemed to be doing very badly. The obvious conclusion was that they were being too heavily preyed upon by lions. Closer inspection, however, revealed that they were being ousted from the best grazing lands by other antelopes – nyala – and being pushed into areas where they succumbed to tick-borne disease. They were not suffering from excessive predation, but from competition. The park managers removed some of the nyala, and the waterbuck recovered.

But why did the nyala not always out-compete the waterbuck, and destroy them? We come back to the difference between islands and continents (◂ page 196). Before nyala were numerous, the waterbuck probably ranged wild all over Africa and their populations possibly fluctuated enormously. On occasions they crashed dramatically through disease and lack of food, possibly abetted by one of Africa's periodic droughts. Then, over the next few decades and centuries, their numbers would build up again. However, the continent of Africa is so huge that even when the waterbuck were at their most numerous they could not consume all its grasses and reeds, and when the numbers crashed they did not fall below the point of no return (◂ page 204). But in a park, even one of hundreds of square kilometers, such fluctuations cannot safely be sustained.

▼ Various populations of eland are threatened. These low-density, highly mobile grazers are vulnerable to any agricultural development, restricted space and hunting.

The plight of the chough and the bighorn

A recent study in Britain revealed that the chough, one of the more elegant kinds of crow, was becoming rare around its native seaside cliffs partly because the cliff tops were no longer grazed by sheep. As a result, the grass on the tops was long. Choughs, like most birds that feed on open ground, do not like long grass. They rely on their vision to escape predators, and seem to feel insecure when their view is obscured. So, if the conservationist wishes to encourage these birds, he should cut the grass. A conservationist that wants to encourage insects needs the long grass.

Such conflicts may occur between professional wildlife managers. There are also conflicts between professional and amateur. Thus in parts of California the dry mountains (◆ page 111) are occupied both by bighorn sheep and by feral donkeys (domestic donkeys that have gone wild) known as burros. The sheep are the native animals and of by far the greatest interest to biologists. They are also shy, and rarely seen except by conservationists. The burros, on the other hand, are descendants of the animals used by the early settlers. They are popular with tourists since they represent a part of American history. Local people and historians are hardly aware of the bighorns' presence, but they are very attached to their burros. Yet the two animals compete, as natives and newcomers often do, and the bighorns are losing out. Which has the greater claim?

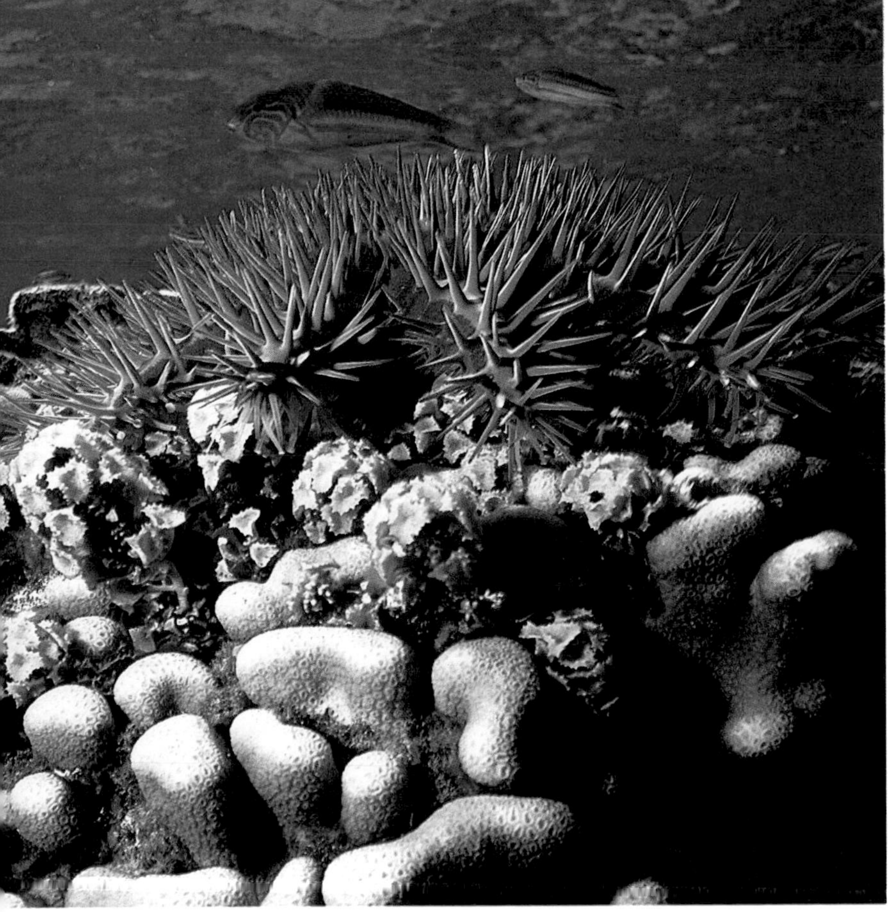

◄ In the Great Barrier reef population densities of 15 adult Crown of Thorns starfish per square meter are known. These result from population explosions that occur every 70 years.

▲ Choughs feed on small insects and nest in cliff crevices. The alpine chough lives at high altitudes and is fairly isolated from humans but not so the more common red-billed chough.

To manage or not to manage?

Generally, wild areas nowadays have to be managed if they are to survive. Sometimes, however, it may pay to leave well alone.

The Great Barrier Reef to the north of Australia is suffering appalling damage from a plague of the Crown of Thorns starfish, which eats many of the corals. Many conservationists suggest that attempts should be made to eliminate the starfish in order to protect the reef.

Other ecologists, such as Dr Tony Underwood of the University of Sydney, urge caution. After all, the Barrier Reef is not a wildlife park – a small area cut out of a larger ecosystem. The Reef represents the entire ecosystem, and there is no good reason to suppose it is different now from how it has been for hundreds of thousands of years. There is no reason to assume that the present plague is unique. Similar outbreaks of starfish could have occurred many times before.

Perhaps then – indeed probably – the reef is adapted to endure such plagues, however harmful they may appear. Intervention in this case may do more harm than good. Conservationists must be guided by the laws and theories of ecology (◆ page 86). But they must remember, too, that every environment is different. They must be prepared just to wait and see how nature proceeds.

Sometimes big is beautiful
When the United States Congress established the world's first ever National Park – Yellowstone, in 1872 – it was "as a public park or pleasuring ground for the benefit and enjoyment of the people".

In general, wildlife areas form a spectrum. At one end, there are small areas devoted exclusively to wild animals or plants (◆ page 225). At the other are the vast National Parks, found in most countries of the world. In between is a whole range of "reserves" – some big, some small, some privately owned, others public, many devoted primarily to wildlife, and a few mainly given over to pleasure but nonetheless harboring many wild species.

Wildlife areas may also be classified in another way. Some, such as the largest National Parks, are intended to contain virtually all the wildlife and ecotypes of a given region. Others are devoted just to one particular kind of landscape or habitat, ranging from sand dunes to "raised bog", with its array of specialist flowering plants and tall mosses.

Finally, some reserves concentrate on the needs of just one or a few species. The eleven National Parks of South Africa show the complete range of intentions (◆ page 230).

When considering habitats, conservationists must decide between several different options. Even when a stretch of land is given over to wildlife, that area, and the organisms within it, must still be looked after and human activity taken into account.

Fair shares in Planet Earth

Conservationists lay claim to land and water but farmers and fishermen want their say too, as do industrialists, miners, town planners and builders. If human beings are to survive on the planet in the company of other species then land must very carefully be apportioned between its claimants.

In general, two principles apply. First, some areas of land and water must simply be used for one purpose only, be it a refuge for wildlife or as farmland. All the conservationist can reasonably ask is that the wildlife reserves are large enough and that land that excludes wildlife really is used to good purpose. Second, whatever land is used for, as much effort as possible should be made to accommodate as many wild species as possible. The important point is that the different areas that each country or continent devotes to wildlife should between them provide suitable habitats for most, if not for all, wild species (◆ page 230).

◀ Norris Geyser Basin in Yellowstone National Park. Currently, almost one-third of the United States is given over to national parks, wildlife refuges and other public lands where flora and fauna are managed. In national forests and areas designated national resource lands, multiple uses of land are permitted but only on the basis that they are compatible with the local environmental status.

▲ Some of the activities of human beings that seem most alien to the needs of wild animals and plants turn out, unexpectedly, to be sanctuaries, as has happened on the Fermilab site. Plants imported from Europe and Asia, such as couch grass, had largely taken over an area that was once prairie. By introducing some of the native flora and fauna, including yellow-winged grasshoppers, an old habitat is being revived.

Strictly from a conservational standpoint, it might be best to exclude humans from wildlife areas altogether, but such purism has shortcomings. No land can be used for anything in the modern world unless people are prepared to pay, and people will not, in general, pay to be kept out. In addition, wildlife cannot survive without widespread public support.

Thus, conservationists must endeavor to bring pleasure to people (so they will in turn give support) and to educate (so that they perceive that conservation is worthwhile). In practice, then, areas devoted exclusively to wildlife – no public admitted – tend to be small and specialized; small areas of woodland or sand-dune, for example, dedicated to nesting birds at specific times of year. Large areas devoted to wildlife must in general accommodate people (➧ page 246).

Whatever a wildlife area is intended for, it cannot fulfill its purpose unless it is big enough. Sometimes "small" really is big enough for the purpose in hand. For example, a small area of "rough" in a golf course in Sussex, England, is conserved as a bog, and harbors cranberries. It is big eneough to ensure that cranberries hang on to their corner of southeast England. But small areas tend to be more vulnerable than large ones. Some ill-informed groundsman could in theory annihilate the Sussex cranberries in an afternoon.

Wildlife in forbidden places

Some 50km west of Chicago, Illinois, the site of the nuclear accelerator Fermilab provides one of the last strongholds of the prairie – the characteristic flora of the American midwest (➧ page 110).

The plants of the prairie include tall grasses, such as the Indian grass and bluestem, and other flowering plants such as yellow and purple coneflowers, golden rod, the wild onion, and the compass plant, whose leaves turn to face east-west. Such collections of largely herbaceous plants would normally be invaded and eventually replaced by trees but fires are common on the plains of the midwest, caused either by lightning or, in the past, by Indians, and trees could not become established. Prairie plants resist fire because of the mass of root below ground. For thousands of years, then, prairie occupied vast areas of the midwest, including much of the state of Illinois. But in 1830 the settlers came, bringing their crops and their European weeds, and the prairie began to disappear. Ten years ago Illinois had only 10km^2 of prairie left, out of its total area of 145,000km^2.

Fermilab was established in 1968 by the US Department of Energy. The structure itself is an underground tube that runs in a circle 6km in circumference. On the surface is a site of 2,700 hectares, where the general public is not allowed. In 1974 Professor Robert Betz gained permission from the Department of Energy to re-establish the original prairie within the Fermilab boundaries. With the help of volunteers (more than 100) he plowed the part of the land and gathered seed of native plants from within an 80km-radius of Fermilab. This they mixed with vermiculite, and sowed in 1975. By 1985 the group had established 184 hectares of prairie, and are now increasing the area, and seeking to introduce more and more species.

In Kruger National Park large-scale irrigation schemes have been introduced which some conservationists see as interference with nature but for many animals are their only salvation

The complicated matter of size

The large game animals of Africa, for example, obviously eat a lot: a single elephant eats as much as a small herd of cattle. Furthermore, wild herbage is usually of poor quality, with a low protein content and very high in fiber (▶ page 108). Thus, whereas a farmer working under ideal conditions in Europe may be able to support two highly productive cows on a hectare of lush, well-fertilized grass, the same sized animals in Africa – for instance eland, kudu, buffalo – might need ten times the area. In addition, Africa in particular is subject to seasonal droughts, and zebra and antelope, such as wildebeest, commonly migrate, from region to region, in search of fresh grazing (▶ page 73). Predators tend to be even more thinly spread than the herbivores; inevitably, there will be fewer predators than prey.

Even when National Parks are very big – Kruger, one of the biggest in South Africa, has two million hectares – they can contain only strictly limited numbers of animals. Some species, such as wild dogs, will be present in tens or hundreds, rather than thousands. The park borders may also cut across migration routes, which means either that the animals tend to leave the park (which puts them in danger) or that they remain, and put extra, unnatural pressure on the grazing.

Two points follow from this. First, even the National Parks are rarely large enough to guarantee the survival of more than a few species in the long term, however well they are managed. Second, for large animals at least, long-term survival is not possible unless the wild populations are constantly bolstered by programs of breeding individuals in captivity (▶ page 235).

Protected areas and living in harmony

Finally, we must acknowledge that a wildlife area is of virtually no use unless it is permanent. There seems little point in preserving a stretch of forest or a species of rhinoceros (▶ page 245) now, if it is simply going to be swept aside in, say, 10 or a 100 years' time. In the evolution of species, 100 years is insignificant. On the face of it, we are too fickle to be in charge of wildlife. The long-term hope is that the human species might, eventually, choose voluntarily to reduce its numbers, when there will again be room for other species. Our immediate task is to ensure that those other species still exist, which might take advantage of such an opportunity.

There are two things that can and must be done to increase the security of wild animals. The first is to give the reserves as much legal protection as possible. One crucial difference between National Parks and other "Nature Reserves" is that the parks are protected by national law, while reserves generally are not. But laws are very difficult to enforce in remote places, as National Parks often tend to be. Many animals in National Parks suffer severely from poaching (▶ page 239). The poachers are not always criminals, however. Often they are merely hungry, and the parks often rob them both of their grazing rights, and of their natural diet.

It is unfortunately true however that only small amounts of land will ever be given the status of National Park. At best, somewhere between 2 and 10 percent of the total area of any one country, the highest figures being reached in countries where human populations are relatively small, as in, for example, Scandinavia. So the survival of many species depends in large part on how well they can adapt to conditions in land that is primarily designed for human purposes. With a little imagination and foresight, we can help them to adapt.

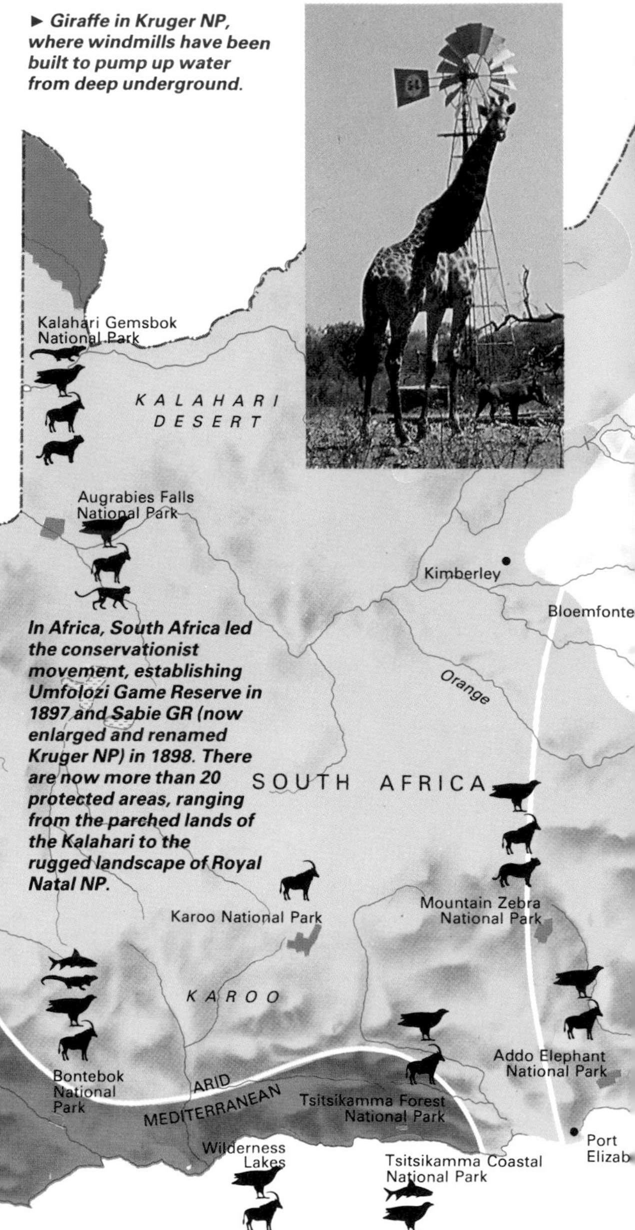

▶ Giraffe in Kruger NP, where windmills have been built to pump up water from deep underground.

In Africa, South Africa led the conservationist movement, establishing Umfolozi Game Reserve in 1897 and Sabie GR (now enlarged and renamed Kruger NP) in 1898. There are now more than 20 protected areas, ranging from the parched lands of the Kalahari to the rugged landscape of Royal Natal NP.

The National Parks of South Africa

The 11 National Parks of South Africa go some way to achieving all the functions of a country's wildlife areas and illustrate the problems of such parks.

Of all the parks only two, the Kruger and the Kalahari Gemsbok, are large enough to contain the full range of the animals and plants in their respective regions. But even Kruger has to be tightly managed, and is bursting at the seams. In particular, the Transvaal region suffers 8- to 10-year cycles of wet and dry weather. When they had the continent to themselves the animals could range far and wide to find grazing but now they must stay in one place. Soon after the Kruger was given National Park status in the 1920s many animals perished in severe droughts. Wells were dug, and now there are windmills to help them through.

In Kruger's game reserve days in the 1890s the hoofed animals, particularly buffalo, were

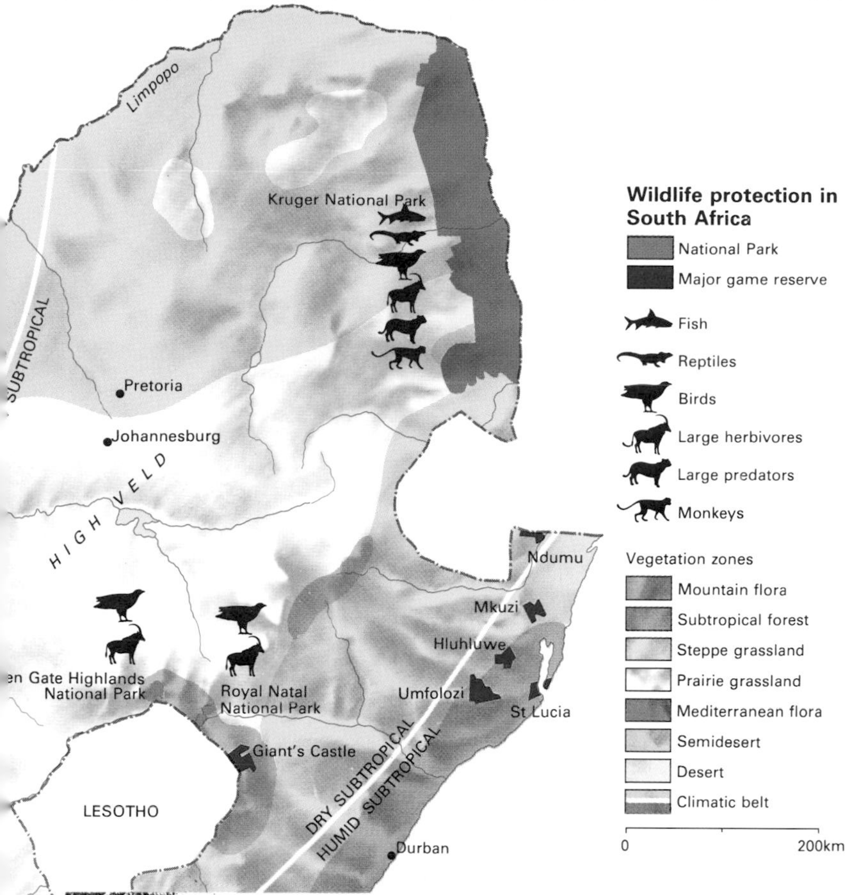

Wildlife protection in South Africa

National Park
Major game reserve
Fish
Reptiles
Birds
Large herbivores
Large predators
Monkeys

Vegetation zones

Mountain flora
Subtropical forest
Steppe grassland
Prairie grassland
Mediterranean flora
Semidesert
Desert
Climatic belt

0 200km

Limpopo, Kruger National Park, Pretoria, Johannesburg, HIGH VELD, SUBTROPICAL, Ndumu, Mkuzi, Hluhluwe, Umfolozi, St Lucia, Golden Gate Highlands National Park, Royal Natal National Park, Giant's Castle, LESOTHO, DRY SUBTROPICAL, HUMID SUBTROPICAL, Durban

◄ *In Umfolozi Game Reserve rhino are captured for relocation to another wildlife area.*

▲ ***Stolsneck Dam, Kruger NP, amid open landscape that allows the animals freedom to roam.***

devastated by rinder pest. Wild animals in general suffer greatly from disease. Again, huge herds spread over a continent can withstand decimation. Smaller groups confined to one place can be wiped out (♦ page 196). Some South African herds of antelope are now vaccinated against anthrax.

Parks, however, can be too successful. Elephants are the main problem. They are slow-breeding, but they have no natural predators, and they rapidly become too numerous (♦ page 238). But there are more positive aspects to management. Thus black rhinoceros disappeared from Kruger in 1936. In 1971 they were reintroduced from Natal, and now there are more than 100. Oribi, gray rhebok and red duiker have also been reintroduced, while surplus animals have been taken to other parks. In general, the translocation of animals from areas of surplus to areas of scarcity plays a large part in modern conservation strategy.

▲ ***A bontebok, the symbol of Bontebok NP, a small sandy depression in a rocky plateau alongside a river.***

Parks devoted to individual species.

The Bontebok National Park is devoted to a single antelope, the bontebok, which in reality is only a subspecies of the blesbok. The park itself acquired National status in 1961, and then had 84 of these animals. Now it has about 300, which is about as many as it can carry. But a species cannot be considered "safe" unless there are at least 500 individuals. The park cannot expand because it is surrounded by farming country which is considered more valuable. So excess animals are transported to other parks, and zoos, with the intention of arranging marriages between the separated animals so they remain a coherent breeding group. In truth, they do breed perfectly well with blesbok, which are reasonably common, but if they were allowed to breed with them, the distinct bontebok subspecies would be lost.

In some parks, the species that the conservationists are mainly interested in serves as a "shield" for many other species. Thus the Addo Elephant National Park, devoted to the last elephants of the Cape Province, also harbors buffalo, vervets, hares, tortoises, and many others. But the bonteboks are in conflict with other creatures – and indeed with the native flora. Bonteboks like grass, but the natural flora of the southern Cape is the highly characteristic fynbos, a marvellous assemblage of plants adapted to dry heat and poor soil, which includes ericas, proteas, serrurias and leucadendrons. The fynbos is threatened, as much as the bontebok. Which should take precedence? The answer may be academic since all the nation's parks are threatened by the rapid increase in the human population

Motorways that, for example, cut through chalklands expose banks that cannot be cultivated or trespassed upon, and are perfect habitats for some threatened butterflies that once flourished on downland

Saving the land

Mining – for coal, sand and gravel, stone, clay, phosphate, and metals such as gold, copper, iron and aluminum – has robbed the world, and in particular the world's wildlife, of millions of hectares: almost a million and a half in the US alone. In the past, waste (spoil) from such mines was simply thrown into heaps. Existing topsoil was buried, while the spoil itself remained extremely inhospitable to plants.

Modern mining operations should not cause such problems. Indeed reclamation can begin while mining or quarrying is still going on. For example, the bottom of a slag heap may be reseeded while the top is still being added to, and one end of a quarry may be landscaped and flooded for waterfowl while the other is still being worked.

The easiest way to restore derelict land is simply to cover it in topsoil and reseed. But the topsoil has to be at least 10cm if plants are to grow at all, while grass needs 25cm, and trees need a great deal more. The cost of transporting such massive quantities is usually prohibitive. In practice, land restorers first analyze the precise shortcomings of the particular area, and seek to remedy them one by one. The problems are of three main kinds: drainage and erosion; toxicity and excess acidity (or alkalinity); and infertility.

Spoil heaps usually drain all too easily. The water runs off them because their sides are so steep, and because they are often compacted by heavy machinery. The water that does sink in, rapidly drains away again because the structure is too porous. The simplest solution to the first problem is to reduce the slope of the heaps and to "rip" the surface with long tines pulled by a heavy tractor, which encourages rain to sink into the heap, instead of running off. If the heap is too free-draining it can be bolstered by adding as much organic material as possible, including chopped straw or wood pulp, and farmyard manure.

Toxic metals, or excess acidity, can sometimes simply be washed, or leached, away provided that the leaching is not allowed to pollute the surroundings (⧫ page 88). Acidity is also countered by adding lime, which is cheap, even in vast amounts.

Finally, all spoil heaps are infertile – too infertile even for wild plants. They lack the essential nutrients, mainly nitrogen, phosphorus, and potassium (NPK) and they usually have such poor structure that they are unable to hold nutrients when they are added. Nitrogen is a particular problem: it is needed in large amounts, and is very soluble, and so very prone to leach away.

One modern solution is hydraulic seeding or hydromulching. Seed of whatever species are required are suspended in a slurry containing thickener (such as alginates, obtained from seaweed) and fertilizer. This is then simply squirted on to the heap surface through hose. Hydraulic seeders can spray up to 60m, and reach places where conventional farm machinery cannot.

When the heap is drained, detoxified, and fertilized, it is ready for sowing and planting. Nitrogen-fixing plants (⧫ page 158) and species able to withstand waterlogging are usually used.

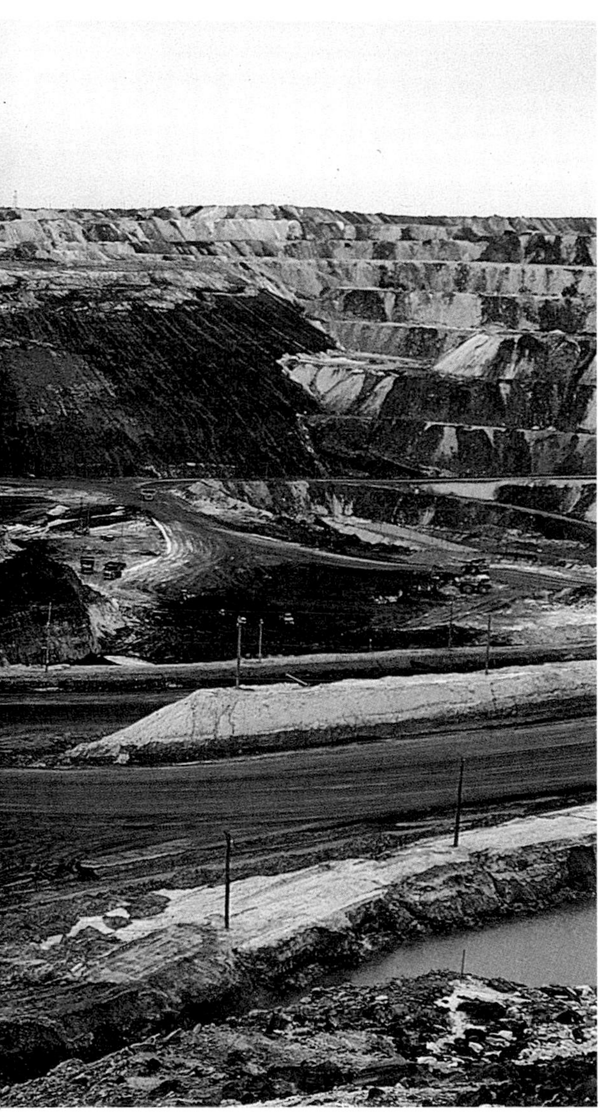

Land for many purposes

Virtually every human activity could be far more compatible with wildlife than is generally the case. For example, many European cities, particularly capital cities, are founded on rivers, the carrier of ancient transport. But waterways, in times gone by, were the source of disease. Accordingly, developers tended to drain or enclose all waterways that were not actually needed for boats. We know now why waterways carry disease, and how to prevent it. We do not need to enclose rivers and drain ponds. If the waterways were restored, the old rivers uncovered, vast and intricate habitats would be opened up.

Even the most bizarre and apparently hostile of human enterprises could be turned to advantage. For example, many conservationists see the motorway as one of the great enemies of wildlife and environment. However, the motorway verge is one of the few places where the general public is not allowed to go, and may be hundreds of kilometers long: a perfect, if narrow, sanctuary.

It is important neither to exaggerate nor underestimate the conservational value of land that is not devoted primarily to wildlife – in which wildlife is simply "hanging on". On the positive side, animals and plants breeding in parks, gardens, cities and on farms do help to boost numbers, which is a vital factor in long-term survival. In addition, the areas between wildlife reserves can provide vital links between them. Thus conservationists argue that hedgerows and motorway verges are important not simply because they provide homes for animals (particularly birds and insects) and plants, but mainly because they may provide natural causeways links between grasslands and woods.

◄ Mute swans once abounded on England's rivers. Now they are becoming rare, or at least extremely localized. They fill their crops with gravel. In so doing they pick up lead weights dropped by anglers and are poisoned. The cygnet on the right shows the characteristic sign of poisoning: kinking of the neck.

▲ Open-cast mining – this is for coal, in Australia – is particularly destructive. On site, habitats are destroyed, while the surroundings are invariably polluted.

► In northern Europe, kestrels thrive in cities. They nest on high buildings (which they regard as "cliffs") and often hunt by the motorways.

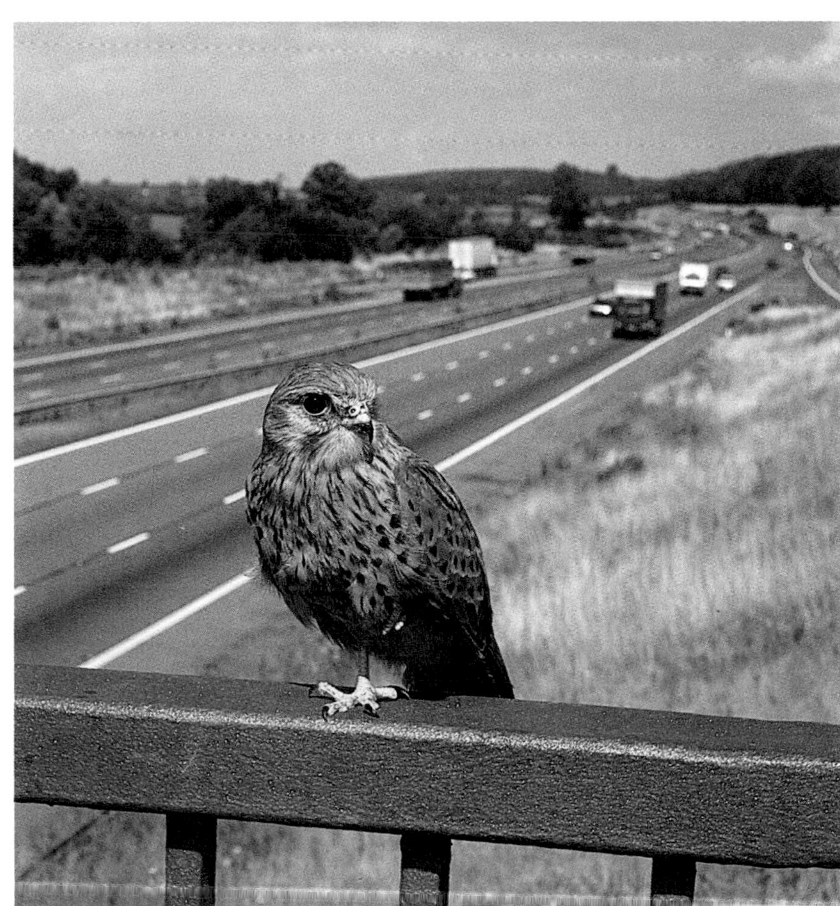

Inadvertent destruction of wildlife

Marine yachtspeople, all unsuspecting, have for many years been interrupting the reproduction of oysters and mussels, and stunting the plants of the plankton (♦ page 122), by covering the hulls of their boats with paints containing tin in organic form. Their object is to kill the sessile adult barnacles that otherwise foul the hull. But minute quantities of tin in the water inhibit everything that grows and, like many such pollutants, organic tin accumulates over time and becomes concentrated in the tissues of those animals that are highest in the food chain. Alternatives to organic tin must be found.

See also
Territory 63-70
Interactions 81-96
An Ape Called Man 129-140
The Shadow of Extinction 195-208
Saving Wildlife 235-248

Human and wildlife compatibility

It has become the tradition, particularly over the past 300 years, to plant towns, parks, and gardens with exotic and highly cultivated plants: European, Australian, and South American species in post-colonial Africa, and Asian and American species in Europe. But exotics are far less hospitable to native insects and birds than are native species. If the unleashed rivers in British towns were planted with willow and alder, and Africa gave back its cities to the acacias (rather than to eucalyptus, from Australia, or jacaranda, from South America), then more of the attendant animals could return. Gardens and parks, too, could be made far more hospitable.

Farms and grazing lands, which now occupy most of the land on Earth that is not simply uninhabitable, could usually be designed to acommodate far more wild species than they do. Possibilities range from the maintenance of hedgerows in Britain to the management of grazing lands in India and the farming of game in Africa. Forestry, too, may be inimical to wildlife, or it can be "sympathetic". Even logging, in tropical rain forests (⧫ page 100), which at present is the most destructive of all human enterprises, can be conducted so as to leave at least some species. Indeed some habitats that humans are constantly disturbing, so that they are always in a state of tension – effectively being pushed rapidly through various states of succession – can be particularly rich in species. The traditional coppice, in which trees are kept cut close to the ground and their branches harvested for stakes, are a case in point. This principle does not unfortunately apply to exploited tropical forest. Nothing, in that case, compares with the pristine state.

The need for planning

The kind of animals or plants that thrive outside but not in nature reserves are not necessarily the ones that are in greatest danger. For example, cities are hospitable and provide useful resources to European foxes (which abound in the suburbs) but are anathema to African wild dogs that are adapted to hunting prey in wide open spaces (⧫ page 56). On the other hand, nature does spring surprises. Britain's three woodpeckers, the greater spotted, the lesser spotted, and the green, all creatures primarily of woodland, now nest in suburban gardens. In general, only a limited number of species will be able to benefit greatly from land that is devoted to human purposes and it is not always easy to predict which ones will benefit.

Similarly, non-reserve areas may be a barrier not a link between reserves. If the "corridor" between reserves is narrow, as hedgerows are, then some species simply will not travel along them. Also, some animals, for behavioral reasons, refuse to cross certain kinds of barrier that a human observer might feel presents no problem. Some woodland butterflies and birds just will not fly across clearings, and a road will cut a population in half. Similarly, a logger who removes just a few trees from a tropical forest so as not to destroy the whole habitat (⧫ page 178) may thereby ensure, inadvertently, that the remaining trees have each become islands, as far as some of the creatures within them are concerned.

Thus there is no room for complacency. We may rejoice that bald eagles in the United States nest in rocket sites and herons fish in London parks but the general rule is that many, indeed most species, even those living under apparently ideal conditions, are likely to be far more vulnerable than they seem. The particular problem is that many wild organisms, both animals and plants, have lost their specialist habitats. The general problem, which applies particularly to large animals, is that there are just too few of each species for comfort.

► In Britain, between 1945 and 1972, 80 percent of hedgerow trees – once the country's richest reservoirs of flora and fauna – were torn up to make way for arable farming. In the 1950s, hedgerows were removed at a rate of 7,250km a year, and nearly a third of Britain's ancient woodlands were cleared. Hedgerows serve as a substitute for woodland, particularly as broadleaved forest has almost disappeared. Garlic mustard, shown here, is among the plants that do particularly well in the slightly exposed, quasi-woodland environment of the hedgerow; and garlic mustard in turn supports orange tip butterflies.

Saving Wildlife

21

The problems of breeding in captivity...Plant and animal reproductive physiology...Pooling resources, genetic and financial...The dangers of breeding new races...The future for conservation...PERSPECTIVE...Reintroduction of rare species...Seed banks and frozen zoos...The elephant situation...Breeding apes in captivity...Big cats, in the wild and in nature reserves

There are three legitimate reasons for breeding wild species of animal and plant in captivity. First, it is a vital part of conservation to keep the numbers of individuals as high as possible – with the caveat that some individuals are genetically far more useful than others (◀ page 202). Second, conservation cannot succeed without the cooperation of the public, and animals and plants in captivity can serve to educate and to show that other species are worth preserving. Finally, wildlife cannot be sustained in the wild except by careful management and this management depends upon knowledge. Captive species provide opportunities for research that can be applied in the wild. But there are problems with captive breeding.

Getting animals to breed

Whereas some animals and plants breed freely in captivity, others do not. The reasons are far from obvious, and some still elude solution. For example, it was many years before cheetahs were bred successfully in captivity. The reason was partly physiological (many male cheetahs are sterile) and partly behavioral (males and females will not mate unless they are kept apart until the female is on heat, which she signals by means of her scent) (▶ page 247).

From captivity to the wild

Two animals that would be extinct by now if they had not been bred in captivity are the Arabian oryx, from the Middle East, bred mainly in the United States, and the Père David's deer, from China, rescued primarily by the Duke of Bedford in England and now breeding in several European zoos. Both now have been returned to their native lands, and illustrate the problems of reintroduction.

The first requirement is to involve the local people, among whom the animals are to live. The Chinese, for example, are anxious to develop their country following the Cultural Revolution. Conservation is one of their causes, and the Père David's deer, which finally disappeared from Eastern China during the Boxer rebellion at the turn of the century, is a symbol of a new age.

The second requirement is to find a suitable location; one that is reasonably protected and self-contained, and providing the right habitat. For the oryx, this was a plateau in Jiddat al Harasis in Central Oman, where the last of the wild oryx were exterminated in 1972. It has richer vegetation than the surroundings, and a natural boundary – the scarp of the plateau itself. Ideal sites in China are still being investigated, but research at Whipsnade Zoo in England indicates that the Père David's deer is extremly tough, and will survive almost anywhere, provided it is not hunted.

The third requirement is to watch over the newly introduced animals, to ensure that they settle down. As Hartmund Jungius, of the International Union for the Conservation of Nature, has pointed out, wild animals rely for survival not only on their instincts – their innate ability to behave in appropriate ways – but also on their knowledge of the local environment.

◀ The Arabian oryx became extinct in the wild in 1972 but survived in zoos around the world – as here at San Diego in the United States. When a herd was first reintroduced into its native land, the Jiddat al Harasis plateau in Oman, some animals wandered off into the desert. Having been raised in Arizona, they did not know the dangers of their homeland. After a few days' misery in the desert the oryx were happy to be herded back up the plateau, and never again broached that natural barrier. Neither will their descendants, who learn from them.

The problems of breeding pandas in captivity – in particular determining when the female is ready to mate – are being overcome by testing for reproductive hormones in the animal's urine

Wild plants in cold storage

"Seed banks" – cold stores of seeds – are already well established. Most of these are concerned with storing old or out-of-fashion varieties of cultivated plants and their immediate relatives (♦page 146). The seed bank of the Royal Botanic Gardens, Kew, in England, is one of the very few international collections devoted to wild species. So far it has samples of 5,000 species of flowering plant in store (out of about 250,000 known species).

The task of establishing a seed bank begins with the field workers. They have to ensure that the seeds they gather, firstly, truly represent the population of a particular species in a particular area and, secondly, reach the bank in good condition. The first task is not as simple as it might seem. All populations of wild plants are highly variable, and it is important to gather as many variants as possible. It is important, too, to avoid simply gathering the most spectacular or unusual individuals, as this would give a badly biased sample. It takes knowledge and skill to gather a truly representative sample – and to do this without decimating the wild population.

To keep the seed in good condition, the gatherer must take it at the right time: when it is mature, but probably a few days before it is due to fall off the plant. The aim then is to keep the seed dry and cool. The moister and warmer it is, the more likely it is to go moldy, to germinate, or simply to die. In general, though, the aim is to post the seed to the central seed bank as soon as possible.

When the seed arrives, physiologists clean it (to remove mold and any surviving insects) and then immediately begin cooling and drying. Most seed will last for a very long time – sometimes for hundreds of years – if its moisture content is reduced to 5 percent, and it is stored at −20°C. Drying is carried out first, by keeping the seed in hessian bags with a relative humidity of 15 percent, at 15°C. In such conditions, most seed reaches 5 percent moisture within a month. Then it can be cooled safely to −20°C for long term storage.

Orthodox and recalcitrant seeds

There are three problems in seed storage, however. The first is that not all seeds will store happily at 5 percent moisture. Apple seed, for example, lasts only 20 to 25 years in store when it is as moist as this. Apple needs to be taken down to 3 percent moisture, if it is to last for hundreds of years. Such important exceptions to the main rule are found not by keeping seed for hundreds of years, but by examining the changes in the seed over short periods, under different conditions of temperature and dryness, and then extrapolating.

Secondly, and in practice more importantly, some seeds do not conform to the simple rules of drying and cooling at all. The ones that do are termed orthodox, and those that do not, recalcitrant. If recalcitrant seeds are dried, they die.

Many tropical trees (♦page 174) have recalcitrant seeds. This is extremely unfortunate, first because some tropical trees are among the most endangered plants in the world, and secondly, because many of them are extremely valuable. The araucarias of South America, for example, are rapid-growing and valuable both for timber and for firewood. Physiologists at Kew are developing methods of storing recalcitrant seeds. Evidently – and against all expectation – they need high levels of oxygen, as well as moisture. They cannot yet be stored for centuries, as is desirable, but at least they can now be kept alive for a few years – rather than months, as in the past.

Inducing germination

The third potential problem is that many seeds enter a state of dormancy, which needs to be broken if they are to germinate. No seed will germinate if it is dry and cold, but most will do so (provided they are still alive) once they are given moisture and warmth. The dormant types, however, also require some additional stimulus, or trigger, to set them off. Many weed seeds, for example, need light before they will germinate. Many legumes (lupin-like seeds) are so tough that their coats must be abraded, simply to allow moisture to penetrate. Part of the task of Kew is to find out precisely what individual seeds require to break their dormancy. Otherwise, like so many Sleeping Beauties, they would remain locked in torpor for ever.

Ideally, seed banks should embrace all plants with seeds: Gymnosperms (pines, cypresses, araucaria) as well as Angiosperms (flowering plants). With limited resources, Kew must concentrate primarily on particularly endangered groups. The tropical trees are one spectacular example. But so too are the grasses and leguminous plants of the Mediterranean, whose habitats are disappearing beneath coastal hotels. These form an important part of the Kew collection, which is expanding each year.

▼ *At Wakehurst Place, scientists from the Royal Botanic Gardens, Kew, conserve seeds from thousands of species of wild plants. Kew concentrates upon plants of potential commercial value, such as legumes, and from threatened areas of the world.*

▼ *European botanists have been gathering exotic plants for their gardens and for medicines and food, for hundreds of years. But collecting-fever reached a peak in the 18th and 19th centuries. Local labor was and is employed to find the plants, as here in the Himalayas of the 1860s.*

The rare Przewalski's horse (◀ page 208), last of the wild horses (apart from zebras) seemed at first sight to pose no such behavioral problems. In the wild, the males gather the females around them, in permanent harems. It transpired, however, that the males like to keep their distance. If the females get too close, the males attack them. Unless enclosures are very large, therefore, they should be fitted with internal barriers, so the females can effectively hide.

With giant pandas, males and females must be brought together when the female is in heat. Females come into heat apparently for only one day a year and neither the zoo keeper nor the male can tell when she is in heat.

Getting plants to breed

Captive plants may suffer from the phenomenon of self incompatibility. Most plants prefer to outbreed; they are better adapted not to mate with their own siblings (which are genetically similar to themselves) or members of the same clone (identical with themselves). Most outbreeders, such as maize and sorghum, will mate with their own siblings if they are forced to (though the offspring of such unions are often inferior because of inbreeding depression (◀ page 204). Others, such as plums and some apples, are physiologically incapable of mating with their own siblings, or clone-mates. In such self incompatible plants the pollen fails to germinate when it lands on the stigma, or aborts soon after germinating.

▲ *Breeding and selection of wild apple* Malus silvestris *has yielded thousands of varieties. Some cultivated types are self-fertile but most require cross-pollination.*

▼ *Spur-thighed tortoises have suffered from the pet trade. That traffic has largely been halted but they face a new danger – their Mediterranean habitats are being plowed for farming.*

In Rwenzori National Park, between 1954 and 1968, elephants reduced the number of trees from over 4,600 to 500

Quantity

The next problem is again one of numbers. Species are not liable to survive in the long term unless they include at least 500 individuals. The breeder is easily deceived. A small population may appear to flourish for several generations, only to crash later. But most zoos feel that they are doing well if they have even a single breeding pair of any one species, particularly large species. Many zoos now specialize, and have large groups of only a few species, rather than one or two of a whole range, as in old-fashioned menageries. Therefore, if zoos are truly to contribute to conservation, they must pool their resources. Animals must be exchanged from zoo to zoo so that the captive animals collectively form one population (➧ page 244).

There are severe limitations, however. First, the magic figure of 500 is rarely reached. Second, transport is expensive, and economics are very important. Third, no country has enough zoos to sustain the required numbers of large animals, so international cooperation is required. But transport of animals across national boundaries is now generally forbidden because of problems of infections such as rabies and foot-and-mouth disease.

The ethics of culling

At the turn of the century there was only a handful of elephants in the reserve that preceded Kruger National Park in South Africa (➧ page 230). In Kruger itself, by 1946 there were 500 elephants. By 1964 there were 2,374. Such a rate of increase would have produced more than 20,000 by the 1980s, if unchecked. Buffalo too were thriving, even though they were almost wiped out by the rinderpest epidemic of the 1890s, and even though they are to some extent preyed upon by lions.

When numbers become excessive the conservationist can do one of three things. First, he can let nature take its course. But too many elephants can destroy a landscape, and then everything starves, and this has happened at times, as at Tsavo in Kenya. Second, surplus animals can be transported to other parks, or to zoos. This is done for animals such as bontebok but nobody wants surplus elephants and buffalo are notorious bearers of disease.

The third alternative is to cull. This began at Kruger in the mid 1960s (using muscle-paralyzing drugs followed by a bullet to the brain, in the case of elephants). Elephant numbers now are kept at around 7,500 and buffalo at around 30,000.

Culling raises several problems. In general, the public do not like it. They see it as an affront to the whole notion of conservation. It is, however, part of grim reality. It can also be turned to advantage. Hunters are prepared to pay to shoot certain animals, which provides valuable income, and the hides and tusks or horns some animals (➧ page 197) fetch a good price on the international market.

A subtler problem, however, is that the best practical way to cull animals such as elephant is to eliminate entire herds or family groups at a time. If individuals are shot from various families – a son here, an aunt there – there is appalling social disruption, as in a human community. However, geneticists argue that it is in the long term interests of the species to keep as many different families as possible (➧ page 202), with each family maintained at the same size. Thus even when the decision is made to cull, there is theoretical conflict: which animals should be killed?

Operation windfall

If conservation is to succeed in the long term, then it must involve local people otherwise they are all too likely to poach endangered species or rob the animals' habitats of trees for firewood. If the people's alternative is to starve, they can hardly be blamed. Only a few reintroduction projects have so far succeeded in obtaining the necessary cooperation however. One is in Zimbabwe, and has been dubbed Operation Windfall.

When Zimbabwe achieved independence in 1980, elephant poaching greatly increased. One reason was for food. Another was because people associated the conservation of elephants with white rule. At first, the government called in the military to curb poaching. But in two areas, they hit on a more enlightened scheme.

In the Chizarira National Park, and the Chirisa Safari Area west of Harare, elephants were becoming too numerous – uprooting trees and leaving the reserves to destroy crops. The wardens recommended a cull of at least 1,500 animals. Dead elephants are valuable, for meat and for ivory. And the government decided, in these instances, to dedicate the earnings to local people.

The cull, in 1981 and 1982, earned almost $1 million, and this went to schools, clinics, and local transport. This included more than $25,000 paid to local people to help in the culling; money paid by tourists for licenses to shoot elephants that were destined to die anyway; and the sale of elephant meat, which was salted, dried, and cut into strips as the traditional "biltong". The biltong was sold expensively in the towns, and cheaply in the local villages. Thus the people benefitted from the local wildlife, and indeed, with luck, will come to regard it as their most valuable resource. At any rate, it is no longer necessary to employ wardens, as the local people now have no reason to poach.

► *When elephants ranged over the whole of Africa their populations probably rose to huge levels and then collapsed. In 1979 there were estimated to be over 1,340,000 elephants in all 35 countries where the species is known to occur. In 28 of these countries, numbers were seriously declining. However, now that elephants are confined to reserves, their populations must be kept within narrow limits by culling. Otherwise they will destroy their habitat during the boom periods.*

Distribution of the African elephant

Range of elephant, 1984

Total numbers, by country

- more than 100,000
- 25,000-100,000
- 10,000-25,000
- 1,000-10,000
- less than 1,000

Countries where elephant populations are not declining

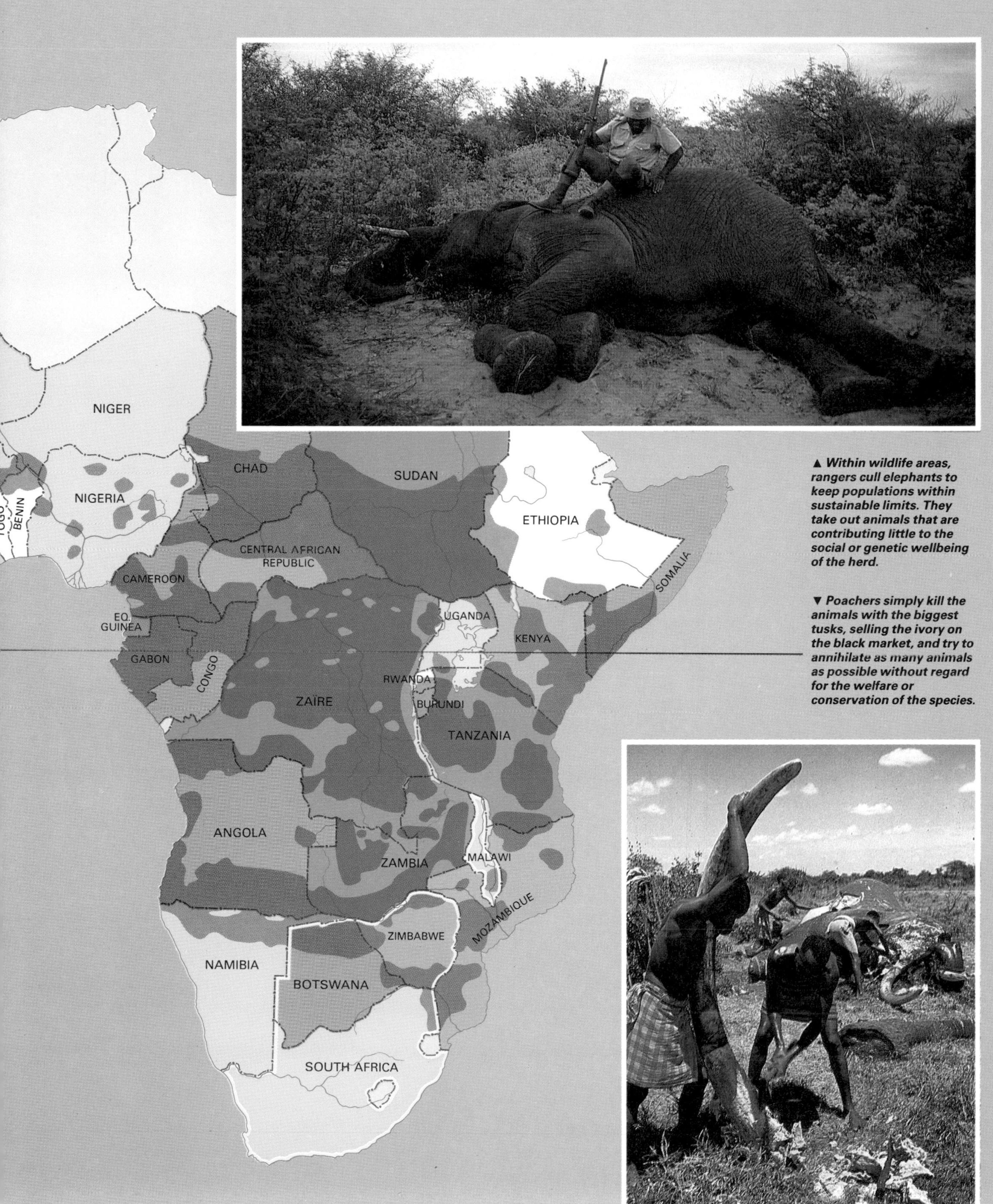

▲ ***Within wildlife areas, rangers cull elephants to keep populations within sustainable limits. They take out animals that are contributing little to the social or genetic wellbeing of the herd.***

▼ ***Poachers simply kill the animals with the biggest tusks, selling the ivory on the black market, and try to annihilate as many animals as possible without regard for the welfare or conservation of the species.***

At least 20 breeds of European farm animals have become extinct during the last 100 years. To counter this trend, conservationists have established rare-breed survival centers as living museums

Some of these problems of breeding in captivity could be overcome by artificial means of reproduction. High numbers of individuals could be maintained, for example, by keeping animals in the form of frozen embryos, rather than as grown animals – the idea of the "frozen zoo" (➧ page 242). And instead of transporting male animals around the world, the biologist may simply take samples of sperm and employ artificial insemination. However, even some of the most basic problems of these alternatives are far from solved. Thus, the sperm of bulls and of men can indeed be frozen very successfully – most cattle are fertilized by artificial insemination, and since 1978, when the world's first "test-tube baby" was born, many babies have been born to infertile couples using in vitro fertilization (IVF) – but this is not yet true for all animals. The sperm of pigs and of many antelopes, for example, have not yet been successfully manipulated in this way and there are many plant seeds that cannot yet be successfully stored.

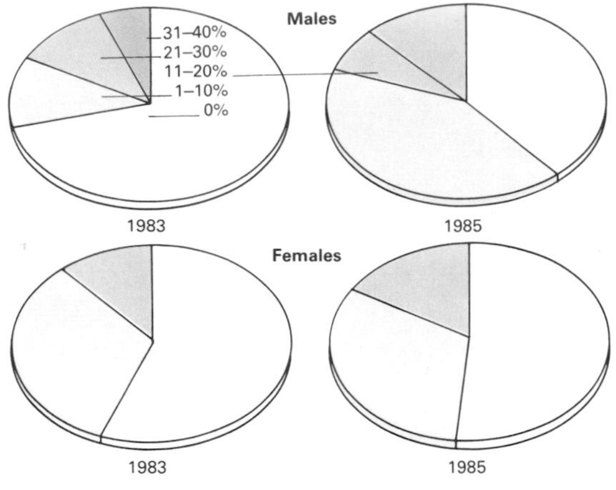

WILD-BORN FOUNDERS

Males: Jambo (b.1961), Kisoro (b.1962), Samson (b.1963), Mumbah (b.1965), Djoume (b.1967)

Females: N'Pongo (b.1957), Shamba (b.1958), Nandi (b.1959), Moula (b.1960), Gogal (b.1961), Caroline (b.1961), Baby Doll (b.1961), Susie (b.1962), JuJu (b.1962), Delilah (b.1963), Lomie (b.1963), Mushie (b.1969), Diana (b.1971)

OFFSPRING

Males: Little fella, Daniel (b.1971), Assumbo (b.1973), Mamfe (b.1973), Tatu (b.1975), Kijo (b.1975), Kisabu, Koundu, Kumba (b.1976), Kambula, Kibabu (b.1977), N'Gola (b.1977), Saul (b.1978), Kakinga (b.1978), Shumbo, Benjamin, Goliath (b.1980), Ruben, Zachary, Kibobo (b.1980), Jomie (b.1980), Kabie (b.1982)

Females: Zaire (b.1974), Rebecca, Kimba, Bamenda (b.1975), Salome (b.1976), Naomi (b.1977), Kishka (b.1978), Killa Killa (b.1978), Kaja (b.1979), Shumba (b.1980), Deborah (b.1981), Kumi, Jouma (b.1981), Leah, Kabinda (b.1982)

Founder representation in 1983

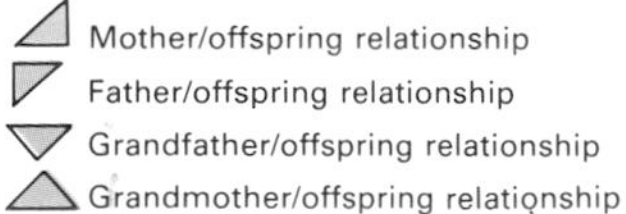

Died before 1983
Died after 1983
Exported before 1983
Exported after 1983

◄▲ Captive-born gorilla population in Great Britain. The genetics of their breeding is illustrated using a matrix (left) in which founder (wild-born) individuals that have bred are set against their captive-born offspring. Each offspring receives half the genes of each parent. Second-generation young possess founder animals' genes through their parents – they have a grandparent/offspring relationship (see key). The shaded portions of boxes for each offspring (50% or 25%) add up to 100% and indicate the founder animal's genes expressed in them. Changes in the population between 1983 and 1985 have resulted in improved founder representation among males, and highlighted which founders must be encouraged to breed.

▲ ***Arctic-breeding snow geese are smaller than their genetically close relatives that live further south. Their own food requirement and that of their young is reduced and they produce smaller eggs which require less time to incubate.***

Keeping the right individuals

A third problem is that breeding for survival is not simply a question of numbers. It matters a great deal which animals are mated together. In general, it is vital to maintain as much genetic variety as possible within a breeding population (◆ page 202). There are several reasons for this. The first is that breeders of wild animals are striving to conserve whole species. Another is that most living things begin to suffer if they are too inbred. In particular, genetic disorders are far more common in inbred animals. It is vital, then, to keep careful records of the pedigree of each animal within all the zoos that are cooperating in breeding programs, to ensure that only the most judicious matings take place. The success of breeding Przewalski's horse in captivity has depended in large part on the records of pedigree maintained since the beginning of this century.

There is, however, a subtle theoretical problem. The curator's desire to arrange precisely the marriages he wants may clash with the desires of the animals. Thus, the aim of the breeder of wild animals is to maintain as much genetic diversity as possible within the limited number of animals he can keep. One way to achieve this is to breed from as many different males as possible – not just use one male as the stud. But curators of modern zoos like to keep animals in natural breeding groups. And many animals, such as antelopes and baboons, naturally form harems – one male, and many females (◆ page 68). So, should the curator allow the animals to do what is natural, or should he instead keep naturally polygamous animals in breeding pairs? One possible solution is to keep the animals (antelope, say) in a harem, but ensure that a good proportion of the females are in fact served by other males (or by insemination, when this is possible).

There are still more problems. Animals that live in the wild are adapted to live in the wild. Animals in captivity must be adapted, to some extent, to captive life. There is a danger of breeding new races of captive animals that in important respects are quite different from the wild species they are supposed to represent.

There can be conflict, too, between the zoo curator's desire to maintain the genetic diversity of his animals, and his need to attract paying customers. Some endeavor to breed only the most attractive animals. This, of course, is what is done by breeders of domestic dogs. But with wild species the philosophy is to maintain as much diversity as possible. The aim with domestic animals is to select rigorously for just a few genes.

Genetic issues – two different stories

The snow goose breeds near the Arctic Circle. Its breeding season is short, perhaps only a few weeks. If the chicks are not born at the correct time of year, they will not grow sufficiently to migrate south before the winter starts (◆ page 76). The birds "know" when to mate and lay eggs because they respond to the particular day-lengths of high latitudes.

If snow geese are bred in zoos at lower latitudes, however, they must become adapted to respond to quite different day-lengths. They would then be incapable of breeding in their natural habitat. Snow geese adapted to breed at low latitudes would be genetically very similar to the wild geese, and they would still be of the same species, but for behavioral reasons they would no longer be capable of mixing with the wild types and would effectively be a new species.

At San Diego Zoo in California, one of the world's greatest zoos, one of the female black-and-white lemurs had a genetic defect: a sunken chest. It did not seem to affect her adversely, but it was a deformity. Should she be allowed to breed and pass on the gene responsible for it? San Diego's director of research, Dr Kurt Benirshke, argued that the gene did her no harm, and that she may well possess a whole number of other genes that were of definite benefit to the species. By rejecting her he might have reduced the colony's long-term chances of survival.

◄◄ ***A lowland gorilla, one of the most popular of zoo animals. The survival of all species of ape – gorilla, orangutan and chimpanzee – depends heavily on successful breeding in captivity. All species have been bred in zoos around the world but careful management is necessary to ensure the correct individuals mate. In most zoos, the ape populations stem from individuals imported from the wild more than 20 years ago. Many of these have failed to breed or now show serious infertility problems. To overcome this, curators have established studbooks providing information for organizing breeding loans and for ensuring animals are not being too inbred (◆ page 202). The data presented in the chart (left) is compiled from such a studbook.***

In male animals with a normal sperm count, abnormal sperm may comprise up to 40 percent of the count without seriously affecting fertility, but if a female's egg is abnormal, total infertility results

Artificial insemination

Many of these genetic and commercial problems are being overcome with the help of reproductive physiologists, who have developed a variety of techniques. First, they can freeze the semen of the animals, and transport the frozen sperm instead of the entire male. The sperm can then be introduced into the females by artificial insemination.

However, the sperm of some animals is highly susceptible to present freezing techniques. All animal cells, including sperm cells, are surrounded by a membrane consisting of protein and lipid (fat). During freezing the lipid has to be protected by adding a preservative. The trouble is that the composition of lipid differs from species to species, and preservatives designed for cattle and humans, for example, do not work on, say, pigs. Physiologists at London Zoo, one of the world centers for such research, are confident that they can devise suitable preservatives for all species, however.

In vitro fertilization

Alternatively, the physiologists can remove ova from the females, and induce conception to take place in a test tube or, more likely, in a petri dish. The conceptus can then, if necessary, be returned to a foster mother. This technique of in vitro fertilization, or IVF, is now almost routine both in human medicine and in the breeding of cattle. But if conservationists decided that some particular female had a particular combination of genes that made her especially desirable to breed from, they could take several ova from her, fertilize them in vitro, and insert them into the uteri of other females. These then become foster mothers, and give birth to the young animals.

IVF could, in theory, be used to generate huge numbers of desirable animals, because the females who produce the desired ova are saved the time and trouble of being pregnant. However, whereas males produce millions of sperm, females release only a few ova – generally only a few hundreds – during their entire reproductive lives. Large, slow-breeding animals such as humans or cattle release only one ovum per month throughout the year, until they become pregnant, and then the release of ova ceases until the pregnancy (9 months or so) has run its course. Thus a cow could at best produce 12 offspring in a year, if all the ova were collected and fertilized in vitro, and the conceptuses were raised by foster mothers. And she can at best produce one calf per year (9 months' pregnancy and 3 months' rest between pregnancies) if IVF is not employed.

For a conservationist, such rates of reproduction may be too slow. But they are developing several techniques to speed things along.

Already, physiologists can increase the supply of ova by injecting females with hormones – gonadotrophics – to encourage release of ova from the ovaries. Thus a cow may be persuaded to release 40 ova in a year, which translates into 40 offspring if these are all raised by foster mothers.

Females can be made far more productive than this, however. They do indeed release only a few ova at a time, but at any one time their ovaries contain hundreds of thousands of eggs. It ought to be possible to remove the entire ovary, or to take the ovary from an animal that is already dead, and extract many thousands of eggs from it. The problem is that most of the eggs are in an immature form – they are not ova but oocytes, the cells that give rise to ova – and these cannot be fertilized. At Cambridge University, England, however, physiologists are devising ways of encouraging oocytes to mature in the test tube.

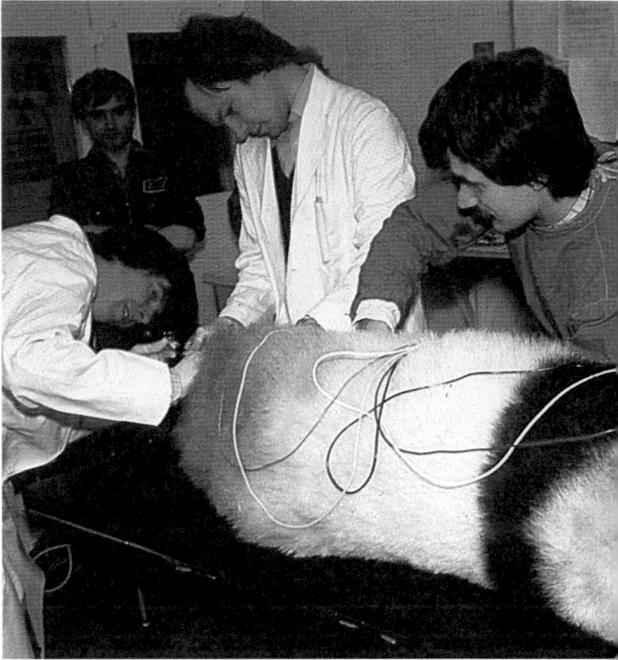

▲ One of the many difficulties of breeding giant pandas in captivity is to detect when the females are ovulating because they show few of the normal signs of being "on heat". London Zoo scientists can now detect ovulation by testing the hormone levels of the urine, and have tried to induce pregnancy by artificial insemination.

▼ This animal is a chimera: roughly half its body cells are those of a goat, and the other half are from the sheep. It was created by mixing the cells of the embryos of the two species, then raising the resulting mixed embryo in the womb of a sheep. This technique could be used to bolster the reproduction of rare species.

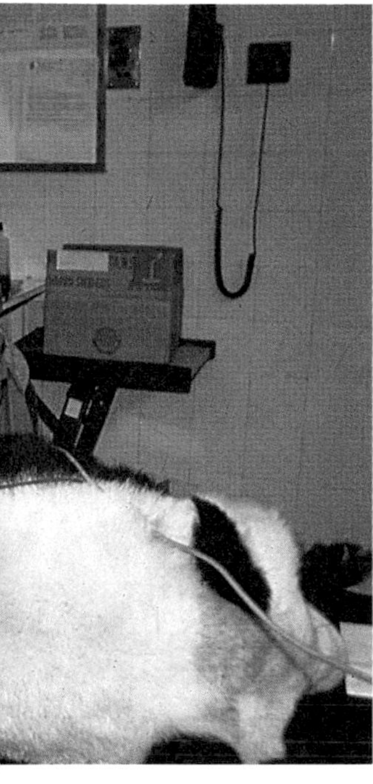

► Reproduction of rare species can be increased by transferring embryos into the uteri of foster mothers. This zebra foal developed in the uterus of the pony.

▼ Eggs may be fertilized "in vitro". Here, the pronuclei of the sperms and eggs of a hamster are about to fuse. Rare species may benefit from the technique.

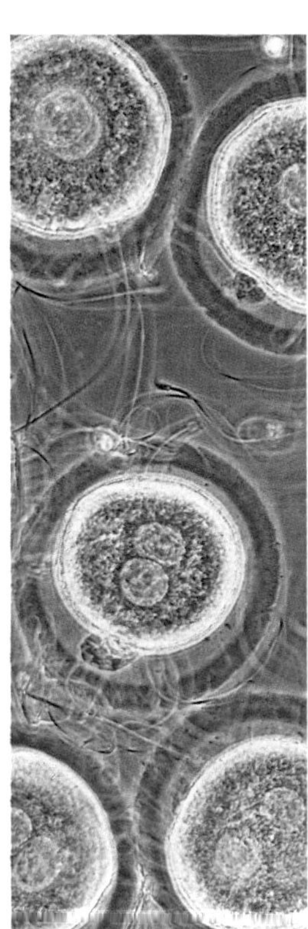

The frozen zoo

In theory, rare, slow-breeding animals could be persuaded to produce hundreds or even thousands of offspring, which is precisely what the conservationist desires. Furthermore, it would be easy to arrange exactly the marriages, between sperm and ovum, that the geneticist decides are most likely to perpetuate the species. These marriages could be arranged even after the parents are dead.

There are snags, though. It is one thing to produce thousands of offspring in captivity, but it is quite another to keep them. Who has room, for example, for 50,000 Sumatran rhinoceroses, assuming artificial reproductive techniques were ever applied to them. This problem is relatively easily solved. The embryos, once produced, can simply be frozen, and kept indefinitely in liquid nitrogen. They can be thawed at any time and inserted into a receptive female. This technique has already been applied successfully both in cattle and in humans. In the future, we will see the development of "frozen zoos", containing huge numbers of embryos from a vast range of species.

Foster mothers for preserved embryos

Neither cattle nor humans are rare. There is no shortage of foster mothers for embryos produced by IVF, whether frozen or not. But when a species is rare, there very obviously is such a shortage. And what is the point of storing frozen embryos from some extinct species of antelope, say, if there are no surviving females to put them into? For it is not yet possible to bring embryos to term outside the womb. The physiology of pregnancy is far too complicated.

Yet there is a solution. If there are no surviving females of the same species as the frozen embryo who can act as foster mother, then the embryo can, in some cases, simply be inserted into a female of a different species.

There are problems. All animals possess an immune system, whose task is to reject invading organisms – parasites. The embryo is genetically different from its mother (after all, half its genes come from the father) and one of the mysteries of pregnancy is why the mother's immune system does not attack it, just as it would attack, say, a bacterium. The fact is, however, that the female immune system does give special dispensation to her unborn offspring, even though it continues, throughout pregnancy, to attack other parasites.

But the female immune system is nonetheless highly discriminating. It will refrain from attacking an embryo of the same species that is inserted into it, but will attack embryos of different species. A sheep, therefore, will not foster the embryo of a goat, even though sheep and goats are related.

The recent decline in rhino populations in the wild is the result of the rapid rise in value of rhino horn which, when ground, is used as an aphrodisiac in India and as a fever reducing agent in the Far East

Mixtures of two species – chimeras

A female sheep will not reject an embryo inside her womb provided only that the outer layer of cells at least of that embryo are from the same species as herself. Provided an embryo of, for example, a goat can be wrapped in cells derived from the embryo of a sheep, then the ewe will carry the goat embryo, and give birth to a healthy goat.

Wrapping goat embryos in sheep embryo cells sounds fanciful, but it has been done, and is not especially difficult. When embryos are very young – just a ball of cells – the cells can be broken apart and mixed with the cells of other embryos. Such a mixture is called a chimera. If such a chimera is put into a sheep then it will develop normally provided the mixed embryo has sheep cells on the outside, and it will develop normally in a goat provided there are goat cells on the outside.

However, animals consisting of a mixture of two different animals are of little use. What the conservationist wants is pure animals. Here, nature is on his side. In the normal course of the development of all mammalian embryos, some cells develop into the body of the animal (skin, muscles, bones etc.) while others develop into the placenta. At the time of birth, the placental cells are detached from the fetus (that is, the umbilical cord is cut) and thrown away.

The placental cells initially form the outside layer of the embryo. With skill, then, the physiologist can mix cells from the embryos of two different species in such a way that he derives "body" cells from one species, and "placenta" cells from the other. Thus he could indeed wrap a pure goat embryo in sheep cells, and a "pure" goat may be fostered by a sheep (◆ page 242).

So the conservationist who wishes to produce a young antelope from a frozen embryo need not despair if there are no female antelopes of the same species left in existence. Instead, he could simply take an embryo from a closely related species, form a chimera with the embryo of the rare species, and insert that into the existing female. If he is skillful, that animal should produce a pure calf of the otherwise extinct species. In truth, this technique has not yet been applied to antelope. But it has been applied to sheep and goats, and there is no theoretical reason why it should not succeed in other species.

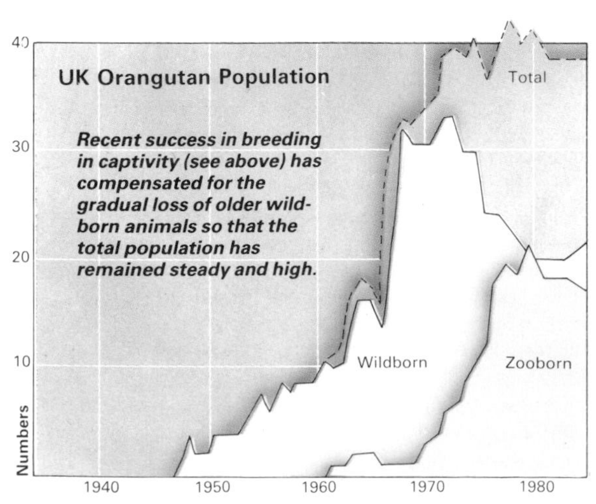

Recent success in breeding in captivity (see above) has compensated for the gradual loss of older wild-born animals so that the total population has remained steady and high.

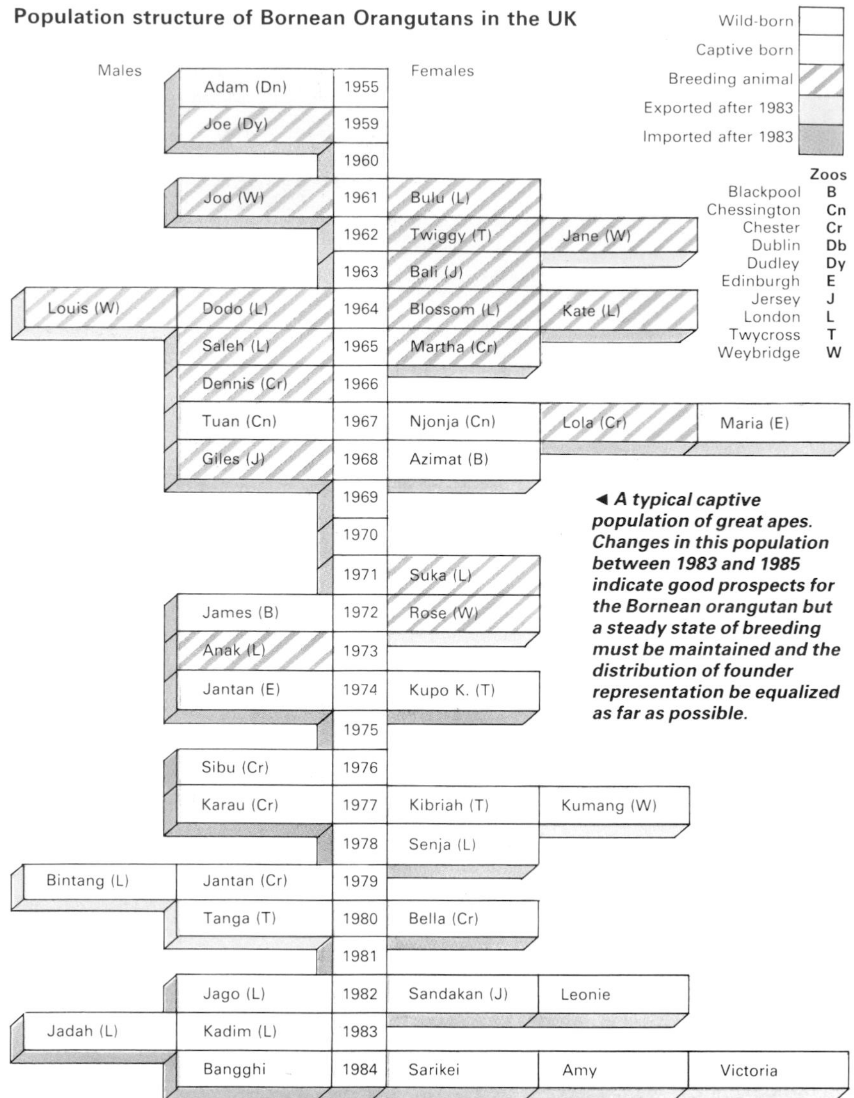

◀ A typical captive population of great apes. Changes in this population between 1983 and 1985 indicate good prospects for the Bornean orangutan but a steady state of breeding must be maintained and the distribution of founder representation be equalized as far as possible.

Preserving subspecies

Even if the zoo curator strives very hard to maintain the greatest possible variety and produce the greatest possible number of animals, he still faces theoretical problems. The greatest, perhaps, is that many species of animal are divided into distinct subspecies (◆ page 197). If the subspecies are too widely separated genetically – that is, on the point of becoming new species – then hybrid offspring produced by mixing the two subspecies may be sexually infertile, or at least partially so, just as mules are sexually sterile. Thus the whole breeding program would be compromised.

The more usual danger, however, is that by mixing different subspecies the breeder loses the variety that exists in nature – the variety he seeks to maintain. Thus Bornean orangutans are different from the Sumatran kind. The Bornean males have huge round faces, and the Sumatrans do not. The two subspecies breed perfectly well together. But if they are mixed (and if they disappear in the wild, which is very likely) then the world will have lost, for ever, the "pure" type. What will remain will be an artificial, manmade hybrid. On the other hand, there is only a strictly limited number of orangutans in captivity. If the two are not mixed, will there be sufficient animals to maintain two separate breeding populations?

The five species of rhinoceros. Indian or Greater one-horned (1), Sumatran or Asian two-horned (2), White or Square-lipped (3), Javan or Lesser one-horned (4), Black or Hooked-lipped (5). The Asian rhinoceroses inhabit rain forests, the African rhinos grasslands.

Rhinos in danger

Rhinoceroses were once a successful and varied group, ranging over much of the world. Now only five species are left, three in Asia and two in Africa, and all are endangered. Perhaps the "safest" of the five now is the white rhinoceros (left). Several zoos, including Britain's Whipsnade, have breeding herds. But hundreds of individuals are needed to guard against extinction. The Sumatran rhinoceros in particular presents a dilemma. In the wild, there are fewer than 800 animals. Some argue that the limited resources available for conservation should be used to support the animals in the wild. Others point out that the pressures that have driven the Sumatran rhinoceros to the point of extinction – primarily deforestation – still exist, and that the wild is too dangerous. In contrast, they say, other species of rhinoceros, including the African kinds, white and black, breed well in captivity. On the other hand, some rhinoceroses may be killed by the attempts to capture them, and whether they are or not, the removal of animals from their native forests will obviously deplete the wild population even further. The pros and cons must be carefully weighed.

Tigers normally avoid contact with people but deliberate man-eating behavior does exist and may be the result of an accidental close encounter that ends with the person being killed

Project Tiger

Project Tiger is an attempt to save a single subspecies in the wild – not just in one place, but at representative sites over its whole range. It is one of the most ambitious conservation projects of all times, and the lessons learned from it are important.

*The tiger in India has suffered badly in this century both from just those kind of biotic and physical factors that have so often driven animals to extinction (**◆ page 195**), and the cause of all those factors is people. It has been hunted – that is, has suffered "predation" – primarily by the members of the British raj, abetted by local maharajahs, and poisoned by villagers. It has suffered competition, as humans have killed its prey – deer such as barasinga, chital, and sambar; antelope such as blackbuck and chinkra; cattle such as gaur, the biggest of all wild cattle, and wild buffalo. Worst of all, its habitat has been destroyed, through the spread of agriculture and the removal of forests (**◆ page 176**), not least to provide fuel for 400 million people (out of a population of more than 700 million) who live in India's commonlands. The estimated tiger population sunk from 40,000 at the turn of the century, to fewer than 2,000, recorded by the All India Tiger Census in 1972.*

1970 was perhaps the lowest point. But in that year the Indian government imposed a national ban on hunting, and in 1972 the Wildlife (Protection) Act came into force. Project Tiger began in earnest in 1973, with the establishment of nine tiger reserves. Two more followed in 1978-79, and another four in 1982-83. The 15 between them are intended to cover as wide as possible a range of habitats, from Corbett in the Himalayas to the forested Bandipur in the south, with semi-desert, swamps, and open grassland all represented.

*The strategy, primarily, depends on excluding people and their cattle from the "core" area of each reserve. But India is crowded, and mostly rural. In many cases villages have had to be relocated, and this has led to much controversy. Around each core area is a "buffer zone", where people are allowed to graze cattle or harvest bamboo. This too has led to controversy; after all, it brings people and cattle into contact with one of the world's most accomplished predators. India, however, needs to be reorganized if its people and wildlife are to flourish, and the short term disruption seems inevitable. What is vital, of course, is to provide local people with some involvement in the project – perhaps through tourism, as in Operation Windfall in Zimbabwe (**◆ page 238**).*

The extent to which Project Tiger is succeeding is still uncertain. Early claims that the population had increased by 40 percent in the first five years (up to 1978) are now considered dubious, but few would doubt that the tigers are increasing. In addition, as is so often the case, protection of one species has a knock-on effect. Many species in India's crowded lands are endangered – including gaur, the great Indian (one-horned) rhinoceros, the wild buffalo, the barasinga, mutjac, swamp deer and hog deer, the mugger (a large crocodile) and the golden langur. The tiger provides a symbol and a focus for Indian conservation.

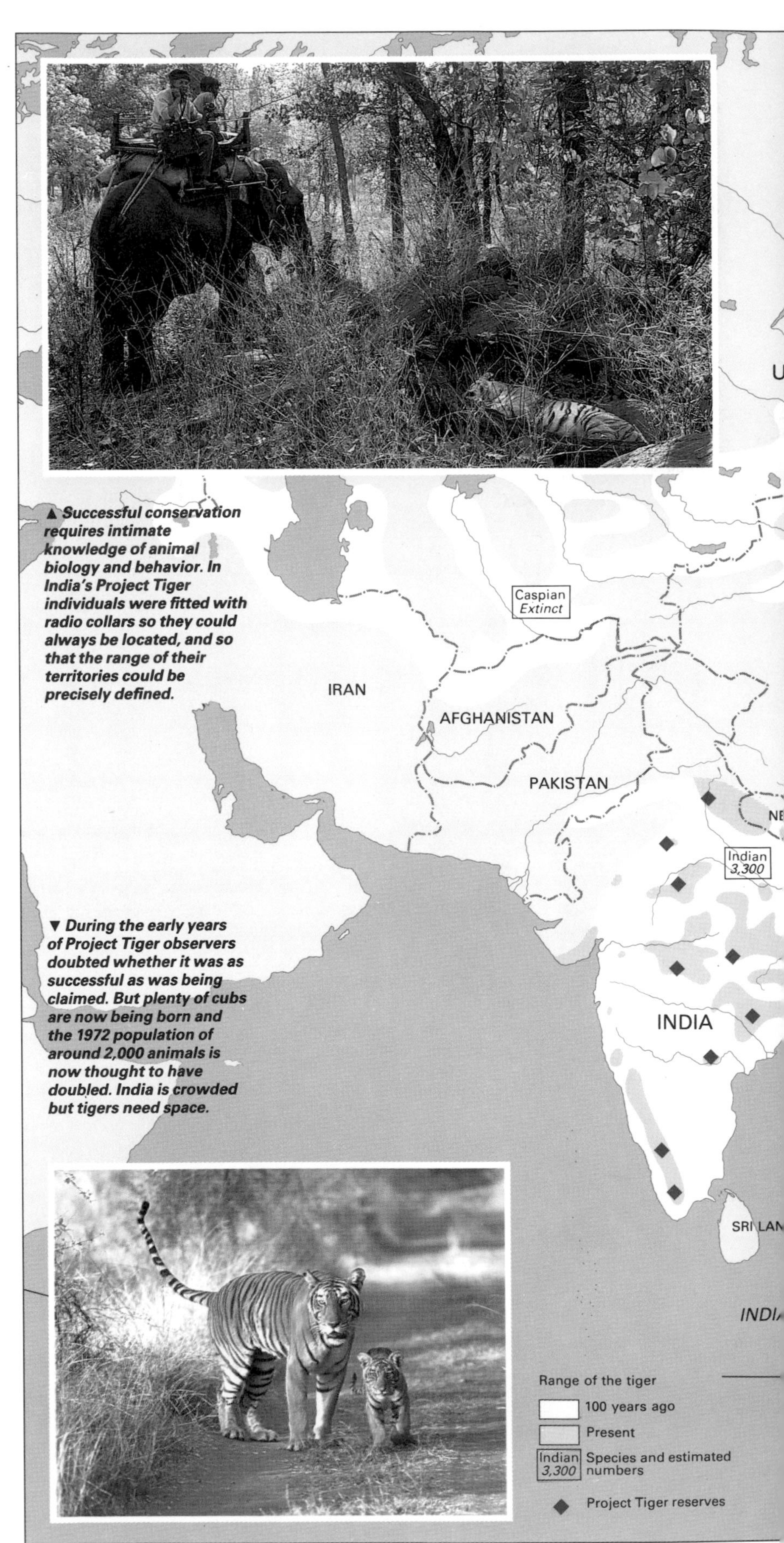

▲ Successful conservation requires intimate knowledge of animal biology and behavior. In India's Project Tiger individuals were fitted with radio collars so they could always be located, and so that the range of their territories could be precisely defined.

▼ During the early years of Project Tiger observers doubted whether it was as successful as was being claimed. But plenty of cubs are now being born and the 1972 population of around 2,000 animals is now thought to have doubled. India is crowded but tigers need space.

▼ **Like most animals with a large range, tigers have become divided into many different races, or subspecies. Of these, the Caspian, Javan and Bali are now extinct. The Chinese until recently regarded their tigers as pests, and less than 40 survive. The few hundred Sumatran tigers share their island with 165 million people, most of whom regard them as vermin. The Indo-Chinese tiger is largely unprotected. But the protected Indian and Siberian seem to be on the increase – showing that well-managed conservation measures can make a difference.**

Breeding the cheetah

In 1969 a study by the Investigation for the Conservation of Nature showed that cheetahs were declining in the wild in Africa. Two institutions in particular then contrived to breed them in captivity: London Zoo, at its country estate at Whipsnade, and the National Zoological Gardens of South Africa, at Pretoria. The latter in 1972, established the Cheetah Research Center at de Wildt, a farm near Pretoria owned by Ann van Dijk and her brother Godfrey. Both Whipsnade and de Wildt have bred hundreds of cheetahs since then, but not without difficulty.

Both groups discovered that cheetahs will not breed successfully in captivity unless the males and females are kept apart until the female comes into heat. They are naturally solitary animals and if they are raised all together they have little sexual interest in each other. One snag is that heat in cheetahs (unlike, for example, in bitches) is often inconspicuous; the females must be watched carefully to spot when they come into season.

A second difficulty is that many male cheetahs are sterile, or at least subfertile. This proved to be the case with 11 out of the 20 male cheetahs that founded the captive colony at de Wildt. What complicates matters is that when males are kept in groups (which is convenient in captivity) some dominate the rest, and only the dominant ones mate with the females. But the dominant ones are not necessarily the most fertile.

*A possible problem with all captive bred animals is that they are liable to become highly inbred (**♦ page 202**). With cheetahs this seems not to be a problem. Even wild cheetahs are highly homozygous, which indicates that they are already inbred. Some time in their history the population must have been reduced to very low numbers and, quite by chance, they survived that genetic bottleneck (**♦ page 204**). They are therefore "accustomed" to inbreeding.*

The relative lack of genetic variety among cheetahs does raise one problem, however. Cheetahs are cats, and extremely susceptible to the infections of domestic cats. Because all cheetahs are genetically similar, they are all equally susceptible, so captive animals are vaccinated.

In addition, many female cheetahs are poor mothers, and neglect or even kill their cubs if they are disturbed – especially their first litters. In the wild, they normally leave their cubs in the open, and shift them frequently to new sites, just as domestic cats re-site their kittens. In captivity, the mothers need quiet, secluded quarters.

The final problem is what to do with the cubs when the programs do prove successful. Both de Wildt and Whipsnade have sent many to other zoos. Attempts to return cheetahs to the wild have so far met only partial success. The animals were able to hunt well enough but were not able to establish territories. Cheetahs already established in the wild drove them out. Within two months of freedom the released animals had covered 500km. Release into the wild will not become vital unless cheetahs become extinct in the wild and if that happens the released animals will be free to establish their own territories.

See also
The Limits of Animals 25-34
Territory 63-70
Interactions 81-96
The Shadow of Extinction 195-208
Threats to the Environment 209-224
Saving Habitats 225-234

▲ ***Inexperience of the wild can put captive bred or escaped domesticated animals at a disadvantage. Lack of skill at predator avoidance has made this horse easy prey for a lion.***

▼ ***North America's whooping cranes have slowly recovered in the wild from 14 individuals in 1941, to about 75 now. But there are another 140 or so in captivity.***

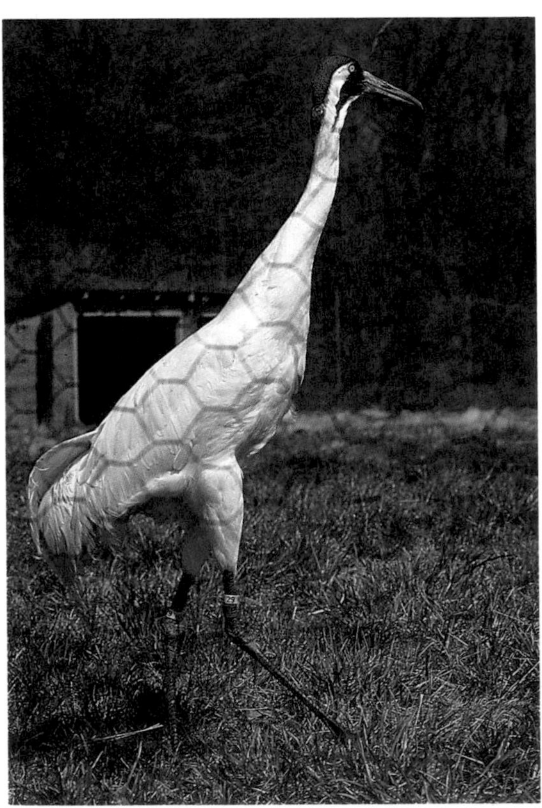

Verdict on captive breeding

Despite all these difficulties, captive breeding is worthwhile. There are clear examples already of species that would by now be extinct had they not been bred in captivity. These include Pere David's deer, from China, the Arabian oryx (◆ page 235) and the Golden Lion Tamarin, from Brazil. But when an animal is clearly or apparently ailing in the wild, it still is not always clear whether it should be left in the wild or brought in. Argument has raged, for example, over whether the remaining Californian Condors should be left to take their chance, or brought into captivity. The trouble is that capture always entails risk – and even when they are captured, animals may fail to breed. In addition, the act of capturing clearly depletes the wild population even more. In this particular case, the captive birds (in Californian zoos) are clearly outstripping the wild ones (now down to single figures), so it now seems positively perverse not to round up the remaining birds.

Because so many animals now are reduced to very small numbers, the ideal, theoretically, would be to combine the wild and captive populations: to bring wild animals into the captive pool to increase the genetic variety (◆ page 202), and to release captive bred animals back into the wild to boost the numbers. In practice there are many difficulties. Taking animals from the wild depletes the wild, and brings problems of disease. Animals returned to the wild from captivity may have behavioral and social problems. Captive-bred predators, for example, may have to be taught how to hunt. This did not prove a problem when captive-bred cheetahs were released in Africa (◆ page 247) but the release still failed because wild cheetahs would not allow the newcomers into their territory.

The hope is that the human species will eventually reduce its numbers – not because of famine or war, but because of a rational and peaceful decision to do so. When and if this happens, a great many more captive-bred animals will be able to return to their homes.

Credits

Key to abbreviations: ANT Australasian Nature Transparencies; BCL Bruce Coleman Ltd; NHPA Natural History Photographic Agency; OSF Oxford Scientific Films; PEP Planet Earth Pictures; PW/KGPM Premaphotos Wildlife/K.-G. Preston-Mafham; SAL Survival Anglia Ltd; SPL Science Photo Library; b bottom; bl bottom left; br bottom right; c center; cl center left; cr center right; t top; tl top left; tr top right; l left; r right

1 Robert Harding **2-3** BCL/B. & C. Alexander **4-5** Dr P. Francis/South American Pictures **7** PEP/D. Gill **8** Frans Lanting **9t** PW/KGPM **9b** Associated Press **10** PEP/Colin Pennycuik **11** PEP/James Hudnall **12** H. Kacher **12-13** BCL/J.D. Bartlett **13** NHPA/M. Danegger **14-15** A. Bannister **14b** Nigel Bonner **15bl** PEP/R. Matthews **15 br** W. Ervin/Natural Imagery **18** Orion Press **19l** Fotomas Index **19r** BBC Hulton Picture Library **20** NHPA/J. Shaw **21** A. Bannister **22** T. Roper **22-23** PEP/K. Lucas **23tl** OSF/A. Shay **23tr** Private collection, London **24t** SAL/J. & D. Bartlett **24b** PEP/A. Kerstich **25** OSF/R. Blythe **26** OSF/G.I. Bernard **27t** T. Roper **27b** BCL/P. Davey **29** NHPA/P. Johnson **30** OSF/Press-Tige Pictures **31** SAL/D. Plage **32-33** Gregory Silber **32b** PEP/A. & M. Shah **33** Bodleian Library, Oxford **34** The Gorilla Foundation/Ronald Cohn **35t** PEP/C. Read **35b** PW/KGPM **36** Biofotos/H. Angel **37t** Michael Fogden **37b** W. Ervin/Natural Imagery **38t** NHPA/A. Bannister **38b** R. Dunbar **39** NHPA/S. Robinson **40** PEP/N. Downer **41l** Jacana/M. Danegger **41r** NHPA/A. Bannister **46** PW/KGPM **47t** NHPA/S. Dalton **47b** BCL/P. & M. Borland **48-49** OSF/G. Merlen **48b** NHPA/E. Hanumantha-Rao **49b** SAL/J.B. Davidson **51** John Deag **52** Agence Nature/Chaumeton **53t** OSF/D. Allen **53b,54** A. Bannister **55l** OSF/D. Curl **55r** PEP/P. Scoones **56t** SAL/J.B. Davidson **56c** ANT/C. & S. Pollitt **56-57b** PEP/J. Scott **58** Michael Fogden **59** BCL/J. Burton **60t** OSF/D. Macdonald **60b** Hutchison Library **61** Agence Nature **62t** PEP/K. & K. Ammann **62b** SAL/J. van Gruisen **63t** BCL/E. & P. Bauer **63b** PW/KGPM **64** Michael Fogden **65t** Aspect Picture Libary **65b** Jacana/F. Gohier **66tl** Jacana/P. Varin **66cl** W. Ervin/Natural Imagery **66cr** NHPA/M. Grey **66b** W. Ervin/Natural Imagery **67t** Doug Wechsler **67cr**, **67bl** NHPA/A. Bannister **67br** PEP/J. Scott **68** Agence Nature/Chaumeton **68-69t** PEP/E. Neal **68-69b** OSF/Pfunder **69r** PW/KGPM **70** A. McGeeney **71t** E. & D. Hosking/D.P. Wilson **71b** OSF/T. Martin **72** SAL/C. Buxton & A. Price **73** SAL/J. Pearson **74** Frans Lanting **74-75** ANT/A. Jackson **75** BCL/J. Foott **76** BCL/J. & D. Bartlett **78,79** Susan Griggs/J. Blair **80t** BCL/B. & C. Alexander **80b** Ardea/I. Beames **81t** BCL/G. Zeisler **81c** E. & D. Hosking **81b** SAL/C. Buxton **82** BCL/S. Kraseman **83t** PEP/Menhuin **83c** Biofotos/H. Angel **83b** Frank Lane Agency **84** NHPA/S. Dalton **84-85** Biofotos/H. Angel **87** PEP/R. Matthews **90** A. Bannister **90-91** PEP/W. Williams **92** Frank Lane Agency/F. Polking **93t** PEP/J. Scott **92-93** Zefa/E. Christian **94t** A. Bannister **94b** OSF **95t** Ardea/F. Gohier **95b**, **96b** Marion & Tony Morrison/South American Pictures **96t** BCL/E. & P. Bauer **98** PEP/J. Lythgoe **100t** NHA/H. Palo **100b** Michael Fogden **101t** Biofotos/S. Summerhays **101c** OSF/D. Thompson **101b** NHPA/H. Palo **102** ANT/G. Fyfe **102-103** NHPA/P. Warnett **103** Jacana/F. Winner **104l** BCL/K. Gunnar **104r** NHPA/S. Dalton **105** E. & D. Hosking **106** OSF **106-107** A. Bannister **107** Agence Nature **108-109t** P. Francis/South American Pictures **108-109b** PEP/E. Neal **109** BCL/J. van Wormer **110** ANT/F. Park **111** SAL/J. Foott **112-113** PEP/R. Salm **112** Frank Lane Agency/M. Newman **113t** BCL/H. Flygare **113b** NHPA/S. Kraseman **114t** ANT/M. Wellard **114b** Frank Lane Agency/Silvestris **115** Biofotos/H. Angel **116t** NHPA/S. Kraseman **116b** A. Bannister **117** NHPA/M. Walker **118l** ANT/P. Horne **118r** Biofotos/H. Angel **119t** NHPA/S. Dalton **119b** PEP/K. Cullimore **120-121t** M.C.F. Proctor **120b** PW/KGPM **121t** Biofotos/H. Angel **121b** NHPA/E.A. Janes **122** PEP/P. David **123** PEP/H. Voigtmann **124t** G.R. Roberts **124b** PEP/J. Scott **125t** Biofotos/H. Angel **125b** NHPA/I. Polunin **126t** PEP/J. Hector **126bl** E. & D. Hosking/D.P. Wilson **126br** A. Bannister **128** Biofotos/H. Angel **130** NHPA/S. Dalton **131t** Hutchison Library **131b** Greg Evans **132t** NHPA/P. Johnson **132b** ANT/G. Claridge **133** USDA **134t** Hutchison Library **134b** Paul Harrison/Panos Pictures **135t** Sally & Richard Greenhill **135b** Aspect Picture Library **136-137t** ANT/G. Wood **137b** Rob Judges/Equinox **138-139** NHPA/P. Scott **140** Greenpeace **141** NHPA/G.I. Bernard **142,143,144** Hutchison Library **144-145** BCL/N. Devore **145** Robert Harding **146** Susan Griggs **146-147** BCL **148inset,148-149** Charles Dowding **148l** Panos Pictures **149** John & Penny Hubley **150** Zefa **150-151** Holt Studios **151** Zefa **152** Holt Studios **154-155** Biofotos/H. Angel **155** Zefa **156** Robert Harding **157tl** Hutchison Library **157tr** L. St Lawrence/B & B Photographs **157cl** Ardea/F. Gohier **157cr** AID Photo **157b,158** Hutchison Library **158-159** Biofotos/H. Angel **159** ZEFA **160-161** Hutchison Library **160** ANT/R. & D. Keller **161** Frank Lane Agency/B. Casals **162l** Holt Studios **162r** Zefa **163t** NHPA/A. Bannister **163b** Robert Harding **164-165** Hutchison Library **171** Zefa/Heilman **172t** PEP/R. Wood **172b** Hutchison Library **173t** Biofotos/B. Rogers **173b** Explorer/A. Perigot **174** PEP/J. & G. Lythgoe **175tr** Explorer/M. Breton **175bl** Robert Harding **175br** Biofotos/H. Angel **176t** PEP/R. Matthews **176b**,**177** Zefa **178-179** Robert Harding **178b** PEP/R. Matthews **179** Holt Studios **180** Edward S. Ayensu **181** Robert Harding **183** Aspect Picture Library/P. Carmichael **184** NHPA/P. Johnson **186** PEP/J. Duncan **188tl** PEP/P. David **188tr** R.B. Mitson, Ministry of Agriculture, Fisheries & Food **188b** Biofotos/H. Angel **189t** Aspect Picture Library/B. Alexander **189b**,**190** PEP **191** Biofotos/H. Angel **193** PEP/P. Capen **194** PEP/K. Scholey **195** Robert Harding **196t** Biofotos/S. Summerhays **196c** Biofotos/H. Angel **196b** Mary Evans Picture Library **197t** Rolf O. Peterson **197b** Aspect Picture Library/T. Nebia **200** Biofotos/H. Angel **201t** BCL **201b** Robert Harding/M. Collier **204** BCL/E. & P. Bauer **204-205** BCL/R. Peterson **205** NHPA/ANT **206** Jacana **206-207** Biofotos/H. Angel **207** Robert Harding **208t** NHPA/L. Lemoine **208b** Biofotos/H. Angel **209t** SPL/G.C. Fuller **209b** Aspect Picture Library **210** Zefa/B. Crader **212** Zefa **212-213** ANT **213** Robert Harding **214-215** Hutchison Library **215t** ZEFA **215b** PEP/W. Williams **216-217** Susan Griggs Agency **217tc** Hutchison Library/J. Egan **217tr** Ardea/A. Weaving **217t** Ardea/L. & T. Bomford **218** Biofotos/H. Angel **218-219** PEP/C. Howes **220** PEP/R. Matthews **221** Holt Studios **222** Robert Harding **223** PEP/D. Gill **224t** Holt Studios **224b** Kent Trust for Nature Conservation **225t** OSF **225b** NHPA/D. Woodfall **226** NHPA/P. Johnson **226-227** PEP/L. Pitkin **227** Frank Lane Agency **228-229** Doug Wechsler **229** Fermilab **230,231tr** National Parks Board, Pretoria **231cl** Natal Parks Board **231b** NHPA/P. Scott **232-233** Robert Harding **232** Chris Perrins **233** Nature Photographers/M. Leach **234** SAL/D. Green **235** NHA **236t** Royal Botanic Gardens, Kew **236b** Mary Evans Picture Library **237t** NHPA/J. Shaw **237b** Nature Photographers/S. Bisserot **239t** OSF/G.I. Bernard **239b** Aspect Picture Library/G. Mullis **240** SAL/J. Foott **241** Ardea/K. Fink **242-243** Dr H. Moore, Zoological Society of London **242** Institute of Animal Physics and Genetics, Cambridge **243t** Zoological Society of London **243b** John Aitken **246t** Ardea **246b** PEP/A. & M. Shah **248t** SAL/J. & D. Bartlett **248b** SAL/J. Foott

Artists Priscilla Barrett; Simon Driver, Chris Forsey; Richard Hook; Alan Hollingbery; Kevin Maddison; Colin Salmon; Mick Saunders
Agricultural advisor Tim Blanchard
Cartographic editor Nicholas Harris
Editorial assistant Monica Byles
Art assistant Frankie Macmillan
Indexer Barbara James
Media conversion and typesetting Peter MacDonald and Ron Barrow

Further Reading

Barnett, S.A. (1981) *Modern Ethology*, Oxford University Press, Oxford.
Boakes, R.A. (1984) *From Darwin to Behaviourism: Psychology and the Minds of Animals*, Cambridge University Press, Cambridge.
Bright, M. (1984) *Animal Language*, BBC Publications, London.
Broom, D.M. (1981) *Biology of Behaviour*, Cambridge University Press, Cambridge.
Chalmers, N.R. (1979) *Social Behaviour in Primates*, Edward Arnold, London.
Dawkins, R. (1977) *The Selfish Gene*, Oxford University Press, Oxford.
Gould, J.L. (1982) *Ethology*, Norton, New York.
Halliday, T.R. (1980) *Sexual Strategy*, Oxford University Press, Oxford.
Krebs, J.R. and Davies, N.B. (1981) *An Introduction to Behavioural Ecology*, Blackwell Scientific Publications, Oxford.
McFarland, D.J. (ed) (1981) *The Oxford Companion to Animal Behaviour*, Oxford University Press, Oxford.
Mace, G. (1983 and 1985) *The Present Status and Future Management of Populations of Great Apes in the United Kingdom*, The Zoological Society of London, London.
Manning, A. (1979) *An Introduction to Animal Behaviour* (3rd edn), Edward Arnold, London.
Marler, P. and Hamilton, W.J. III (1966) *Mechanisms of Animal Behavior*, Wiley, New York and London.
Messent, P. and Broom, D. (1986) *The Encyclopedia of Domestic Animals* Grolier International.
Miller, T. and Armstrong, P. (1982) *Living in the Environment*, Wadsworth, California.
Moran, J., Morgan, M., Wiersma, J. (1986) *Introduction to Environmental Science* (2nd edn), W.H. Freeman, New York.
Owen, J. (1980) *Feeding Strategy*, Oxford University Press, Oxford.
Slater, P.J.B. (1985) *An Introduction to Ethology*, Cambridge University Press, Cambridge.
Trivers, R. (1985) *Social Evolution* Benjamin/Cummings, Menlo Park, California.

Glossary

Many words are explained in the text, and if they are not listed here, the definition can be found by use of the index. Words in CAPITALS are glossary terms.

Acid rain
Precipitation, both rain and snow, made acidic in reaction through the chemical POLLUTION of air by waste gases, such as oxides of sulfur and nitrogen, from industry and car exhausts.
Adaptability
The ability of an organism to alter its mode of behavior, or even its physiology, when placed under a new type of stress.
Adaptation
The process whereby, under the influence of NATURAL SELECTION, an organism gradually changes genetically in such a way that it becomes better fitted to cope with its environment.
Aggression
This is a controversial term, best used as a loose categorization of attack and threat behavior but sometimes taken to include a much broader group of activities. A cat attacking a mouse (predatory aggression), a bird singing (a behavior that tends to repel rivals) and a human speaking assertively, have all sometimes been described as aggressive. As the causes of these actions probably have little in common with fighting within a species, such broad usage is probably best avoided.
Altruism
Behavior that brings advantage to others than those who perform it. The term is usually taken to include a reduction in the performer's own FITNESS.
Amino acids
Organic compounds containing both basic amino (NH_2) and acidic carboxyl (COOH) groups. Amino acids are fundamental constituents of living matter – some hundreds of thousands of amino acid molecules are combined to make each PROTEIN molecule.
Ancestor
An individual from which an animal is descended. In EVOLUTION, it refers to a living or fossil species from which a present-day species is believed to have descended. Most domestic animals are derived from wild ancestors whose wild descendants are considered to be of the same species as the domesticated descendants.
Arable crop
A crop grown on tilled land.
Artificial selection
The process whereby the animals that produce offspring are chosen by humans, in contrast to NATURAL SELECTION where biological fitness determines the success of reproduction.
Atmosphere
The gaseous envelope that surrounds the Earth.
Behaviorism
A school of psychology founded by J. B. Watson which rejected introspection and stressed the importance of objective observation. Their studies were largely concerned with experiments on learning carried out in highly controlled and simplified laboratory situations.
Benthic
Of or relating to the depths of the ocean.
Biological control
The use of one species (usually a PREDATOR or PARASITE) to control the population of another.
Biomass
The total quantity of organic matter associated with the living organisms of a given area at a particular time.
Biome
A major global ecological unit, or type of flora and fauna formation (e.g. savanna grassland, boreal forest).
Biosphere
That part of the Earth which is capable of supporting life: includes part of the atmosphere, lithosphere and hydrosphere.
Breed
To produce offspring. Also, to cross selected individuals to produce offspring with desired characteristics. Also (noun) a race or strain whose members, when crossed, produce offspring with the same characteristics as the parents.
Breeding
The rearing and crossing of animals or plants so as to change the characteristics of future generations.
Carbohydrates
Organic compounds of general formula $C_x(H_2O)_y$: e.g. sugars, starches, cellulose. Carbohydrates play an essential part in the metabolism of all organisms.
Carnivore
An animal which eats the flesh of another animal, usually involving the death of the latter.
Cereal
A general term for grains produced by members of the grass family, e.g. wheat, rice, barley, millet, which are commonly consumed by people.
Climax
Stage in development of an ECOSYSTEM when there is no further net growth in BIOMASS; relative to other stages, climax flora and fauna are rich and their interrelationships complex.
Colonization
Invasion of a new habitat by plants and/or animals.
Community
A collection of populations of a number of species interacting together.
Competition
The interaction between two or more species, or between individuals of a single species, in which a required resource is in limited supply and consequently one or both of the competitors suffer in their growth or survival.
Conditioning
Learning by association. In classical conditioning two stimuli are associated so that the second comes to elicit a response formerly only elicited by the first.
Conservation
The rational management of and care for the BIOSPHERE, in order to avoid the creation of imbalance resulting in the destruction of habitats and the extinction of species.
Cultivation
The management of an ECOSYSTEM with the specific intention of channeling energy into human beings.
Decomposer
An organism which relies upon the dead tissues of other organisms as an energy source; in using this energy it liberates nutrients from those tissues into the ENVIRONMENT.
Detritivore
An individual which feeds upon the dead remains of other organisms, both plant and animal.
Display
Movement pattern used in communication. These are often striking, stereotyped and species-specific, especially during courtship and aggressive behavior. Displays may also function between species.
Diversity
A measurement of the richness of species in a given area, sometimes also incorporating the evenness of COMMUNITY, i.e. the degree to which certain species dominate the community in numerical terms.
Domestication
The process of altering the behavior and physiology of initially wild populations through ARTIFICIAL SELECTION for such characteristics as docility, high yield of desired products or ease of breeding.
Dominance
The situation in which one animal dominates another in fights or in access to resources such as sitting positions, food and mates. Dominants are often older and stronger, but fighting may not be involved in encounters because subordinates frequently defer. In some social groups relationships may be consistent and clear enough for animals to be placed in a linear dominance hierarchy in which each is dominated by those above it and dominates those below.
Ecology
The study of organisms in relation to their physical and living environment.
Ecosystem
A unit which includes all the living organisms and the non-living material within a defined area, the size of which is relatively arbitrary.
Endangered species
A species whose population has dropped to such low levels that its continued survival is insecure.
Energy
The ability to do work. It may take many forms, such as light, heat and chemical: it changes form as it flows through the ECOSYSTEM.
Environment
The surroundings of an ORGANISM, including both the non-living world and the other organisms inhabiting the area.
Enzyme
A PROTEIN which is a catalyst of biochemical reactions. There are many different kinds, each kind directly promoting only one or a very limited range of reactions.
Epiphytes
Plants that grow on other plants, but do not derive any nutrients or water from them.
Ethology
The biological study of behavior.
Eutrophication
The enrichment of a HABITAT (often aquatic) with nutrient elements such as nitrogen or phosphorus.
Evolution
The process by which species have developed to their present appearance and behavior through the action of NATURAL SELECTION in determining the survival of those individuals most suited to their ENVIRONMENT.
Extensive farming
A system of farming in which animals are kept in fields or open country rather than in buildings or yards with little space per animal.
Extinction
The complete elimination of a population: often used in a global sense, but can be used of local population.
Fat
A substance which can be extracted from tissues by organic solvents such as ether but not water, and stored in adipose tissue. True fat is a compound of glycerol and fatty acids. Fats form a potential energy source.
Fitness
A measure of the degree to which a given genetic type succeeds in reproducing itself. An organism's fitness is related to its ecological adaptedness and sexual proficiency.
Fodder
Feed for domestic HERBIVORES.
Food web
The complex feeding interactions between species in a community.
Free ranging
In a domestic animal, having access to a more varied environment than a cage or pen. In wild animals, a lack of territoriality and a wide individual range of movement or dispersal.
Gene flow
Interchange of genetic factors between and within populations as a result of emigration and immigration of individuals.
Genes
The units of inheritance which are transmitted from generation to generation and control the development of an individual.
Genetic drift
The occurrence of random changes, irrespective of selection and MUTATION, in the genetic make-up of small isolated populations.
Genus (plural genera)
A taxonomic division superior to SPECIES and subordinate to family.
Grazing
The action of a HERBIVORE which feeds upon herbaceous vegetation close to the ground.
Greenhouse effect
The accumulation of gases, such as carbon dioxide, in the atmosphere which prevent infra-red radiation leaving the Earth and hence cause increasing global temperature.
Habitat
The locality within which an organism is found, usually including some description of its character.
Hay
Grass that has been cut and dried as feed for livestock.

Herbivore
An animal which feeds exclusively on living vegetable matter.
Heterotroph
An ORGANISM that needs preformed organic molecules, produced by other living things, in its diet.
Home range
The area which an animal or group of animals occupies or visits. As it is not necessarily defended from others, it is distinguished from TERRITORY.
Hominid
A man-like animal belonging to the evolutionary line leading to Man.
Hormone
Organic substance produced in minute quantity in one part of an organism and transported to other parts where it exerts a specific effect, e.g. stimulating growth of a specific type of cell.
Hybrid
The offspring of parents which are not genetically identical.
Imprinting
A process whereby young animals learn the characteristics of other individuals, normally parent, early in life. In filial imprinting they come to devote their social responses to that individual. Learning about parents and siblings may also influence mate choice through sexual imprinting, the animal seeking a partner similar to those with which it was reared, but not usually identical with them.
Inbreeding
Breeding with close relatives, sometimes associated with reduced offspring survival.
Intelligence
That capacity which enables an individual to learn tasks, reason and solve problems. Such capabilities being based on many attributes, testing intelligence in humans is open to numerous biases. In animals, intelligence is also hard to assess: they may find superficially difficult tasks to which they are adapted easy and other, apparently simpler, tasks much more difficult.
Kin selection
Selection acting on an individual in favor of the survival not necessarily of that individual but of its relatives (which carry the same GENES). An example is the ALTRUISM between genetically related members of a social insect colony.
Metabolism
The chemical processes occurring within an organism. Including the production of PROTEINS from AMINO ACIDS, the exchange of gases in respiration, the liberation of energy from foods and innumerable other chemical reactions.
Microbe, microorganism
Organisms of microscopic or ultramicroscopic size, such as bacteria, some fungi, viruses.
Mutation
A structural change in a GENE which may give rise to a new heritable characteristic if it occurs in one of the germ cells.
Natural selection
The process by which those organisms which are not well fitted to their environment are eliminated by predation, parasitism, competition, etc, and those which are well fitted survive to breed and pass on their genes to subsequent generations.
Neo-Darwinism
The modern theory of evolution which combines both NATURAL SELECTION and genetics.
Niche
In ecology, the totality of interactions of a given ORGANISM. It includes the HABITAT it lives in, its food sources, PARASITES and PREDATORS, its special requirements for a burrow, nest or other living space, and all other factors affecting its survival.
Nutrient cycle
The movement of elements around an ECOSYSTEM between living and non-living components.
Omnivore
An animal which is prepared to consume both plant and animal material in its diet.
Opportunist
An organism which as a result of its capacity to migrate rapidly, to breed fast, or to survive in a dormant state, is able to take advantage of an opportunity, such as ECOSYSTEM disturbance, to expand its population.
Organism
A living creature, animal, plant or MICROBE.
Outbreeding
The mating of an individual with one that is unrelated, whether of the same or a different breed; the opposite of INBREEDING.
Parasite
An organism which is totally dependant upon another organism (the host) for its energy. It is usually very closely associated with its host and often causes its reduced growth or reproduction but only rarely kills it.
Parental investment
Any investment by a parent in an individual offspring which increases the offspring's chances of surviving at a cost to the parent's ability to invest in other offspring. On the basis of this theory parents should not invest too much time and effort in one offspring but should allocate their investment to maximize the number of their young that survive to breed.
Pasture
The growing herbage eaten by cattle. Also a piece of land covered with this. Also (verb) to feed animals on this.
Pedigree
Genetic origin, line of succession.
Pelagic
The upper part of the open sea, above the BENTHIC zone.
Pest control
The reduction of pest populations by various means, including chemical and BIOLOGICAL CONTROL.
Photosynthesis
The synthesis of organic compounds, primarily sugars, from carbon dioxide and water using sunlight as the source of energy, and chlorophyll, or some other related pigment, for trapping the light energy.
Plankton
Mostly very small animals and plants of sea or lake which float or drift almost passively. Of great ecological and economic importance, providing food for fish and whales.
Pollution
The disruption of a natural ECOSYSTEM as a result of human contamination.
Polygamy
The mating system in which one individual has two or more mates, either simultaneously or successively (the latter is also known as serial monogamy). In polygyny one male has several females. In polyandry one female has several males. The latter is rare because the reproductive success of females is usually limited by the number of eggs they can produce rather than by the number of mates they can have.
Population
A more or less separate breeding group of animals. Also, the total number of individuals counted in a given area.
Predator
An animal which feeds upon populations of other animals (the prey); sometimes also used of herbivores, where the plant is the prey.
Productivity
The amount of weight (or energy) gained by an individual, a species or an ECOSYSTEM per unit area per unit time.
Protein
A complex bio-molecule, made up of one or more chains of AMINO ACIDS. Where made of several chains, each of these is known as a polypeptide chain.
Reflex
An automatic and involuntary response which is the simplest form of reaction to an external stimulus. Reflexes may involve as few as two or three nerve cells. Examples are a dog scratching when its side is irritated and the constriction of the pupil of the eye when light shines into it.
Ruminant
A mammal with a specialized digestive system typified by the behavior of chewing the cud. The stomach is modified so that vegetation is stored, regurgitated for further chewing, then broken down by symbiotic bacteria.
Saprophytes
A term used to describe the nutritional characteristics of those fungi and bacteria that live on dead matter or excretory products. The word has now largely been replaced by saprophyte or saprobe; the ending -phyte reflected the mistaken idea that fungi were plants.
Scavenger
An animal which relies upon other animals to kill or collect food and then takes advantage of the unwanted remains.
Scent mark
Site where the secretions of scent glands, or urine or feces, are deposited and which has communicative significance. Often left regularly at traditional sites which are also visually conspicuous. Also the "chemical message" left by the means; and (verb) to leave such a deposit.
Sessile
Of animals; fixed to the seabed, riverbed or other substrate, either permanently or for most of the time. (Also used in botany, to describe leaves with no stalk.)
Silage
FODDER preserved in a silo or pit, without previous drying.
Sociobiology
That branch of behavior study concerned with the social behavior of animals, its ecology and evolution.
Species
Basic division of biological classification, subordinate to GENUS and superior to subspecies. In general a species is a group of animals similar in structure and which are able to breed and produce viable offspring.
Strategy
Mode of behavior adopted by an individual when it could, either actually or theoretically, behave in a different way. In theoretical discussions, an evolutionary stable strategy is that which, if adopted by the whole population, cannot be bettered by immigrants adopting a different strategy.
Stress
Physiological state induced in animals by conditions they are unable to tolerate and cope with, such as pain or overcrowding.
Subspecies
A recognizable subpopulation of a single species, typically with a distinct geographical distribution.
Symbiosis
A close relationship between different species.
Territorial behavior
Behavior used by an animal to delineate or defend an area from intruders either on behalf of itself or for a social group. Includes birdsong, scent marking and visual display.
Territory
An area occupied by one or more individuals of a species and defended against the intrusions of others.
Trace element
A chemical element which must be available to an organism for its normal health although it is necessary only in minute amounts.
Trial and error
A form of learning, also called instrumental CONDITIONING, in which an animal comes to associate performance of a behavior pattern with its consequences so that, for example, a response which by chance yields reward will be repeated more often.
Variation
Differences between individuals which may be caused either genetically or environmentally. The differential survival of genetic variants results in the process of NATURAL SELECTION.
Vitamin
Any of several organic substances, distinguished as vitamins A, B, etc. occurring naturally in minute quantities in many foodstuffs and regarded as essential to normal growth, especially through their activity in conjunction with ENZYMES in the regulation of METABOLISM.
Wild
Not domesticated: intractable, opposite of tame. Also a habitat that has had little or no disturbance from the activities of Man.

Index

Page numbers in *italics* refer to illustrations, captions or side text.

A

B

C

D

E

F

G

H

I

J

K

L

M

N

O

P

Q

R

S

T

U

V

W

Y

Z